T0328311

Renewable Energy Finance

Renewable Energy Finance

Theory and Practice

Santosh Raikar

Seabron Adamson

Academic Press is an imprint of Elsevier
125 London Wall, London EC2Y 5AS, United Kingdom
525 B Street, Suite 1650, San Diego, CA 92101, United States
50 Hampshire Street, 5th Floor, Cambridge, MA 02139, United States
The Boulevard, Langford Lane, Kidlington, Oxford OX5 1GB, United Kingdom

Library of Congress Cataloging-in-Publication Data
A catalog record for this book is available from the Library of Congress

British Library Cataloguing-in-Publication Data
A catalogue record for this book is available from the British Library

ISBN: 978-0-12-816441-9

For information on all Academic Press publications visit our website at https://www.elsevier.com/books-and-journals

Publisher: Brian Romer
Acquisition Editor: Brian Romer
Editorial Project Manager: Michael Lutz
Production Project Manager: Selvaraj Raviraj
Cover Designer: Mark Rogers

Typeset by TNQ Technologies

To our daughters, who represent the future generations that must be protected from the perils of the global climate change....

Anaya and Isha
- *Santosh Raikar*

Georgia
- *Seabron Adamson*

Contents

The URL for the companion site when it goes live should be https://www.elsevier.com/books-and-journals/book-companion/9780128164419.

Foreword

For those of us in the renewable energy field, these are heady times. Energy policy, green energy, climate change – these are no longer specialist topics, but are being discussed every day in the news, in politics, and around the dinner table. There is a sense of change in the air, and, as in all moments of profound change, one of the key questions is "How are we going to pay for all of this?"

No one knows for sure, but a transition to a global, low-carbon, renewable energy sector will likely require trillions of dollars in capital investment. Most of that money must be raised in private capital markets. So, the solution to the great environmental challenge of the 21st century will almost certainly involve ensuring that the capital markets can be harnessed to fund renewable energy investments on a colossal scale.

One of the key mechanisms for funding new large-scale renewable energy projects is project finance, which is non-recourse to the sponsor, and hence dependent on the future cash flows available to the project from its operations. In this book, Santosh Raikar and Seabron Adamson, noted project finance practitioners, describe the concepts of project finance and how it can be applied to investments such as wind farms and solar facilities. They also describe the detailed financing structures used by sponsors and lenders to manage the risks and uncertainty inherent in these projects. In doing so, they make accessible decades of practical experience, which is not often seen described clearly in print. Raikar and Adamson also introduce some relevant areas of public policy, engineering, energy economics, law, and other topics, and how these interact to make large-scale renewable energy investments viable.

I believe this book can be a primer for a broad range of people focused on the renewable energy sector: finance students, project developers, policymakers, academics, and bankers. Given the scale and urgency of the energy financing challenges ahead, the clear description of how and why renewable project finance works that is available in this volume will be useful in a range of contexts – not just closing the next deal.

Raymond S. Wood, MBA
Senior Investment Banker – Energy and Power
MIT Sloan School of Management, Class of 1990

Preface

If climate change is one of the greatest challenges of the 21st century, then one of the biggest questions in contemporary finance should be how to pay for the measures needed to control it. Since switching to renewable energy resources is at or near the top of most policymakers' list of measures for reducing carbon emissions, as a practical matter the financing problems arising in new renewables projects should have substantial commercial and public policy interest.

While there is a large body of literature on renewable energy economics, I know of no other text that explains how a large-scale renewable energy project – such a wind farm or solar photovoltaic facility – actually gets financed in detail. This book seeks to address this gap, providing the analytical frameworks and cash flow modeling tools used by project sponsors and lenders in real-life projects. For example, the chapter on tax equity is the first of its kind, providing a complete breakdown of the different complex tax equity financing structures widely used in renewable energy projects. In writing it, we have drawn upon several decades of joint experience as renewable project finance practitioners. The book has benefitted in particular from my co-author Santosh Raikar's long career in investment banking, which has focused on the development and application of project finance techniques to renewable energy.

One of the first things I learned in project finance is that it requires an understanding of far more than financial theory and spreadsheet modeling. Never having studied law, I found one of my first assigned tasks in joining the professional workforce to review and then model complicated power off-take agreements. Anyone who has worked on renewable energy in the real world knows that no project developer or banker can succeed without some knowledge of the legal, economic, engineering, public policy, and tax issues associated with these projects. There is no other way to understand the full risks to the project and how these are allocated in the contractual and financing structure.

For this reason, our approach to the subject of renewable energy project finance is interdisciplinary. We examine the different power market structures and transmission interconnection concepts that affect how and where projects get built. We also offer perspective on the valuation of projects during the development and operation phases. We discuss debt sizing, the public policy environment for renewables, and the complex web of contracts that support a project. For this reason, students of energy policy and law may also find this book useful in understanding the elements needed to make renewable projects financeable.

We also emphasize how to think about projects as a set of risks. Some risks are amenable to quantitative analysis, such as the distribution of revenues over a year given expected wind speeds. Others must be considered more holistically. Project risks can be managed and allocated, but someone always bears the risk. Experience soon teaches that it is better to have thought through these risks before the event.

Project finance does depend on models of future project cash flows. Our approach in writing this book has been to provide examples based on real-world experience, and to also provide the reader (through access to the book website) with the actual financial models used. These provide not only a way for students to understand how changes in inputs and structure affect cash flows and risks, but also could be used by professionals as a starting point for developing their own custom models for their own projects.

Many energy analysts believe that any deep decarbonization of the global energy sector will require trillions of dollars of investment, much of it directed at new renewable energy generation projects. Governments can set the agenda and shape the policy environment, but much of this capital must be raised from private capital markets. Project finance is a widely used financing structure in many countries that will be critical to make renewable energy projects financially viable if carbon policy goals are to be met. It is our hope that this book makes some small contribution towards advancing the shift to clean energy, by providing insight into the financial structures used to make large-scale renewable energy a reality.

Seabron Adamson
Boston, Massachusetts
September 2019

Acknowledgements

A lot of people directly or indirectly contributed to this work. We offer our thanks to them and many others for their contributions and support.

Some of the original material on project finance was developed for a series of informal lectures at MIT. We would like to thank John Parsons for his input (and for suggesting the cost of capital for pharmaceutical R&D as an analytical framework for the valuation of renewable energy projects under development), as well as Frank O'Sullivan, Antje Danielson, and Rowan Elowe of the MIT Energy Initiative. Akshar Wunnava, Yichen Du, and Mustafa Ali (presidents of the MIT Energy Club) and Jason Jay and Bethany Patten (of the Sloan Sustainability Initiative) helped organize some of the original talks. MIT Sloan students Lisa Khanna and Lydia Li provided valuable research assistance.

Much of the material was developed for our class on renewable energy investments at the Carroll School of Management at Boston College (BC). We would like to thank the BC Department of Finance for their encouragement in the class who provided the motivation to write a book for which no real text was available. We would also like to thank Ronnie Sadka and Elliott Smith of the Department of Finance at Boston College for their encouragement and support.

We offer our thanks to Todd Glass and Scott Zimmerman from Wilson Sonsini Goodrich and Rosati ("Wilson Sonsini") for spreading our academic stint to the West coast. They hosted Santosh twice at the Law School at University of California, Berkley, where they teach a law course on Renewable Energy Project Development and Finance. Their guidance was instrumental in developing the course outline and reading materials for our course at BC.

The finance practice is incomplete without the legal analysis and documentation. Santosh is fortunate to have had outstanding support from Sean Moran and Michael Joyce from Wilson Sonsini since 2013. These two provided the legal term sheets included in the appendix. Sean has been at the forefront of developing the tax equity structure for renewable energy projects since the early 2000s. Most of the tax analysis and structuring discussed in the book Santosh learned by working through various transactions. Lauren Collins provided outstanding expertise to review Chapter 6 from a tax law perspective and provided helpful case law research.

Norton Rose Fulbright has always been proactive in disseminating the latest developments in the renewable energy industry. The Project Finance Newswire published by the firm is a valuable first stop for detailed analyses of legal developments influencing the industry. Keith Martin deserves a special mention for advancing the industry in the right direction through advocacy and education. His colleagues – Michael Masri (currently at Orrick), David Burton, Roger Eberhardt, and John Marciano (currently at Akin Gump) – supplied materials that were very helpful in designing the course and this book.

Tax equity modeling is highly complex. We were fortunate that Rubiao Song from JP Morgan readily shared a simplified spreadsheet model to be used in our class. While Santosh developed most of the book's spreadsheet examples, Rubiao's spreadsheet example provided a reliable resource to ensure the calculations were consistent with the industry practice. Dennis Moritz from Advantage for Analysts provided an additional layer of scrutiny of Chapter 6.

Our current and former employers who supported this project are also due some heartfelt thanks. This includes Jacob Rosenfeld and State Street for Santosh and his current group at Silverpeak. He would especially like to thank Colleen Floberg, Antonio Giustino, Erica Nangeroni, Harshal Mohile, Michael Alexander, and Winston Chen who contributed in multiple ways. Erica was very helpful with her research assistance on energy storage. We would both like to thank Alex Margolick for his substantial editing and organizational efforts. Santosh would like to thank his family for outstanding support during his time away from home.

Seabron would like to thank his colleagues at Charles River Associates, especially Chris Russo, Derya Eyilmaz, and Billy Muttiah. Please note that this book reflects the personal views of the authors and does not reflect the views of CRA, its employees, or its clients. He would also like to thank Karsten Neuhoff and Nils May of DIW Berlin for their helpful suggestions on the German experience, and Mark Russell for his help on graphics. The discussion of energy markets and transmission risks has benefitted from conversations over the years with many people in industry and academia, but especially with Richard Tabors, who helped develop the theory of spot pricing and who started Seabron's work on energy and climate issues while still a graduate student at MIT. As always, he thanks his wife Ali for her support.

We thank the Lawrence Livermore National Laboratory, DSIRE, PJM, and Bridge to India Energy Private Limited for granting permission to use their graphics in the book.

Finally, we would like to thank Elsevier and its team, including Tommy Doyle, Michael Lutz, J. Scott Bentley, Indhumathi Mani, Selvaraj Raviraj, and Joseph Hayton. They have provided valuable guidance and support through the publication process.

Financing the new energy economy

1

For most of history, almost all energy used by humans was renewable. The iconic windmills of the Netherlands date as far back as 1200 AD, and water and wind driven mills were used before that era to grind grain. As late as 1800, almost all of the energy consumed in Paris – then one of the largest and richest cities in the world – came from firewood and charcoal, although by the end of the 18th century firewood had to be transported over 100 miles on average to the French capital to meet the demands of a growing urban population.[1] Only around 1830 would coal, the first fossil fuel to play a major role in the global energy economy, start to replace wood as the dominant energy source for Paris, London, and the other great European cities.

The evolution of renewable energy

The renewable energy derived from flowing water powered the early mills of the Industrial Revolution, but the growth of the modern world is heavily linked with the use of fossil fuels. Oil, coal, and natural gas power much of the man-made world around us, and are indirectly embodied in almost every product and service we consume.

In the United States and much of the rest of the developed world, the availability of cheap oil and other fossil fuels was taken largely for granted until the oil embargo of 1973, in which some members of the Organization of Petroleum Exporting Countries (OPEC) imposed an embargo on sales of oil to the United States. When combined with other supply and economic factors, the price per barrel of oil quadrupled over time after the embargo, bringing to the wider public recognition of the growing Western reliance on imported oil, and stimulating the belief among many that oil was a finite resource which could be depleted in a relatively short period.

The United States responded over the rest of the 1970s with a mixture of policies, from encouraging greater domestic production to conservation. The potential finiteness of reserves of oil and other fossil fuel resources became engrained in the minds of political leaders and the public imagination. In April 1977, US President Jimmy Carter addressed the nation on the need for a comprehensive national energy policy, with a strong focus on conservation. Carter famously referred to the energy crisis as the "moral equivalent of war," and stated that "because we are now running out of gas and oil, we must prepare quickly for a third change, to strict conservation and the use of permanent renewable energy sources, such as solar power."

[1] Kim, E., & Barles, S. (2012). The energy consumption of Paris and its supply areas from the eighteenth century to the present. *Regional Environmental Change*, 12(2), 295–310. http://doi.org/10.1007/s10113-011-0275-0.

Renewable Energy Finance. https://doi.org/10.1016/B978-0-12-816441-9.00001-5

Carter's initial energy policy speech was criticized as being short on specific policy details. One of the concrete policies to emerge was the *Public Utilities Regulatory Policy Act of 1978* (PURPA). In the first major stimulus for renewable energy in the modern era, PURPA, among its other provisions, required electric utilities to purchase energy from certain "qualifying facilities", which included renewables. The PURPA era brought forth the first major wave of new renewable energy projects in the United States.

Other countries also responded to the energy crisis. Energy research and development expenditures rose in the European Union in the late 1970s and early 1980s but various major European countries responded differently to the energy shocks. France concentrated on its nuclear industry. Germany, which had a large coal industry, was less dependent on imported oil for its total energy consumption, but also became after the 1986 Chernobyl accident more skeptical of nuclear power. Green ideas had taken hold in Germany even in the 1970s, and these provided an initial basis of public support for renewable energy research and development. Denmark, with few energy resources of its own, was highly dependent on imported oil. In 1976 two reports from the Danish Academy of Technical Sciences played a major role in shaping Danish renewable energy policy and the country's support of wind energy, including tax credits and investment subsidies.[2]

Renewable energy in the global energy economy

Despite these developments, the share of renewable energy in total energy consumption has generally remained low in most countries, except for a few countries with abundant hydroelectric resources such as Norway.

At the global scale, renewable energy resources made up less than a fifth of all energy consumption in 2017, as shown in Fig. 1.1.[3] Fossil fuels accounted for almost

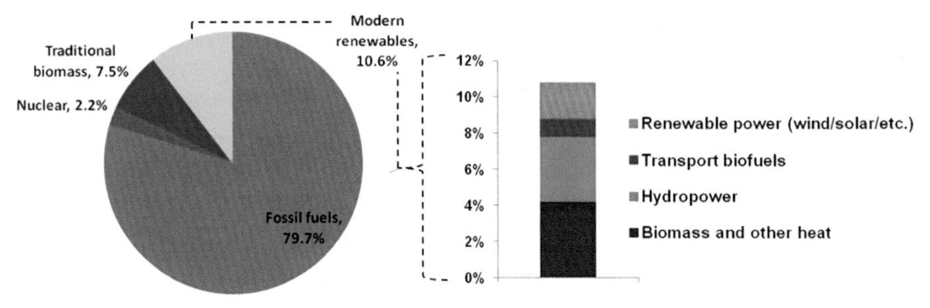

Fig. 1.1 Global share of renewable energy in final energy consumption.
Source: REN21 Data.

[2] Aklin, M., & Urpelainen, J. (2018). *Renewables: the politics of a global energy transition*. Cambridge, MA: The MIT Press.

[3] Data from REN21, *Global status report 2019*, 2019. Available from www.ren21.net.

80% of total final global energy consumption in 2017. Another 7.5% came from traditional biomass sources (firewood, charcoal, etc.), which remain important energy sources for many people in some developing countries. The "modern" renewable energy sector accounted for only 10.6% of total final energy, and of this fraction only 2% of total final energy came from renewable power generation such as wind farms, solar photovoltaic (PV) and other sources.

A relatively low share for these renewable energy sources also holds true in the United States. While wind and solar power have made significant gains in the United States, it has grown from a low base. Fig. 1.2 illustrates the sources and uses of energy in the United States in 2018, based on analysis done by the Lawrence Livermore National Laboratory.[4] The left-hand side shows sources of energy, and total production in "quads" in the year.[5] On the right-hand side are consuming sectors (residential, commercial, industrial, and transportation). The electricity sector is an intermediate sector - other primary energy sources are used to generate electricity - but final consumption of electricity is in the four final sectors. Similar "Sankey" diagrams are available for other countries — we focus here on the US only as one example.[6]

A brief review of this annual energy flow diagram illustrates some important aspects of US energy consumption. First and foremost, only just over a third of energy used actually provides useful energy services to customers and businesses — most of the energy produced is lost as "rejected energy" or waste heat. For example, consider the transportation sector, shown in the lower right-hand box. The transportation sector used 28.3 quads of energy inputs in 2016, but only achieved 5.95 quads of useful services (moving cars and trucks down the road or planes through the sky, for example). The rest was dissipated in heat, and reflects the relatively low thermal efficiency of combustion engines, aircraft turbines, and other transport power sources.

Second, it is easy to see that the transportation sector is one of the largest end-use segments, and is almost entirely based on petroleum at present, with small amounts of biofuels (chiefly from ethanol and biodiesel), natural gas, and (minimal) electricity used.

Third, the largest total user of energy is the electric generation sector, which used approximately 38% of total energy. Over half of the input energy for generating electricity came from coal and natural gas fossil fuels, with nuclear being the next largest contributor. Despite their rapid recent growth, wind, geothermal, and solar energy made up only about 3% of the total energy mix in the United States. Renewable hydro

[4] Graphic used by permission of Lawrence Livermore National Laboratory. Available from https://flowcharts.llnl.gov/.

[5] A "quad" represents one quadrillion British Thermal Units and is a unit used in large scale energy analyses. To give some perspective, one quad of crude oil will fill about 80 supertankers. Note that total U.S. consumption in the year was approximately 101.2 quads, so the figures in the boxes are close to a percentage value as well. The Appendix includes a full discussion of the energy units and conversions useful in energy project finance.

[6] The International Energy Agency has an online tool for generating these diagrams for various countries and regions at https://www.iea.org/sankey/

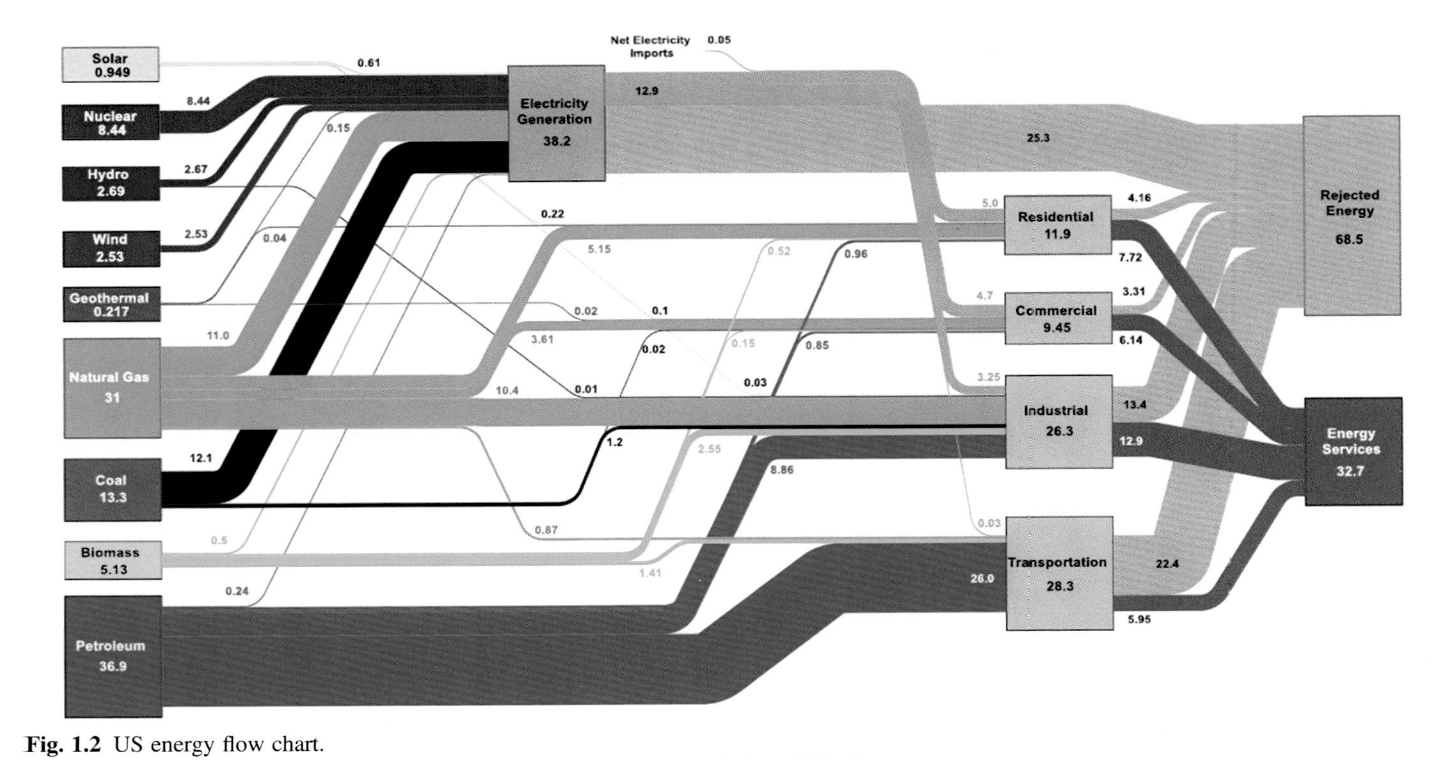

Fig. 1.2 US energy flow chart.
Source: Lawrence Livermore National Laboratory. Reproduced by permission of LLNL.

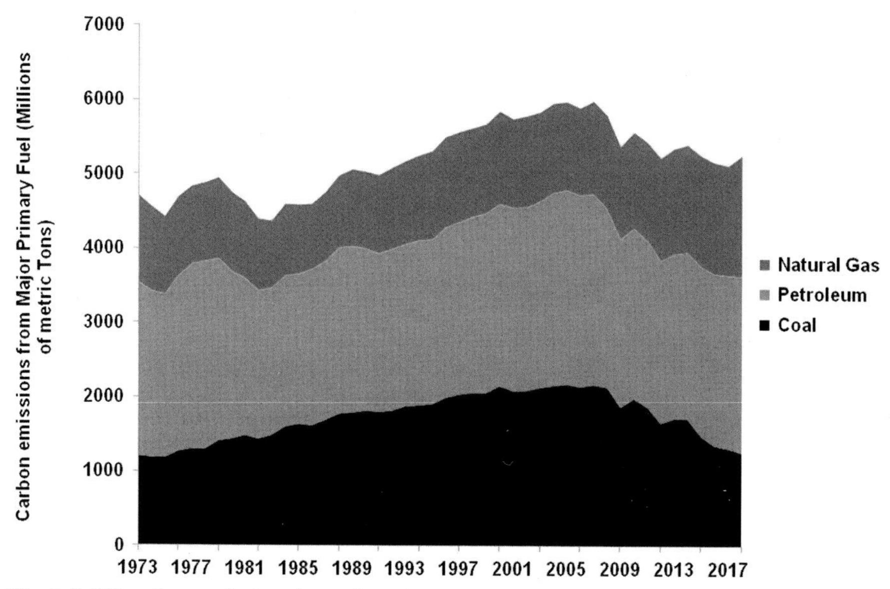

Fig. 1.3 US carbon emissions by major primary fuel.
Source: EIA Data.

power from dams made up another 2.5%, but few expect that percentage to grow in the future due to the lack of new sites on which to build large dams.

While in the 1970s many feared the exhaustion of fossil fuel resources, a primary focus of renewable energy policy now is the reduction of carbon emissions and other pollutants while maintaining similar standards of energy production.

Fig. 1.3 shows trends in carbon emissions from the US energy sector by primary fuel (coal, petroleum and natural gas).[7] Not all fossil fuels are the same in terms of carbon intensity. Of the major fuels, coal has the highest carbon content per unit of energy, while natural gas (which is primarily methane) has the lowest. Emissions peaked in the mid-2000s, and declined with lower coal use. Recently, emissions have started to creep up again, driven substantially by the transportation sector.

There are numerous energy and environmental policy studies examining future scenarios for transitioning to a low-carbon energy economy. While a full exploration of this complex topic is outside the scope of this book, analysis of a few common threads gives insight into the likely needs for new renewable energy projects:

- Most scenarios share the deep decarbonization of the power generation sector, switching from coal and other fossil fuels to renewables such as wind and solar, often coupled with storage. Most recent US analyses focus on expanding renewables over other technologies such as new nuclear or carbon sequestration, based on current economics.
- As seen in Fig. 1.2, transportation is one of the largest end-use sectors. Many analysts predict the large-scale conversion of much of the ground transportation fleet of cars and trucks to

[7] Data from U.S. Energy Information Administration. Available from www.eia.gov.

plug-in electric vehicles. This would lower direct carbon emissions, but would in turn raise electricity demand, further increasing the need for new renewable power generation.
- Finally, the industrial sector, which currently uses large amounts of natural gas and oil, may also need to be significantly electrified.

These lessons are not limited to the United States. A key finding across a wide range of future global energy scenarios is that demand for low-carbon renewable energy, such as wind and solar, will likely increase dramatically in coming years. This will require tremendous capital investment.

Financing a low carbon energy future

The energy sector is highly capital intensive, and the capital stock often lasts for decades. Global investment in renewable energy has already increased substantially, as shown in Fig. 1.4.[8] Solar and wind projects dominate the current investment landscape.[9] The largest markets for renewable investment in recent years have been China, Western Europe, and North America.

While current global investments are large, the scale of renewable energy investment needed to meet long-term carbon emissions targets is much larger.

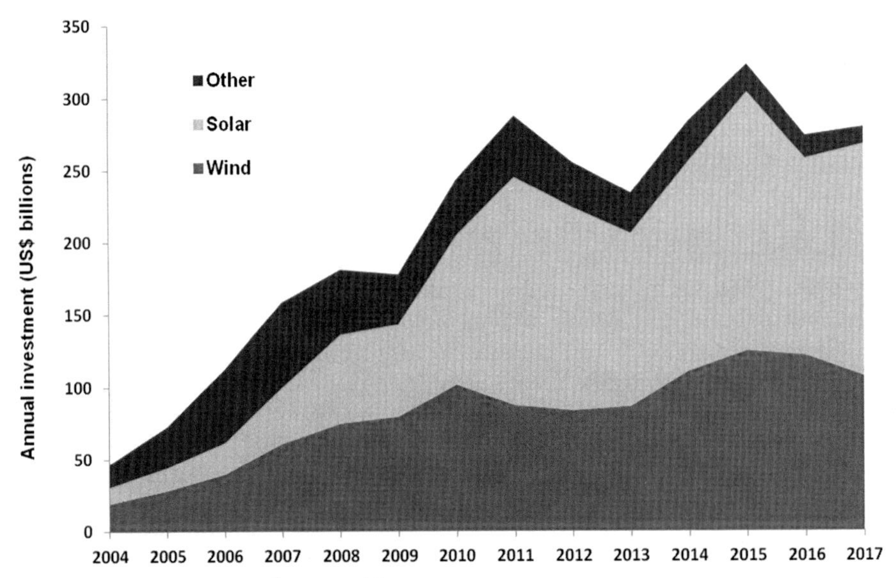

Fig. 1.4 Global investment in renewable energy.
Data from IRENA.

[8] Data from IRENA, 2019. Available from www.irena.org.

[9] The "Other" category in Fig. 1.4 includes liquid biofuels, biomass and waste-to energy projects, small hydro, geothermal and ocean energy, and other related investments.

Most scenarios aiming at keeping warming at 1.5° Celsius (the target set in the Paris Agreement) would require very substantial cuts in emissions from the power sector. The International Renewable Energy Agency (IRENA) estimates that the share of renewables would need increase to 65% of global energy supply by 2050 – more than triple the current level – to meet the Paris targets.[10] As will be discussed in Chapter 13, some analysts suggest that capital expenditure of over $2 trillion per year for several decades would be required to meet even a more modest 2° Celsius target. A significant fraction of that investment would be directly in renewable energy generation projects.

The role of project finance

In most market economies, which make up a large share of total global output, new energy infrastructure is financed primarily by the private sector, and not directly by government. Private capital finances over 90% of new renewable energy investment globally.[11] To make the investments in green energy many anticipate are needed on a global scale will require harnessing the capital markets to raise the trillions of dollars required.

A critical method for getting renewable energy infrastructure built is project finance – that is, finance backed not by the balance sheet of a corporate investor but rather by the expected operating cash flows from the project itself. Substantial experience has been built up in adapting project finance structures to the challenges posed by renewable energy, such as the variability of output from one period to the next given the unpredictability of wind and solar resources.

Two core themes are woven throughout the successive chapters of this book. First, project financed renewable projects are not standalone investments, but are deeply embedded in local support policies, legal and energy market structures. To understand how project finance works for a large renewable asset requires understanding these contexts. For this reason, this book touches on a wide set of topics not usually found in a finance book such as public policy, law, economics and tax regulations.

Second, the project finance structures used for renewable energy projects must manage the complex risks associated with large sunk cost investments which will be generating power for decades to come. A financing is thus supported by a set of contracts that shift risks between the project participants in an acceptable way, to ensure a bankable project. The sponsor and the lenders are just two of the parties needed to reach financial close, and each party has its own interests and risk tolerances to consider.

Organization of this book

The focus of this book is on project finance in the renewable energy sector, and subsequent sections will describe the detailed mechanisms of project finance and how they

[10] International Renewable Energy Agency (IRENA), *Renewable energy: A key climate solution, 2017.* Available from www.irena.org.

[11] IRENA, *Global Landscape of Renewable Energy Finance*, 2010. Available from www.irena.org.

have been adapted to the needs of the green energy sector. Given the nature of the subject, the text is necessarily interdisciplinary in nature.

The organization of the book is structured around the concept of flexibility. Chapter 2 describes the renewable support policies that have underpinned most renewable project development around the world. As such, it provides some key background for the material to follow. The subsequent three chapters (Chapters 3 to 5) make up a core unit that describes the covers the economics of renewable energy projects, the basics of renewable project finance, and how financing and contractual structures allocate risk. While not every structure can be discussed, the emphasis is on providing the reader with a core understanding of the basic concepts to help with the analysis of any future financing.

Chapters 6 and 7 make up a second unit, focusing on more specialized project finance structures. This includes the tax equity structures commonly used in the United States and the structures commonly used in financing distributed generation assets, such as rooftop solar.

Chapters 8 to 10 make up a third unit, focusing on how renewable energy projects function in the context of electricity markets. These chapters discuss managing transmission costs and risks for projects, and a few of the alternative off-take structures used for managing merchant price risks in these markets.

Chapters 11 and 12 explore the valuation of renewable energy projects and development companies, and the issues associated with financing energy storage assets, a rapidly growing class of projects closely associated with renewable energy.

The extended final chapter describes the growth in global project finance for renewables, and examines the lessons from four case study countries: Chile, Germany, India, and China.

Public policy mechanisms to support renewable energy

2

Large scale renewable energy projects such as wind farms and utility-scale photovoltaic facilities require large capital investments. To make such investments feasible in a project finance context, investors need to know that the price received for the power generated on average will be high enough to make the investment economic, and that revenues will be stable enough to allow the project to cover its operating and interest costs and debt repayment on schedule. As described in Chapter 4, large scale renewable projects often have **off-take arrangements,** such as **power purchase agreements** (PPAs) which fix the price of energy sold for a considerable period. From a public policy perspective, policy mechanisms to support renewable energy have been required, as renewable energy has generally been more expensive than fossil fuel-derived sources. Without these policy supports, the development of renewable energy sources would have been much slower or impossible.

This may be changing – at least in some jurisdictions. Renewable energy costs have been falling rapidly and in some cases are approaching or are below the costs of fossil fuel generation, such as gas-fired generation. In 2018, for example, the Dutch government announced that two large offshore wind farms would be built in the Netherlands with no direct subsidy, although the government did support the grid connection.[1] Unsubsidized projects have also been announced in other countries. If costs continue to fall, renewable electricity generation may become cheaper than fossil generation in a wide variety of cases.

Support renewable energy or tax conventional energy?

Renewable energy support mechanisms exist to support policy goals such as avoiding carbon emissions derived from electricity generation. From an economic perspective, the problem is often not that renewable energy is too expensive, but rather that fossil-fired generation such as coal and gas is too cheap. That is, the cost to customers of generating power reflects only some of the costs of production (the *private costs*), and does not reflect the full *social costs* of providing this power, which should, even in theory, incorporate the damages created by emissions of carbon dioxide and other pollutants on surrounding stakeholders.

Pollution is a classic case of an externality – a cost to others not making the economic decision. The owners of a factory emitting a pollutant, for example, might see only a small impact on their own wellbeing, but their pollution could have large

[1] Wentworth, A. (March 20, 2018). *Subsidy-free offshore wind farm to be built in the Netherlands*. Retrieved from http://www.climateaction.org/news/subsidy-free-offshore-wind-farm-to-be-built-in-the-netherlands.

Renewable Energy Finance. https://doi.org/10.1016/B978-0-[illegible]

impacts on the local inhabitants. Since the factory owners see only a small proportion of the total consequences of their emissions decisions, their decisions will be inefficient from the perspective of society as a whole. Economists generally see externalities as a classic form of "market failure" in which the welfare-maximizing properties of the "invisible hand" of the free market may fail to hold.[2]

Fig. 2.1 represents the externality problem in a simple economic diagram. The production of electricity has a rising (private) marginal production cost curve as shown by the upward sloping solid blue line. The more electricity is required, the higher the marginal cost to generate the last unit. These are private costs to the generators, and do not reflect any costs imposed by their emissions on others (these are referred to as "external costs").

Electricity consumers have a declining demand curve, shown as the downward sloping red line. The demand represents the marginal benefits consumers see for using an additional unit of electricity. In the absence of external costs, the output of electricity at Q at price P (where marginal cost equals the marginal benefit of consumers) would be the socially optimal production level. Producing more or less than Q would in general make society worse off.

However, consider a situation where external pollution costs exist. In this case, the **marginal private cost** (MPC) and the **marginal social costs** (MSC) differ, due to the costs (such as health and environmental impacts) that production of electricity imposes on everyone else. Marginal social costs are substantially higher than marginal private costs. In Fig. 2.1, the marginal social costs of generating electricity are shown by the upper dashed blue line. This line is the sum of the increasing marginal private cost of generating electricity, plus the external cost created by carbon pollution (fixed at a constant rate in this stylized example).

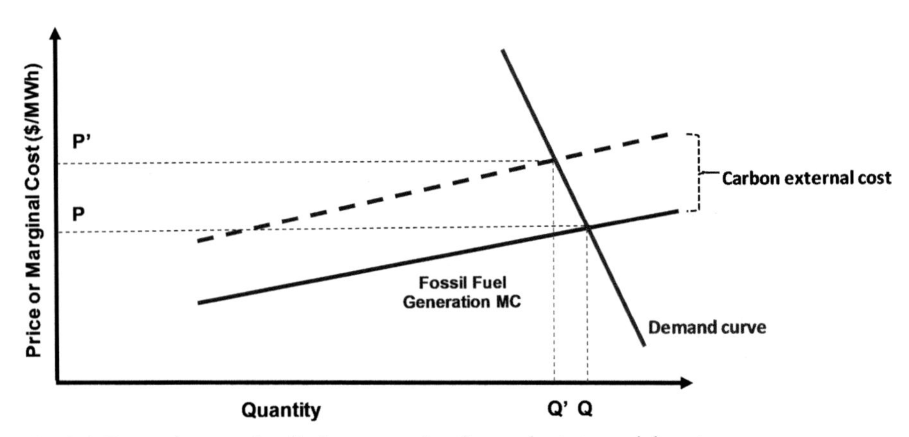

Fig. 2.1 External costs of pollution – moving from private to social costs.

[2] Romer, D. (2012). *Advanced macroeconomics*. New York. McGraw Hill/Irwin.

Where there are external costs (from carbon pollution for example), matching supply and demand at the private costs of supply is inefficient, as fossil generators are not including the costs they impose on others in their production decision. The socially efficient outcome is found at Q', where the marginal social cost curve intersects the marginal benefit curves. At this point of equilibrium, as shown in the diagram, there is still production of electricity (although less by the amount of $Q - Q'$) and still some carbon pollution. The resulting market price (P') is also higher, reflecting the addition of the marginal costs of pollution into the market outcome.

Carbon taxes

If market prices for energy do not reflect the total costs (e.g., including environmental impact costs), then an obvious policy mechanism is to correct the prices. In so doing, renewable and other clean energy resources should also be better able to compete in the market.

Many economists favor a system of carbon taxes as a public policy mechanism to control carbon and other greenhouse gas (GHG) emissions. Under an optimal carbon tax system, a tax is levied on emissions of carbon that is equal to the marginal incremental social costs of those emissions (for example, the marginal costs imposed on everyone else through climate change). In this way, energy prices can reflect not only the marginal private cost (MPC) in Fig. 2.1, but the marginal total social cost (MSC). Carbon emissions can be reduced to an efficient level, and the rise in energy prices also provides better incentives for consumption.[3]

Fig. 2.2 illustrates how carbon pricing can help support the development of renewable energy. The horizontal axis shows the quantity of energy supplied in a period

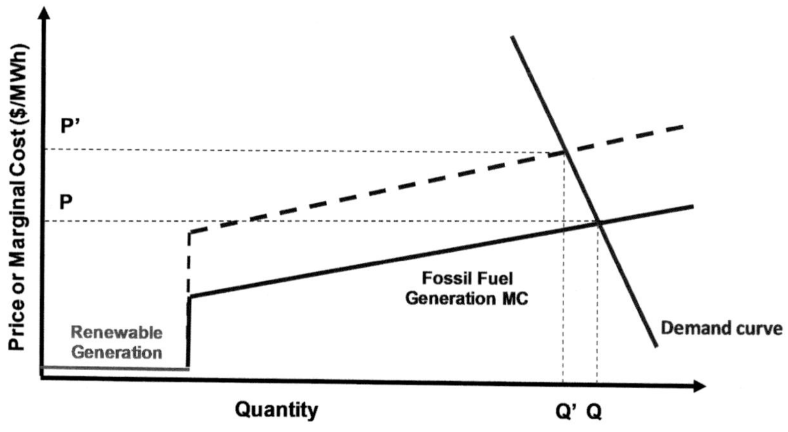

Fig. 2.2 Carbon pricing can help support renewables.

[3] Such a system of taxes was proposed by the economist A.C. Pigou as early as the 1920s. See Pigou, A. C. (2013). *The economics of welfare*. Palgrave Macmillan (Palgrave Classics in Economics edition published 2013).

(e.g., 1 hour). The vertical axis represents the marginal cost of supply. Renewable resources such as wind and solar often have high capital costs, as they are expensive to build initially, but very low marginal costs (no fuel costs). Conventional fossil generation such as gas and coal-fired plants have higher marginal costs, even without considering their environmental impacts, as they have to pay for fuel. The solid blue supply curve (marginal private cost curve) is thus at almost zero for the segment that can be supplied solely from renewable generation, and then increases with output at higher levels. The demand curve for electricity is again illustrated as the downward sloping red line, which shows that consumers want to buy less electricity at higher prices. Without a carbon tax applied, the market clears where the supply curve and the demand curve intersect. This gives an equilibrium quantity of Q at a price of P, as before. However, as was illustrated in the previous figure, this is not a socially optimal outcome as the costs of carbon pollution from fossil-fuel generation must also be considered.

When a carbon tax is imposed on emissions (here shown as equal to the external cost in the previous example), it raises the marginal cost of supply electricity from fossil fuel sources, raising the marginal supply curve to the upper blue dashed line. At this price, slightly less power will be demanded, but the market equilibrium price rises from P to P', as shown in Fig. 2.2. This increases margins on sales for renewable generators – the market price is higher and their costs have not gone up, as their generation does not create carbon emissions. Economically, a perfect carbon tax could eliminate the need for any renewable energy support mechanisms limiting carbon emissions, as energy producers themselves would have the incentive to make the right production decisions. Energy consumers would also see "correct" prices, and hence carbon reductions could be made at lowest total cost.

Carbon taxes on fossil fuels are relatively easy to implement in theory, as the carbon content of natural, gas, oil, and other fuels is well known. Implementing the carbon tax therefore implies levying a tax on fossil fuels at their point of production or import. Since many fuels (such as gasoline) are already taxed, there is often already an existing administrative mechanism for collecting such taxes.

Tradable permit systems

Carbon taxes are simple in concept, but it is not so simple to set the level of the tax. In theory, a tax to control carbon emissions optimally should set the tax equal to the marginal damages such emissions create, but these are not easy to estimate.[4] In practice, policymakers may be left trying to guess what quantity of emissions will result from a given carbon tax.

In economics, prices and quantities are, in some ways, different sides of the same coin – changes in prices affect market quantities, and vice versa. Under a **tradable permit system**, instead of imposing a fixed additional cost on emissions, a fixed

[4] Metcalf, G., & Weisbach, D. (2009). The design of a carbon tax. *Harvard Environmental Law Review, 33*, 499–556.

number of permits is allocated, each allowing the emission of a fixed amount of carbon (e.g., one permit is needed for one ton of carbon emitted). An important feature of these emissions permits is that they can be traded in a market – so a permit holder can either use their permit in order to emit a ton of carbon, or sell it to someone else. In this way, a market price for carbon emissions can emerge which reflects the overall cap on emissions set by regulation (these policies are often also referred to as "cap-and-trade" measures). Under some limiting economic assumptions, the market price for permits (like in a carbon tax) provides incentives to emit less carbon at the lowest total social cost.

Tradable permit systems have a fairly long history for sulfur dioxide and other acid rain pollutants in the United States, and a number of New England and Mid-Atlantic states have banded together to implement such a system for the power sector under the Regional Greenhouse Gas Initiative (RGGI).[5] The best-known tradable permit system for carbon operates in the European Union. The EU Emissions Trading System (ETS) implements a cap on greenhouse gases (including carbon) covered by power plants and some other industrial facilities.[6]

One problem for tradable permit systems is that prices can be quite volatile, especially when government policies change or other factors impact the value of carbon permits. This in turn can reduce the willingness of investors to rely on tradable permit prices to support their investments. For example, Fig. 2.3 shows ETS carbon prices

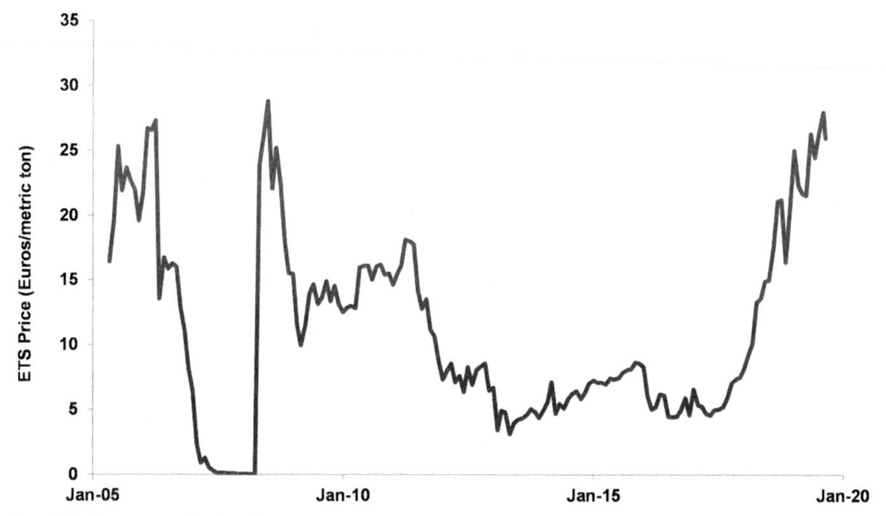

Fig. 2.3 European ETS prices.

[5] For more information see www.rggi.org.

[6] For more information on the EU ETS see the description available at https://ec.europa.eu/clima/policies/ets_en

from April 2005 to August 2019.[7] Prices were heavily affected by allocation issues in the early period, which drove the price to almost zero in 2007. In later years ETS prices have again been quite low. This is at least partially explained by the existence of other policies (discussed later) to support low carbon renewable energy in Europe and excess permits accumulating during the economic crisis. Prices have again risen in 2019, after implementation of a market stability reserve plan in January 2019. If high ETS prices continue in Phase 4 of the ETS (covering 2021–30) this could be a positive for new renewable projects in Europe over this period.

Carbon tax and tradable permit systems have many attractions in theory, but have not been widely used in many countries (including the United States at a national level). Price and policy stability, as noted above, has also been an issue. Imposing new taxes has often proven politically difficult, and like all taxation systems the distributive effects on different groups of citizens can be quite large. So, while carbon taxes and related tradable permit systems remain widely discussed, most renewable energy development in practice has been supported by other mechanisms.

Direct subsidies

If renewable energy is too costly in the market in comparison with conventional sources, one potential policy solution is to directly subsidize investment in the sector. While in theory this could involve direct government transfers, in the United States this has often been accomplished through the tax system. While the impact may be the same, tax credits are less likely to attract political opposition than a direct subsidy, which would require federal appropriations each year.

Two forms of tax subsidy have commonly been employed in the United States. The production tax credit (**PTC**) provides an inflation-adjusted corporate tax credit for each kilowatt-hour (kWh) of electricity produced from qualifying facilities, including wind and geothermal. The PTC is paid in addition to any other revenues received for the power sold. The tax credit lasts for the first 10 years of facility operation. In 2016 Congress mandated that the PTC will be gradually be phased out.

The investment tax credit (**ITC**) is typically used for solar facilities and may be used in lieu of the PTC. The ITC provides a tax credit of up to 30% of the installed costs for certain technologies including solar and small wind turbines. While a PTC is paid out over time, based on production, the ITC is a one-off tax credit, paid when the facility goes into service.

While the implementation of the PTC and ITC are complex, the tax credits are in effect economically similar from the facility owner's perspective. As will be discussed in more detail in Chapter 6, to receive the tax credit, the corporate owner has to own equity in the facility itself, a legal restriction that has created a class of "tax equity" investors and complex financial structures to ensure that the tax incentives are efficiently used within the project finance structure. These tax subsidies have often

[7] Prices shown in Fig. 2.2 are for monthly EUA futures. Data from Bloomberg.

been combined with other policy support mechanisms (discussed in the next section) to finance renewable energy projects in the United States.

In terms of creating good incentives for investment in renewable energy, the design of the ITC and PTC both leave something to be desired. The ITC rewards investment, which may not directly reflect renewable energy produced or emission avoided. In fact, the ITC may encourage over-investment as the more costs claimed, the higher the tax credit received.

The design of the PTC does encourage greater renewable energy production, since the more energy generated, the larger the tax credit received. However, the PTC has the weakness that the tax credit is the same regardless of location or time of generation – no incentive is provided to generate power where or when it is most needed or where renewable energy can best avoid emissions from conventional fossil fuel generation sources. Nor do the PTC or ITC subsidies provide any incentive for consumers to use electricity more efficiently.

Quantity-based mechanisms

One way to support renewable energy has been to require some proportion of total consumption to be purchased from qualifying renewable suppliers. This purchasing is usually done by the utilities or other load-serving entities (such as electricity retailers, where these exist). In the US, other than the federal tax credits discussed previously, the most common support mechanism used in the United States has been renewable portfolio standards (**RPS**). There is no federal RPS mechanism, but many states have their own standards.

An RPS is effectively a quantity-based mechanism, in which the utilities or retailers of electricity are required to supply a defined percentage of their customer supply from qualifying renewable energy resources. For example, under a 30% RPS in the state, a local utility would have to generate or purchase 300 megawatt hours (MWh) of renewable generation for every 1000 MWh sold to its customers. Not all states have an RPS, as illustrated in Fig. 2.4.[8] To add to the complexity, many US states have more than one "class" of RPS, with different requirements for different types of renewable generation (e.g. solar, offshore wind). The result is a complex patchwork of policies across the different states. Several states have announced plans to shift to 100% renewable electricity generation.

Many RPS rules include a renewable energy credit (**REC**) system. A REC is an attribute correlated with one unit of renewable energy generation (such as one MWh of wind power) in a location (e.g. New Jersey). A utility which generates or purchases more renewable energy than it needs to meet its RPS requirement can sell the associated RECs to another utility that is short against its requirement. This is designed to allow utilities to more easily and cheaply meet their RPS requirements by creating a market for specifically-qualifying renewable energy. Outside of the United States,

[8] Information shown in Fig. 2.4 from NCSL and state websites.

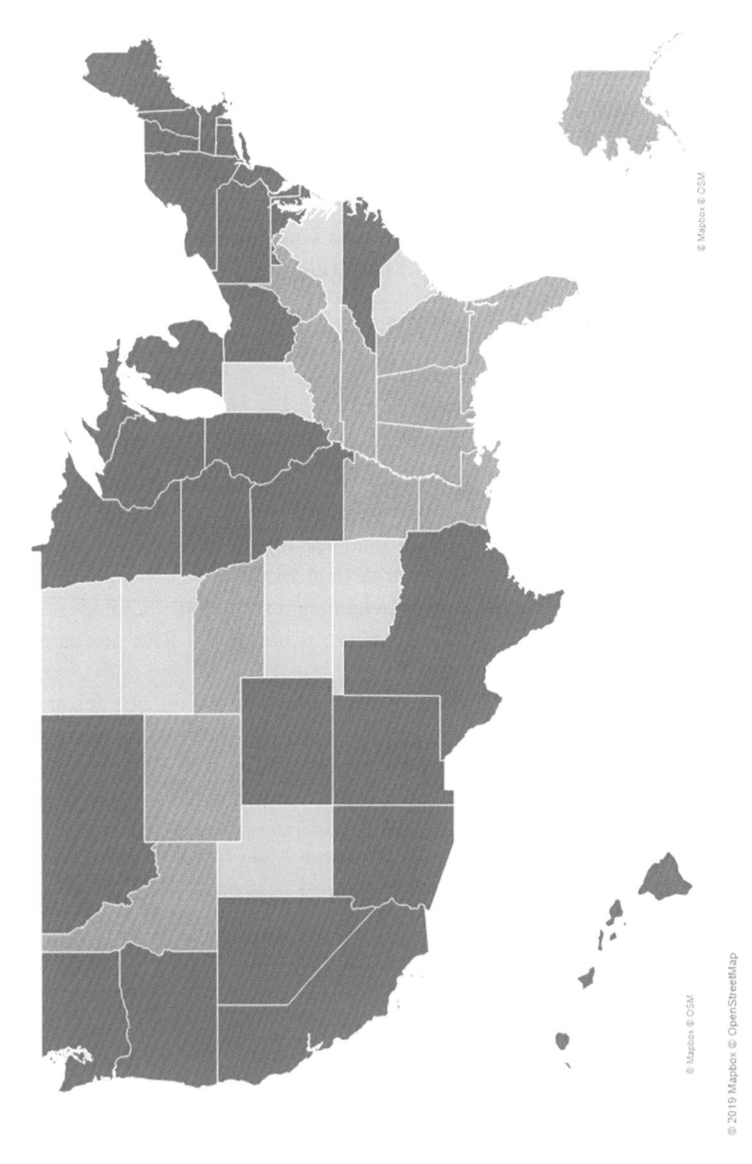

Fig. 2.4 Renewable portfolio standards (dark blue) and voluntary goals (light blue) by state.

policies of this type are often referred to as tradable renewable energy credit (TREC) schemes.

Like tradable permit systems (to which they are economically similar mechanisms), REC prices can be highly volatile, dependent as they are on the supply and demand of renewable energy generation in narrow (often state-level) markets. For example, from 2012 to 2019 the price of compliance RECs (those needed to comply with state-level RPS rules) has varied in some states by a factor of ten or more.[9] REC prices are also heavily influenced by state policies, which have often shifted over time.

In practice, RECs are often sold bundled with the output energy under PPAs or other long-term contract, and many renewable energy projects will have little or no direct exposure to volatile REC prices. Despite these issues, RPS has been a major driver of investment in renewable energy in the United States. In practice, a substantial percentage of new renewable energy projects are still driven by RPS requirements, and backed by a PPA with a utility. Indirectly, RPS rules have therefore been a major driver of renewable project investment in the United States.

One major criticism of RPS rules is their complexity and instability. Since the rules are set by individual states, there is little scope for ensuring that renewable generation investment is made where it is most cost effective (solar power in sunny Arizona, for example, instead of New Jersey). The rules have also often been subject to change, which makes investing and developing new projects more difficult. Finally, the wide variety of state regulations may require project developers to plan projects across many states, and to track the local legislative and regulatory developments which could make an otherwise attractive project infeasible overnight.

Other types of fixed quantity mechanism are **renewable energy auctions or competitive tenders**. These typically take the form of a reverse auction, in which renewable developers bid to supply a quantity of qualifying renewable energy (for example, an amount per year of wind-generated electricity) at a specified price. There are numerous designs for these mechanisms.[10] Around 50 countries use these auctions to procure renewable energy, and use of these mechanisms around the world continues to grow in popularity. By defining the specific product to be procured, national regulators can flexibly define the quantity, type and characteristics of the renewable energy needed, to ensure that the renewable energy contracted meets local policy objectives.

Experience with these auction mechanisms in several countries (notably Chile, but also in Germany and India) is discussed in Chapter 13. Latin America has been a regional leader in the use of auctions to support renewables development, with both Chile and Brazil being noteworthy examples.

[9] O'Shaughnessy, E., Heeter, J., & Sauer, J. (2018). *Status and trends in the U.S. voluntary green power market (2017 data)*. National Renewable Energy Laboratory. Retrieved from https://www.nrel.gov/docs/fy19osti/72204.pdf.

[10] Hochberg, M. & Poudineh, R. (April 2018). *Renewable auction design in theory and practice: lessons from the experiences of Brazil and Mexico*. Oxford Institute for Energy Studies, OIES Paper EL 28.

Price-based mechanisms

Another way to create incentives for investment in renewable energy is to set a price for new renewable energy, separate (and usually higher) than the prevailing electricity price. While price-based mechanisms have not been widely used in the United States in recent years, **feed-in tariff** (FIT) programs have been widely used in many countries including Germany, Canada (especially Ontario), and China. The international experience with FIT programs in Germany is discussed in Chapter 13.

In simple terms, under a FIT program, the government sets a price under which it will purchase qualifying renewable energy, typically guaranteeing the price for a long period to allow investors to have some certainty of recovering their costs once they build a new project. The costs are usually recovered from grid charges or from other customer surcharges. FIT programs are popular with renewable energy project developers since they provide a sure market for new projects that can meet the price requirement.

FIT programs have created incentives for large investments in some jurisdictions, but are not without their critics. Governments may have difficulty in setting the tariff prices correctly: if the FIT price is too low no one will invest in new projects, and government policy objectives will remain unmet. Set the FIT too high and the number of new projects may be unexpectedly high, with large cost implications for consumers. In practice, some FIT systems have limited the amount of new capacity that can qualify by limiting interconnection into the transmission grid, making some projects impractical.

A variation on the FIT mechanism is a **feed-in premium** (FIP). Under a FIT, the price received by the project is independent of the electricity market price – similar to a long-term contract. Under a FIP, the project receives a premium to the prevailing market price for electricity. This gives project developers some incentive, for example, to create projects where they are most needed by location, or that can generate the most during peak demand periods. FIP mechanisms can be designed using a fixed premium or a sliding scale which varies with the underlying price. Various caps and floors can also be incorporated into these mechanisms to help projects manage risks which could otherwise affect investment. Like FIT programs, FIPs have also been widely used in Europe.

Impact of public policies on renewable energy finance

Investments in renewable energy projects are risky: much of the costs have to be sunk before a single unit of energy is generated, and once built, the facilities and equipment cannot be easily redeployed. Given these risks, renewable energy project finance has largely been built on a base of long-term fixed price contracts. These may be in the form of PPAs between the project company and a creditworthy off-taker such as a utility, or long-term fixed payment mechanisms such as FIT or auction mechanisms paid for by a government agency.

What has been important in almost every jurisdiction is that a locked-in price for generated renewable energy is paid for a period long enough for the project to have a reasonable expectation of earning a return on its upfront capital investment. While mechanisms have varied significantly over countries and time, the need for off-take price stability at a reasonable rate is a common feature of renewable energy project finance in general.

Basic project finance concepts

3

Project finance structures are widely used inside and outside the renewable energy sector. Before exploring the specific application to renewable energy projects, this chapter introduces the basic concepts of project finance and how it is used.

Project finance defined

Project finance may be defined as "the financing of one or more projects on a non-recourse basis through a specially-created single purpose entity." The definition highlights two important characteristics that distinguish project finance from traditional corporate finance, the source of most corporate funding.

The first characteristic is the nature of the recourse available to the lender to secure payment. In simple terms, from a lender's perspective, the recourse defines what assets are available to back the loans. A recourse financing provides the lender(s) the right to make claims against the assets of the entity that provides the recourse. Recourse may be established either through the loan documents or through an explicit guarantee agreement. Thus, in the event a borrower cannot meet its obligations under the financing contracts, a lender can pursue its claim with the entity that provided the recourse.

A typical example of a recourse financing is one in which a corporation guarantees the obligations of its subsidiaries. This is sometimes the only way a corporate subsidiary can access funds from a financial institution at reasonable interest rates. It should be noted that the recourse provided under the financing contracts is often limited to an amount agreed upon in the loan documents. Recourse-based financings are widely used, but have the drawback that the amounts contractually guaranteed by the parent corporation are typically recorded as **contingent liabilities**, and therefore are reflected into the guarantor's (e.g., the parent corporation's) financial ratios. This can, for example, affect the parent corporation's ability to access capital in the future.

A non-recourse financing, on the other hand, prohibits the lenders from making claims against the project sponsor(s), and their recourse is limited to the specific entity they financed. This distinction is important, as a non-recourse financing essentially offers the project sponsor a "walk-away" option, since its liability in a failed project is limited to the equity it has contributed. Banks and other financing institutions, however, are exposed to more risk (relative to traditional corporate financing) because the repayment and recovery of the debt in the event of default are therefore limited to the assets of the specific project company they are financing. The lenders are thus

Renewable Energy Finance. https://doi.org/10.1016/B978-0-12-816441-9.00003-9

directly exposed to the economic risks of the project and cannot seek relief from the sponsor if the project does not perform as expected.[1]

The second characteristic that distinguishes project financing from traditional corporate financing is the existence of a company with the specific purpose of owning and operating the project. Project financing is geared toward a single project (or a portfolio of similar projects) with a finite useful life, a pre-specified purpose, and a well-defined operational plan. Consequently, these projects are typically owned by a special purpose company (SPC). For US-based projects, SPCs are typically organized as a limited liability company (LLC), but numerous variations exist around the world.

Historical development of project finance

Project finance has been used for centuries to finance a variety of projects. The first documented project finance transaction dates back to 1299 when the English Crown used a loan from the Italian merchant bank Frescobaldi to finance the exploration and development of the Devon silver mines.[2] In more recent history, project finance has been used in the development of the Panama Canal, North Sea oil field infrastructure, the Trans-Alaskan pipeline, and a range of other projects.

Today, project finance encompasses a broad range of sectors including not only energy but also other core infrastructure projects such as bridges, tunnels, and toll roads. Project finance has also been used in financing long-term health care facilities such as hospitals. Furthermore, technology and telecom companies have discovered project finance as a vehicle to finance data centers, cell phone towers, and other investments. Project finance as a discipline has developed into a variety of sub-categories such as infrastructure finance, public project finance, and others, each with their own eco-system of advisors, sponsors, consultants, and financial institutions.

Fig. 3.1 illustrates the annual global project finance volume between 2013 and 2016, and the breakdown by sources of capital (equity vs. debt, with debt divided into loans and bonds). As shown, project finance deals are on average highly leveraged or geared (with a high ratio of debt to total capital invested).

Fig. 3.2 depicts the breakdown of the project finance volume by sector for the years 2013 to 2016 for North America. A few trends can be easily observed in the two graphs:

[1] The recourse/non-recourse financing distinction should not be confused with consolidations from an accounting perspective. Special purpose companies (SPCs) created for project finance purposes need to be consolidated on the sponsor's balance sheet based on the variability of its interest and voting rights according to the so-called Variable Interest Entity (VIE) rules. For example, if a sponsor undertakes non-recourse financing and if the accounting rules determine that the sponsor's economic interest is "variable" enough to justify consolidation; the sponsor will need to consolidate the equity interest as per the rules.

[2] Kensinger, J. W., & Martin, J. D. (1988). Project finance: raising money the old-fashioned way. *Journal of Applied Corporate Finance, 1*(3), 69–81. https://doi.org/10.1111/j.1745-6622.1988.tb00474.x.

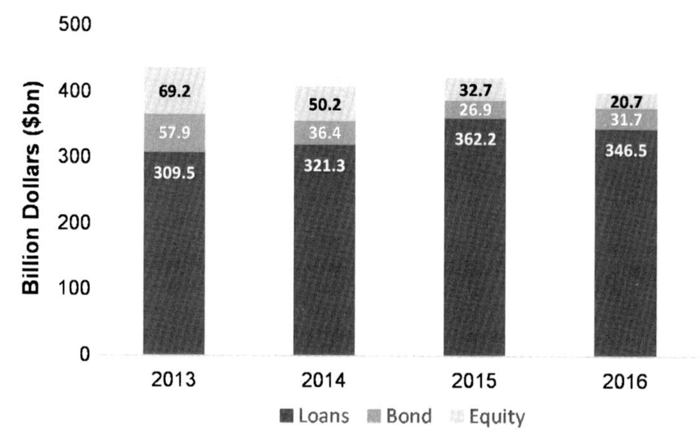

Fig. 3.1 Global project finance volume.
Source: Dealogic data.

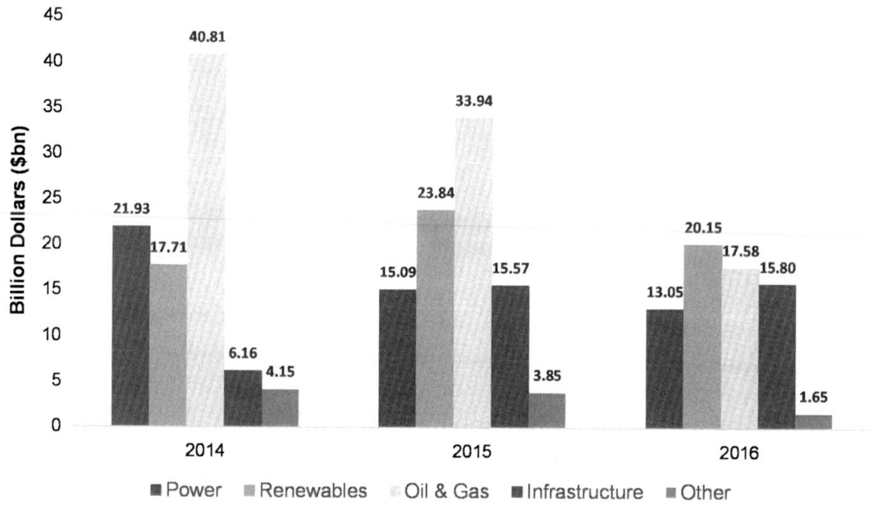

Fig. 3.2 North American project finance volume by sector.
Source: Dealogic data.

- Project finance transaction volume often exceeds $400 billion globally, making it an important component of global financial markets and a critical method of supporting capital investment.
- While the overall project finance transaction volume has been stable over the past few years, renewable energy projects form a larger share of the overall transaction volume, especially in North America.
- Bank loans are the primary capital source providing the bulk of financing. Overall, if consistent with the underlying project economics, the **advance rate** —or debt-to-total capitalization— for project finance transactions can exceed 85%.

Comparing project finance with traditional corporate finance

Due to the non-recourse nature of the financing, its limited purpose, and the useful life of the projects involved, project finance has complexities that are not usually encountered in traditional corporate financing. The project sponsors and financial institutions providing equity and debt financing are forced by the nature of the non-recourse structure into evaluating and assessing idiosyncratic risks associated with each project. Risk allocation in project finance will be discussed further in Chapter 5, but it is clear that the complex risk assessment and allocation processes inherent in project finance make transactions time-consuming and relatively expensive to finance. The universe of potential investors is also generally limited to parties who have extensive experience in such transactions.

Table 3.1 provides a comparison of project finance with traditional corporate finance.

Table 3.1 Comparison of project finance with corporate finance.

Dimension	Corporate finance	Project finance
Financing vehicle	Multi-purpose firm	Single-purpose entity
Recourse	Yes	No
Type of capital	Permanent with indefinite equity	Finite-time horizon matching life of project
Dividend policy and reinvestment	Corporate management decides	Pre-defined dividend payout; no reinvestment
Capital investment decisions	Opaque to creditors	Transparent to creditors
Transactions costs	Relatively low - fairly standardized transaction forms	Relatively high for complex transactions
Leverage ratio	Generally low but varies from sector to sector	High leverage (anywhere from 50%–90%)
Basis for credit evaluation	Overall financial health of corporate entity or guarantor entity	Technical and economic feasibility of project
Investor/lender base	Typically broader/deep secondary markets	Relatively highly sophisticated investor base and thinner markets
Financing costs	Relatively low owing to deeper market liquidity	Relatively high owing to complex risk allocation and thinner markets

As shown in Table 3.1, project finance, which typically lacks the deeper markets for the debt and equity issued, differs from traditional corporate finance in almost every way, due to the standalone nature of the entity being financed. This in turn affects how and when project finance can be used, if at all.

Project finance limitations

While project finance is an important vehicle for financing power and other infrastructure projects across the world, it has notable limitations and downsides. These include:

- **Longer execution timeframe:** While project development and construction can span over years, the financing process itself may be extensive. The due-diligence process alone may take months, even for the simplest projects. The longer execution time may imperil the economics of the project, as financing costs may be dependent on market conditions.
- **High transaction costs:** Each project is different, with its own set of risks, contracts, and technical features. Therefore, there is little scope for standardization. The customized nature of each project financing can make transaction costs higher compared to traditional corporate financing.
- **Higher debt costs:** As discussed before, due to the non-recourse nature of the financing, the lenders are exposed to idiosyncratic risks specific to a project they are financing. As a result, the lenders are required to devote significant resources to understand the project's technical aspects, construction process, important contracts, etc. There is often heightened uncertainty during the construction phase leading to higher risk of default and prospects for lower recovery. As a result, the lenders will demand higher compensation for the risk they underwrite. The risk premium depends on market conditions and other factors, but can often vary anywhere between 25 and 200 basis points (bps), depending on the nature of the project being financed.
- **Higher insurance requirements and costs:** Since the projects are financed on a non-recourse basis, the lenders require that each project has its own insurance policies to ensure that the insurance proceeds can cover the outstanding financing amounts. The insurance markets for certain types of perils may be very thin, which may drive up the insurance costs. The insurance costs can be contained if the sponsor has an institution-wide insurance policy program covering a broad portfolio of projects. However, such a program is accessible only for deep-pocketed sponsors. Moreover, despite institution-wide insurance programs, the lenders may still insist on project-specific sub-limits for each peril, which can be costly.
- **Stringent lender reporting requirements:** The highly structured nature of the transaction as well as significant oversight requirements result in stringent reporting requirements for projects, which may require additional administration resources and increase the overall cost of operations.
- **Significant lender oversight and delays in decision-making:** The non-recourse nature of the financing implies that project lenders are exposed to the financial risks arising from non-performance or sub-par performance. Therefore, lenders want to have a say in matters that may be critical to the overall health of the project, and ensure that any issues or problems are dealt with at an early stage. Lenders maintain asset management teams who receive periodic updates through operations reports, financial statements, etc. As a result, the sponsor rarely has a free hand in operations of the project. The contractual requirements to receive consents or votes on important matters may also delay decision-making, which can create financial ramifications.

- **Intricate risk allocation resulting in complex documentation:** Most of the negotiation and documentation for project finance revolves around risk allocation among various parties. Given the sheer number of counterparties involved and their conflicting objectives, finding an acceptable risk allocation can be quite difficult. Nevertheless, the legal documentation that governs various contractual arrangements must accommodate the final negotiated outcome in terms of risk allocation, and hence the documentation can be very complex. Furthermore, the documentation alone can drive up the execution time and transaction expenses. Since the legal documentation typically covers the lifespan of the project the counterparties need to institute operational procedures to ensure compliance with the documents.
- **Greater disclosure of proprietary information and business strategies from the sponsor:** The lenders need access to a significant amount of confidential information about technology of the project, construction outlay, permitting, internal legal analysis, market assessment, and other items from the point that a project is presented for the financing proposal to the financial close. The stringent reporting requirements cited above may also require disclosure of private information to the financial institutions on a regular basis. Consequently, the sponsor may be forced into divulging a significant amount of information about its business strategy to third parties. To help mitigate this business risk, lenders are obligated to abide by confidentiality restrictions for a specific period of time and sponsors usually work with certain "relationship banks", both of which may help limit the spread of confidential information.

Why use project finance?

Given all of these disadvantages, it might seem that project finance would rarely be used, since conventional equity and debt markets allow for investment funds to be raised in deep, liquid markets with relatively low transaction costs. The transactions costs of project finance deals, with their complex structures and contractual arrangements seem to be a particular hindrance, as these costs fall on the project sponsor. Project finance, however, is thriving, as discussed previously. Economically, there must be some countervailing advantages that outweigh the considerable additional costs.

Financial economists have identified a number of theoretical structural advantages that can help make project financing a better model than traditional corporate finance in some cases by reducing total financing costs and/or avoiding specific transaction-related problems.[3]

First, the contractual structures inherent in project finance can help mitigate some of the agency risks inherent in large-scale, sunk cost project investments. Remember that in a conventional debt or equity fundraising, the funds and later returns may typically be used for "general corporate purposes", at the discretion of corporate management. But investors, having put up the money to build a large project, may worry that management will not be properly motivated to return future cash flows to them, and instead might embark on other projects or "empire building" at their expense. The tight

[3] This section draws heavily on the analysis of Benjamin C. Esty in Esty, B. C. (June 2004). Why study large projects? An introduction to research on project finance. *European Financial Management, 10*(2), 213–224.

constraints on how cash flows are distributed inherent in single-purpose project finance structures can help mitigate this risk, to the benefit of investors.

Project finance structures can also help limit agency conflicts and opportunistic behavior between project parties. "Hold up" behavior may be anticipated when a project has inputs or outputs that are not freely traded, or where government action might allow for expropriation. A tightly structured project company in which interactions between parties are governed by strict contracts (and available cash flow already has pre-determined uses such as debt service) can help limit these risks. When the scope for opportunistic behavior is large, joint ownership of the project may also be used as a further means to limit such behavior which could scuttle a project.

Second, project finance may provide a way to avoid underinvestment. A sponsor may have a potentially attractive (e.g., an expected positive NPV) project but may not be able to undertake the investment without breaching its own debt covenants or other constraints. By moving the investment into a project-financed structure, the sponsor could raise most of the funds necessary against the project's own economics, without risking the integrity of its own finances and imperiling the returns of its existing investors.

Third, large projects are often highly risky from a sponsor's point of view. Consider a sponsor with a set of existing successful projects, all conventionally financed, but which is considering a large new project. This project is NPV-positive and otherwise attractive, but, like all projects, has inherent risks. If the project underperforms, it could create incremental distress costs (e.g. bankruptcy, other restructuring costs) and drag down the entire firm. Project financing this large project provides a tool for managing the risks of financial "contagion" between the existing projects and this new opportunity, again providing a method for avoiding underinvestment in an otherwise attractive project.

Project finance loans versus bonds

Debt in project finance, as in the traditional corporate finance sector, can be in the form of loans (from a bank or another lender) or in bonds. As was illustrated earlier, debt in the project finance market is dominated by banks, and loans are the lending instruments of choice. There are several reasons why bank loans are much more common than bonds in project finance.

First, banks have financed infrastructure and other similar projects over many years. As a result, the banking community has accumulated significant experience with project finance and the underlying assets and businesses. Banks likely have more experience with project finance than typical bond investors.

The second reason relates to the evolution of project risk over time. According to a study published by Moody's Investor Services, project finance loans have a relatively high default rate during the initial years (primarily due to inherent construction risks).[4]

[4] Moody's Investor Service. (March 5, 2014). *Default and recovery rates for project finance bank loans, 1983–2012*.

As projects gain a track record, the default risk often goes down considerably – sometimes even lower than a typical A-rated corporate loan. Therefore, with a well-diversified portfolio of project finance loans, a bank may partially mitigate construction risks and earn a better risk-adjusted return on its investment relative to the risk of the loans. As a result, by accessing cheap financing available from saving deposits, the banks can provide competitive financing and still earn a handsome return on regulatory capital.

Third, banks are willing to underwrite construction loans and can provide a flexible financing mechanism that automatically converts into a term loan once certain conditions are met. Bond market participants are not really interested in construction loans – bond investors are not equipped to provide financing in installments as the construction milestones are met, and in the absence of such installment-based financing, the project sponsors are unwilling to bear the cost of "**negative carry**" that up-front financing entails. Bond investors generally do not have access to the engineering and technical resources needed to assess construction risks, and typically are also not comfortable with these risks.

Fourth, the bond underwriting process is complex, time consuming, and requires up-front investment by the sponsor in terms of marketing the bonds, roadshow presentations, etc. Bond issuance also comes with stringent securities law requirements not affecting bank loans.

Fifth, due to the diffused public ownership, restructuring and renegotiation of bonds may be very difficult if a project runs into trouble. Bonds often come with expensive pre-payment penalties as well. Banks typically have longer-term relations with project sponsors and may be more commercially reasonable when disputes arise or the project runs into trouble.

Sixth, bonds often need to be rated by the credit rating agencies; this process is time consuming, expensive, and may be quite conservative. Traditionally, for example, the credit rating agencies have taken a conservative outlook toward debt sizing for renewable energy projects. As a result, the bond financing may yield a lower financing amount relative to project finance loans.

Against these disadvantages, bond-based project financing does have several advantages over bank loans:

- Bonds are fixed-rate instruments. Therefore, the project sponsor does not have to worry about interest rate risks.
- Bond investors are generally insurance companies, pension funds, and other institutions that are looking for long-term investments and are often willing to go much longer in terms of tenor (term) than banks.
- Bond investors often have a more relaxed governance standard than banks and there is less interference with the operations and decision-making. Bond investors may have less stringent reporting requirements too.
- Bond financing may help to keep open valuable credit lines that can be used for general corporate purposes. Because project finance loans are typically held on the bank's balance sheet, the credit envelope available to a sponsor recedes as the bank's aggregate debt exposure to the sponsor rises. Especially in the advent of deteriorating market conditions,

a sponsor may require access to short-term capital, such as a revolving credit facilities, and find it difficult to find liquidity from bank lenders who are already "full" on the sponsor.

Considering the complexities, costs, time, and resource commitment, bond financing is seldom justified unless the financing amount to be raised exceeds $1 billion. In such situations, by using a combination of bank loans and bond financing, a project sponsor may achieve the lowest financing costs and an efficient execution.

Despite the obvious advantages of bank loans over bonds, a private placement market has sprung up to cater to insurance companies that offer competitive financing for small projects. The insurance companies are often looking for longer-dated assets to match their insurance liabilities. Consequently, project finance bonds serve as an attractive asset class. By providing flexible financing similar to bank loans (e.g. with construction draws, delayed draws, etc.), insurance companies can be competitive. Furthermore, the National Association of Insurance Commissioners (NAIC), which helps oversee insurance companies, provides certain flexibilities in terms of credit ratings for the instruments that insurance companies can invest in. As a result, insurance companies can reduce credit rating costs and time, thereby making them more competitive to banks.

Modeling project cash flows and debt service

Financial modeling is an integral component of project finance. Financial models are used in structuring deals, analyzing project risks, and sizing of debt for projects. Since, in a non-recourse financing, lenders have no access to collateral to secure their loans to the projects, understanding a project's cash flows and its ability to repay debts and interest is critical.

The financial modeling process typically starts at the inception of the project by the sponsor, and is refined and fine-tuned at every step to ensure that the project remains viable and the project economics meet the sponsor's investment criteria.

Other parties participating in the financing process have their own models and these may differ materially in terms of assumptions, based on their objectives and role. For example, a lender might have a conservative set of assumptions for expected electricity generation, power prices, and other parameters, and these might differ significantly from the assumptions used by the sponsor, which would result in a conservative revenue and cash flow forecast and a lower debt sizing. In many cases, there is substantial back-and-forth between the lenders and the sponsor to settle the assumptions that go into the lender's financial model. In some cases, the project will have an independent engineer or other advisor who effectively works with both parties to help get to a commercial resolution regarding the input assumptions that determine the financing amount.

Modeling project cash flows

In this section, we will review a simple project *pro forma* model in conjunction with analysis of debt sizing. For simplicity, the analysis will be limited to a simple project finance loan and we will use the same set of assumptions used in the lender's model to calculate the sponsor's internal rate of return (IRR).

As a first step of the analysis, we will review the inputs to the *pro forma* financial model for a wind power project. Modeling for solar photovoltaic projects is quite similar, with some differences that will be noted along the way.

Project generation

One key aspect of renewable energy project finance is that output is often inherently uncertain – it depends on patterns of weather (wind speeds, cloud cover, etc.) that cannot be known completely when the project is planned. A substantial component of renewable energy financing deals with the uncertainties in future project output.

Renewable Energy Finance. **https://doi.org/10.1016/B978-0-12-816441-9.00004-0**

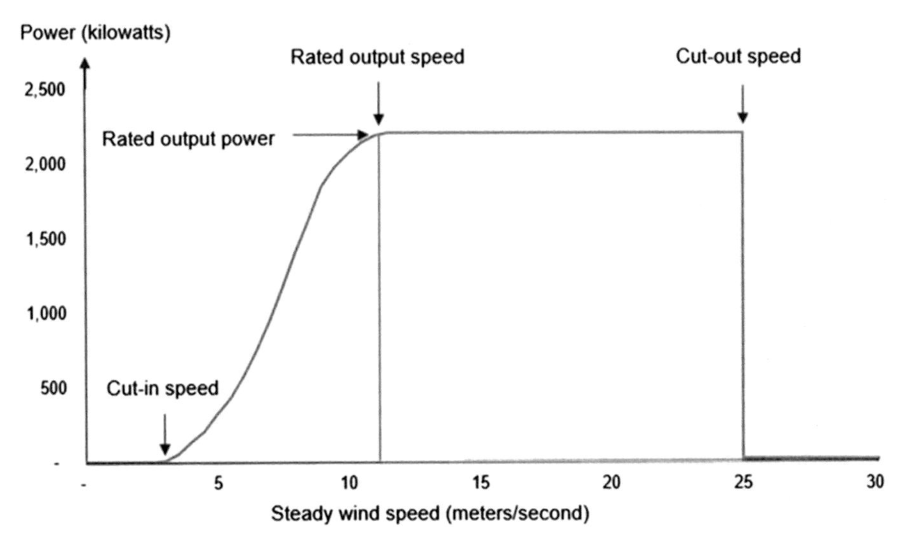

Fig. 4.1 S-curve for a wind turbine.

To model future revenues requires the modeling of expected future generation. For a wind project, as discussed in more detail in Chapter 7, projections for electricity generation for a project are developed using wind data from the region, often using data from weather stations in the vicinity of the project. The wind projections are commonly adjusted for the topography of the project, the type and siting of wind turbines, and any neighboring wind projects. Independent consultants are often hired to forecast future wind patterns and calculate future generation.

The wind resource profile is then combined with the "**S-curve**" for the wind turbine to produce a forecast of power output. The "S-curve", as illustrated in Fig. 4.1, is named for its shape. The curve shows the relationship between the power output of a wind turbine and the wind speed. As shown in the figure, at very low wind speeds, the turbine produces no electricity, as the wind is insufficient to turn the blades at all. Power output increases with wind speed until the maximum output power level is reached, and then levels off. At very high wind speeds, wind turbines are designed to cut-out to avoid damage, and power output again returns to zero.

The project sponsor or consultants combine wind pattern data with the S-curves of the turbines to model forecast electricity output over time. This is typically done using a Monte Carlo simulation model, which yields the "Gross Generation" of the project in MWh. The simulation model results also provide an analysis of the statistical variation of the generation at various confidence intervals. Once the gross generation profile is available, the numbers are adjusted for various inefficiencies and losses present at the project. Each project is different and some energy losses are unavoidable. The loss factors are themselves of probabilistic nature. Consequently, the Monte Carlo simulation needs to be re-run with various stochastic loss factors modeled in order to arrive at the "Net Generation."

The spread between Gross Generation and Net Generation can vary from project to project. It can be minimized with superior turbine technology (e.g. a cold climate

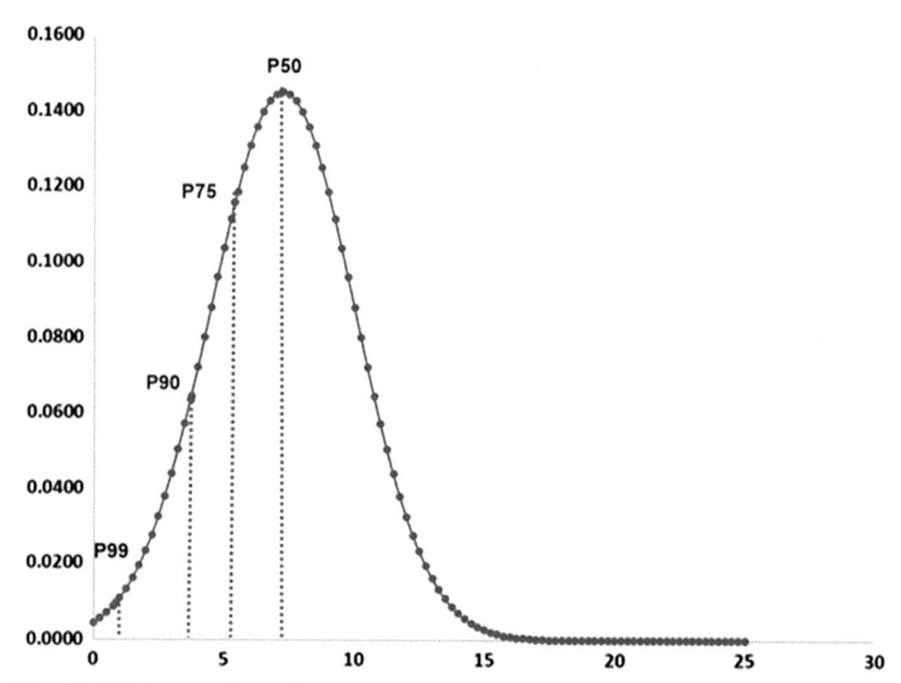

Fig. 4.2 Wind generation probability distribution and p-values.

package that allows the turbines to operate in deep freezing temperatures), which can be costly. The Net Generation has its own probabilistic distribution, as illustrated in the blue curve in Fig. 4.2.

It is inconvenient to model financial outcomes for all of the possible different levels of output that could happen in a year (or another period). Therefore, the financing parties usually rely on confidence intervals for expected generation based on the simulated net generation from the project. The p-50, p-75, p-90, p-95, and p-99 values are the most frequently cited confidence intervals.

The p-values are effectively percentiles, and can be thought of as how confident one can be (if the underlying data is correct) that the generation over the period will exceed the value given. P-50 is the median of the modeled data – 50% of the time the project should generate at least this much electricity in a year. P-90 is a more conservative value – 90% of the time the project should generate at least that amount. As shown in Fig. 4.2, the p-90 value will therefore be lower than the p-50 value. Finally, the p-99 value is even lower and more conservative – 99% of the time the project should generate this level of electricity in a year.

The values can be calculated over different time periods, with p-values calculated over 1 year or 10 years' worth of data being the most frequently cited numbers. It should be noted that the 10-year p-values for all confidence intervals higher than the p-50 confidence interval are higher than those for the 1-year values. This is because the longer the time period, the more predictable the wind resource. The mean of the distribution is generally stable over the 1–10 year period. Therefore, 1 year p 50 values are often similar to the 10-year p 50 values.

The p-values are calculated after considering the various weather events and equipment degradation a project is likely to experience through its useful life. Therefore, it is not unusual to have one constant number used to estimate electricity generation through the life of the project. Sometimes, depending on the methodology being used, or if there are circumstances unique to a project, the independent engineer may provide wind generation estimates that vary from year to year.

Solar projects follow a similar process to establish the electricity generation profile. The profile estimation is somewhat easier because solar data has been tracked for several decades, so the solar resource (i.e. sunshine) is usually more predictable than the wind resource. As a result, several desktop programs have been developed that can estimate electricity generation for a given location based on the solar modules in use, array configuration, inverters, tracking devices, etc. Since the standard deviation for the solar resource is narrower (e.g., the equivalent blue curve in Fig. 4.2 would show a much narrower distribution), the p-values for electricity generation generally fall within a narrower range.

For solar projects, it is customary to produce the electricity generation estimate for the first couple of years and use a constant degradation factor to reduce the electricity output year-over-year to account for equipment degradation. The degradation factor that goes into the financial model, while in fact a technical quantity, may also be subject to negotiation, like everything else. It can typically vary between 0.5% and 1.0% per year. However, as solar photovoltaic (PV) technology matures, financing parties are typically getting more and more comfortable with an annual degradation assumption of approximately 0.5%.

Power prices

If the project company has entered into a PPA, the power prices agreed upon in the PPA are used in the financial model, as the PPA specifies the price paid to the project for its output. In many cases, the sales price of electricity specified in the PPA is linked to the inflation rate. In this scenario, the parties must agree on forecast inflation data for modeling purposes.

While it is often beneficial to have 100% of a project's electricity output contracted through its useful life with a PPA to provide cash flow stability, this is not always the case. For example, the term of the PPA may not cover the full useful life of the project. In this scenario, for the portion of electricity output subject to merchant risk – the risk that future power prices are subject to variation – project modelers often have to use a forecast of future local power prices. However, these are always uncertain and subject to substantial volatility.

The sponsor and lenders will generally need to agree upon a merchant price forecast. One possible proxy for future merchant power prices is the forward price curve at a location closest to the project. However, forward price data may not be acceptable to the parties for various reasons. First, forward price data is often not readily available and may be limited to certain locations. Second, forward price data will generally not cover the full useful life of a project. Finally, the forward prices may be affected by liquidity constraints. Other alternatives include purchasing forecasts from data providers or engaging specialist power market consultants to generate price forecasts to be used in the models. These are typically based on bottom-up fundamentals modeling, and can be complex and expensive.

Due to the fact that power prices vary from one location to another (caused by power grid constraints), the power price forecast must account for **transmission basis** (as discussed in Chapter 9) to the extent the power prices agreed in the PPA are based on a location other than the project's location. This too is complex, and requires specialist modeling skills.

Modeling project revenues and costs

Project revenues are typically estimated by multiplying net generation from the project (at a given p-value) by the expected price (from the PPA, for example). Adjustments for basis, other losses, and/or PPA price escalation may be required, depending on the project details. This provides the forecast revenues in each period (e.g. a year or quarter) in the financial model.

Operating expenses comprise the various categories of expected expenses necessary to keep the project in operation once it is built. These will typically include items such as turbine operations and maintenance (O&M) costs, balance of plant O&M, land lease expenses, property taxes, sales taxes (if any), O&M for communication and transmission equipment, regulatory and professional fees, etc. Detailed budgets for these amounts will usually be created based on project O&M contracts.

Forecast earnings before interest, depreciation, and amortization (EBITDA) is calculated as the difference between expected revenues and operating expenses in each period. EBITDA is a very common basic measure of project cash flow.

Debt sizing for a fully contracted project

Lenders are rightly concerned about the cash flows from the project available to meet debt service obligations – especially in a non-recourse project financing where project cash flows are the only significant means of debt repayment. To quantify the cash flows available for debt service relative to the mandatory debt service, the lenders use a concept called Debt Service Coverage Ratio (DSCR). The **DSCR** is defined as the ratio of "Cash Flow Available for Debt Service" (**CFADS**) to mandatory debt service, which is the sum of the interest and mandatory amortization or repayment. The metric provides a quick assessment of how much cushion the borrower has to service the debt. The CFADS is typically calculated as EBITDA minus various recurring financing fees, working capital adjustments, adjustments to debt service reserve accounts (a topic covered in subsequent sections), and similar quantities.

Lenders typically want to ensure that there is enough cushion in both base case and downside scenarios. Accordingly, lenders usually back-solve for a borrowing amount based on a DSCR they are comfortable with for a base case and downside case. Given the variable nature of the renewable energy resource, the downside cases for a renewable energy project can be subject to interpretation. As discussed earlier, energy generation from wind and solar projects can be estimated using probabilistic measures. Consequently, lenders have become comfortable with using the energy generation estimates at various confidence intervals to size the debt. Lenders generally consider both p-50 and p-99 production levels in their debt sizing. Some lenders may consider a lower confidence interval, but generation estimates using the 1-year p-99 confidence interval has become almost an industry standard. The 1-year p-99 level of energy generation implies there is a 1% chance that the energy generation in a given year would be below this value. Of course, as a matter of conservatism, the lenders usually use their own adjustments to the energy forecasts and *pro forma* cash flow projections to arrive at the base case and downside scenarios.

Once the numbers are finalized, the underwriters will typically run the model using a generation forecast based on 1-year p-99 production values and calculate a debt size such that the lenders are assured of the repayment of mandatory debt service. Some lenders may desire more cushion, and consequently size the debt with a margin of 1−10% (in addition to using a conservative p-99 production forecast). In other words, the lenders would size the debt with a DSCR of 1.01x to 1.10x. So, for example, if the DSCR was forecast to be 1.00x in a year, the forecast CFADS would be equal to the sum of interest and debt repayment due in that year. At a DCSR of 1.10x, CFADS would be 10% greater than interest and debt payments due. The higher the DSCR required, the smaller the debt that can be supported by the project, all other things being equal.

Since the project is likely to operate closer to the mean generation, lenders will also want to ensure there is enough coverage at 10-year p-50 values. Consequently, the debt underwriters will usually run the model again, replacing the 1-year p-99 values with 10-year p-50 values. Other model assumptions will generally remain the same as there is little, if any, relationship between a project's operating expenses and its energy production in a year. In the 10-year p-50 production scenario, lenders are often comfortable underwriting at a DSCR of 1.30x to 1.50x.

The underwriters also typically calculate debt size using the 10-year p-50 value and use the lower of the two values. Therefore, debt sizing will be based on the more conservative of the 1-year p-99 and the 10-year p-50 production scenarios.

Other debt sizing parameters

The DSCR is a key debt sizing parameter, but other parameters also affect the quantity of debt that a project can support. For example, the term of the debt is generally decided by the useful life of the project and the term of the off-take agreement. With a longer debt term, repayment of the loan principal can be spread out over more years, providing a higher level of possible debt for the same level of cash flows in each year. Lenders are often comfortable attributing a useful life of 25 years for wind farms and up to 35 years for solar projects. The term of the off-take contract can extend up to the useful life of the project. Accordingly, the lenders may consider a debt term extending up to the term of the off-take contract. Other lenders may consider a slightly shorter term (e.g. by one to five years) for the sake of conservatism.

The size of the potential debt for the project also depends on the interest rate, as with higher interest rates, interest payments each year will increase. Project finance loans provided by banks usually reflect a floating underlying interest rate. Banks may quote the interest rate linked to LIBOR (the London Inter-Bank Overnight Rate, a common short-term dollar interest rate benchmark), or similar inter-bank rates applicable to the currency of the debt financing. For example, a project finance bank loan might be quoted as LIBOR + 250 bps, or the current LIBOR rate plus 2.5%.

Although variations exist, the LIBOR term is generally linked to the debt payment frequency: banks will use the one-month LIBOR index for a monthly debt repayment schedule, the three-month LIBOR index for a quarterly repayment schedule, etc. Projects are generally unable to bear substantial interest rate risks, and banks or other lenders are keen to ensure that the project can service its debt even if interest rates (as reflected in the LIBOR index) increase. The project financing loan documentation will therefore typically require that the project company hedge at least 75% of the mandatory amortization through the life of the debt.

As the project finance loans are floating rate instruments, banks are generally willing to accept low pre-payment penalties. The project company, however, may face "swap-breakage" costs if the loans are prepaid ahead of schedule. The swap-breakage costs arise because the interest rate swaps are financial derivatives. If such instruments are terminated prior to the stated maturity of the contracts, the investment bank (serving as the counterparty) may be owed a mark-to-market payment (depending on the interest rates at the time of the termination).

Bonds are an alternative to bank loans for some project financings. Bonds have fixed interest rates, and interest rate hedges are not required for bond financings. However, bond indenture documentation typically includes stringent make-whole provisions which may be triggered by various specified conditions, including change of control. The make-whole provisions are incorporated in order to protect the financial investors from any change in interest rates, which may affect the return on the reinvestment of the debt repayment proceeds.

Debt sizing example

The following example illustrates debt sizing for a 200 MW wind project with an annual electricity generation of 700,800 MWh/year (implying a **capacity factor** of 40%) in the p-50 production scenario and 525,600 MWh/year in 1-year p-99 production scenario (implying a capacity factor of 30%).[1] The wind project has a PPA with a creditworthy utility for a price of $50/MWh escalating at 2.0% per annum for the term of 25 years matching the useful life of the project. The project has operating expenses of $8 million for the first year (approximately $40/kW-year), while operating expenses

[1] The capacity factor is the projected generation of a project in a year divided by the maximum possible output, if the project ran at full capacity for 24 hours a day, 365 days a year (8760 hours = 24 hours/day × 365 days/year). So the capacity factor for this example project is 700,800 MWh/[200 MW × 8760 hours] = 40%. For more details on energy units and basic calculations, see Appendix A.

are expected to increase with the inflation rate. The lender has specified an inflation rate of 2.5% per annum to be used in the projections.

For simplicity, it is assumed that there are no financing fees or adjustments to working capital or the debt service reserve account so that EBITDA equals CFADS. The debt financing accrues interest at 4.5% per annum. The project has a capital cost of $2100/kW or $420 million in total. The lenders require a sizing DSCR of 1.30x in the p-50 production scenario and 1.0x in the p-99 production scenario. Finally, we have assumed that the lender is willing to provide financing matching the PPA term of 25 years.

The calculations in Table 4.1 show that the project has debt capacity of $371.72 million in the p-50 production scenario, and those in Table 4.2 has $324.12 million in the p-99 production scenario. Therefore, a lender is likely to extend the lower of the two quantities or $324.12 million in debt financing.

It should be noted that the debt amortization is sculpted to the cash available from the project such that the debt sizing DSCR is maintained.

Fig. 4.3 shows the amortization profile for the project, extending over the life of the loan.

The equity IRR for a project changes over the project's lifetime. Fig. 4.4 illustrates how the pre-tax IRR to the equity sponsor changes over time for the sample project.

The debt sizing exercise also illustrates that the debt sizing is fairly robust with an average DSCR of approximately 1.49x and a minimum DSCR of 1.48x in the p-50 productions scenario. Intuitively, this suggests that the project's cash flows could potentially withstand a reduction of 32.43% in any given period before the project company would be at risk of defaulting on any mandatory debt service payments.[2]

In conjunction with the debt sizing, lenders typically perform sensitivity analyses to understand the robustness of the cash flows to various stresses to electricity output, increased operating expenses, and other factors, with the goal of ensuring that the project generates enough cash flows to service debt through the life of the loans or bonds.

Debt tranching

For large project finance transactions, it is possible to have several different classes of debt securities. Different tranches or classes of debt may be used in order to increase the marketability of the debt financing, or to provide additional leverage.

Prioritization to assets

Different classes of debt securities are often segregated by their priority to the collateral security in the event of liquidation. In this framework, there are multiple classes of debt that are secured by the same collateral (i.e. substantially all of the assets and income of the project company), but there is a prioritization as to how the assets are divvied up if the projects face a financial restructuring (e.g. bankruptcy or liquidation). Specifically, the first-lien debt has top priority to the assets and cash, while the second-

[2] (1-(1.00x/1.48x)) = 32.43%.

Table 4.1 Debt sizing in the p-50 scenario.

Assumptions

		Unit
Plant Capacity	200	MW
P-50 Capacity Factor	40.0%	%
Operating Expenses	$40.00	per KW-Year
PPA Price (year 1)	$50.00	per MWh
PPA Escalation Rate	2.0%	%
Inflation Factor	2.5%	%
Sizing DSCR	1.30x	Ratio
Interest Rate	4.5%	%
Hours in Year	8,760	Hours
Conversion to USD 000	1,000	USD

Debt Sizing

Pro Forma

Period	0	1	2	3	4	5	6	7	8	9	10	11	12
Electricity Production		700,800	700,800	700,800	700,800	700,800	700,800	700,800	700,800	700,800	700,800	700,800	700,800
PPA Escalation Factor		1.00	1.02	1.04	1.06	1.08	1.10	1.13	1.15	1.17	1.20	1.22	1.24
Inflation Factor		1.00	1.03	1.05	1.08	1.10	1.13	1.16	1.19	1.22	1.25	1.28	1.31
PPA Rate		$50.00	$51.00	$52.02	$53.06	$54.12	$55.20	$56.31	$57.43	$58.58	$59.75	$60.95	$62.17
Electricitiy Revenue		35,040	35,741	36,456	37,185	37,928	38,687	39,461	40,250	41,055	41,876	42,714	43,568
Less: Operating Expense		(8,000)	(8,200)	(8,405)	(8,615)	(8,831)	(9,051)	(9,278)	(9,509)	(9,747)	(9,991)	(10,241)	(10,497)
EBITDA		27,040	27,541	28,051	28,570	29,098	29,636	30,183	30,740	31,308	31,885	32,473	33,071
Cash Flow Available for Debt Service (CFADS)		27,040	27,541	28,051	28,570	29,098	29,636	30,183	30,740	31,308	31,885	32,473	33,071
Mandatory Debt Service (Sizing)		20,800	21,185	21,577	21,977	22,383	22,797	23,218	23,647	24,083	24,527	24,979	25,439
Debt Schedule													
Beginning Principal		371,723	367,651	363,010	357,768	351,891	345,343	338,087	330,083	321,290	311,665	301,163	289,736
Less: Interest Expense		(16,728)	(16,544)	(16,335)	(16,100)	(15,835)	(15,540)	(15,214)	(14,854)	(14,458)	(14,025)	(13,552)	(13,038)
Less: Amortization		(4,072)	(4,641)	(5,242)	(5,877)	(6,548)	(7,256)	(8,004)	(8,793)	(9,625)	(10,502)	(11,427)	(12,401)
Ending Principal	371,723	367,651	363,010	357,768	351,891	345,343	338,087	330,083	321,290	311,665	301,163	289,736	277,335

Period	13	14	15	16	17	18	19	20	21	22	23	24	25
Electricity Production	700,800	700,800	700,800	700,800	700,800	700,800	700,800	700,800	700,800	700,800	700,800	700,800	700,800
PPA Escalation Factor	1.27	1.29	1.32	1.35	1.37	1.40	1.43	1.46	1.49	1.52	1.55	1.58	1.61
Inflation Factor	1.34	1.38	1.41	1.45	1.48	1.52	1.56	1.60	1.64	1.68	1.72	1.76	1.81
PPA Rate	$63.41	$64.68	$65.97	$67.29	$68.64	$70.01	$71.41	$72.84	$74.30	$75.78	$77.30	$78.84	$80.42
Electricitiy Revenue	44,439	45,328	46,235	47,159	48,102	49,064	50,046	51,047	52,068	53,109	54,171	55,255	56,360
Less: Operating Expense	(10,759)	(11,028)	(11,304)	(11,586)	(11,876)	(12,173)	(12,477)	(12,789)	(13,109)	(13,437)	(13,773)	(14,117)	(14,470)
EBITDA	33,680	34,300	34,931	35,573	36,226	36,892	37,568	38,257	38,959	39,672	40,399	41,138	41,890
Cash Flow Available for Debt Service (CFADS)	33,680	34,300	34,931	35,573	36,226	36,892	37,568	38,257	38,959	39,672	40,399	41,138	41,890
Mandatory Debt Service (Sizing)	25,908	26,385	26,870	27,364	27,866	28,378	28,899	29,429	29,968	30,517	31,076	31,644	32,223
Debt Schedule													
Beginning Principal	277,335	263,908	249,399	233,752	216,907	198,802	179,369	158,542	136,248	112,411	86,952	59,789	30,835
Less: Interest Expense	(12,480)	(11,876)	(11,223)	(10,519)	(9,761)	(8,946)	(8,072)	(7,134)	(6,131)	(5,058)	(3,913)	(2,691)	(1,388)
Less: Amortization	(13,428)	(14,509)	(15,647)	(16,845)	(18,106)	(19,432)	(20,827)	(22,294)	(23,837)	(25,459)	(27,163)	(28,954)	(30,835)
Ending Principal	263,908	249,399	233,752	216,907	198,802	179,369	158,542	136,248	112,411	86,952	59,789	30,835	0

Table 4.2 Debt sizing in the p-99 scenario.

Assumptions

		Unit
Plant Capacity	200	MW
P-99 Capacity Factor	30.0%	%
Operating Expenses	$40.00	per KW-Year
PPA Price (year 1)	$50.00	per MWh
PPA Escalation Rate	2.0%	%
Inflation Factor	2.5%	%
Sizing DSCR	1.00x	Ratio
Interest Rate	4.5%	%
Hours in Year	8,760	Hours
Conversion to USD 000	1,000	USD

Debt Sizing

Pro Forma

Period	0	1	2	3	4	5	6	7	8	9	10	11	12
Electricity Production		525,600	525,600	525,600	525,600	525,600	525,600	525,600	525,600	525,600	525,600	525,600	525,600
PPA Escalation Factor		1.00	1.02	1.04	1.06	1.08	1.10	1.13	1.15	1.17	1.20	1.22	1.24
Inflation Factor		1.00	1.03	1.05	1.08	1.10	1.13	1.16	1.19	1.22	1.25	1.28	1.31
PPA Rate		$50.00	$51.00	$52.02	$53.06	$54.12	$55.20	$56.31	$57.43	$58.58	$59.75	$60.95	$62.17
Electricity Revenue		26,280	26,806	27,342	27,889	28,446	29,015	29,596	30,187	30,791	31,407	32,035	32,676
Less: Operating Expense		(8,000)	(8,200)	(8,405)	(8,615)	(8,831)	(9,051)	(9,278)	(9,509)	(9,747)	(9,991)	(10,241)	(10,497)
EBITDA		18,280	18,606	18,937	19,273	19,616	19,964	20,318	20,678	21,044	21,416	21,794	22,179
Cash Flow Available for Debt Service (CFADS)		18,280	18,606	18,937	19,273	19,616	19,964	20,318	20,678	21,044	21,416	21,794	22,179
Mandatory Debt Service (Sizing)		18,280	18,606	18,937	19,273	19,616	19,964	20,318	20,678	21,044	21,416	21,794	22,179
Debt Schedule													
Beginning Principal		324,117	320,422	316,235	311,529	306,275	300,441	293,997	286,909	279,142	270,659	261,423	251,392
Less: Interest Expense		(14,585)	(14,419)	(14,231)	(14,019)	(13,782)	(13,520)	(13,230)	(12,911)	(12,561)	(12,180)	(11,764)	(11,313)
Less: Amortization		(3,695)	(4,187)	(4,706)	(5,255)	(5,833)	(6,444)	(7,088)	(7,767)	(8,483)	(9,236)	(10,030)	(10,867)
Ending Principal	324,117	320,422	316,235	311,529	306,275	300,441	293,997	286,909	279,142	270,659	261,423	251,392	240,526

Period	13	14	15	16	17	18	19	20	21	22	23	24	25
Electricity Production	525,600	525,600	525,600	525,600	525,600	525,600	525,600	525,600	525,600	525,600	525,600	525,600	525,600
PPA Escalation Factor	1.27	1.29	1.32	1.35	1.37	1.40	1.43	1.46	1.49	1.52	1.55	1.58	1.61
Inflation Factor	1.34	1.38	1.41	1.45	1.48	1.52	1.56	1.60	1.64	1.68	1.72	1.76	1.81
PPA Rate	$63.41	$64.68	$65.97	$67.29	$68.64	$70.01	$71.41	$72.84	$74.30	$75.78	$77.30	$78.84	$80.42
Electricity Revenue	33,329	33,996	34,676	35,369	36,077	36,798	37,534	38,285	39,051	39,832	40,628	41,441	42,270
Less: Operating Expense	(10,759)	(11,028)	(11,304)	(11,586)	(11,876)	(12,173)	(12,477)	(12,789)	(13,109)	(13,437)	(13,773)	(14,117)	(14,470)
EBITDA	22,570	22,968	23,372	23,783	24,201	24,625	25,057	25,496	25,942	26,395	26,856	27,324	27,800
Cash Flow Available for Debt Service (CFADS)	22,570	22,968	23,372	23,783	24,201	24,625	25,057	25,496	25,942	26,395	26,856	27,324	27,800
Mandatory Debt Service (Sizing)	22,570	22,968	23,372	23,783	24,201	24,625	25,057	25,496	25,942	26,395	26,856	27,324	27,800
Debt Schedule													
Beginning Principal	240,526	228,779	216,106	202,459	187,787	172,036	155,152	137,077	117,750	97,107	75,082	51,605	26,603
Less: Interest Expense	(10,824)	(10,295)	(9,725)	(9,111)	(8,450)	(7,742)	(6,982)	(6,168)	(5,299)	(4,370)	(3,379)	(2,322)	(1,197)
Less: Amortization	(11,747)	(12,673)	(13,647)	(14,672)	(15,750)	(16,884)	(18,075)	(19,327)	(20,643)	(22,025)	(23,477)	(25,002)	(26,603)
Ending Principal	228,779	216,106	202,459	187,787	172,036	155,152	137,077	117,750	97,107	75,082	51,605	26,603	0

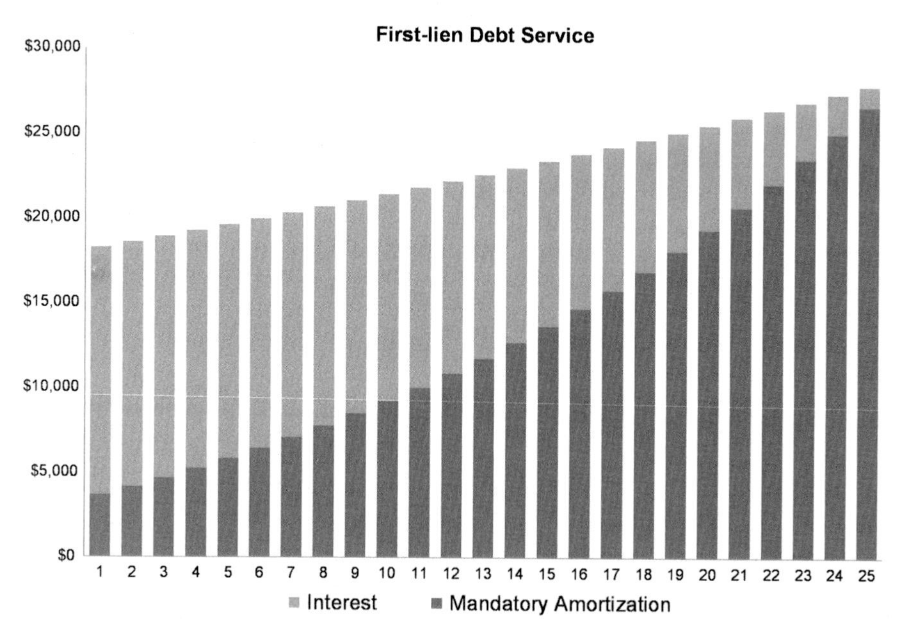

Fig. 4.3 Project amortization profile.

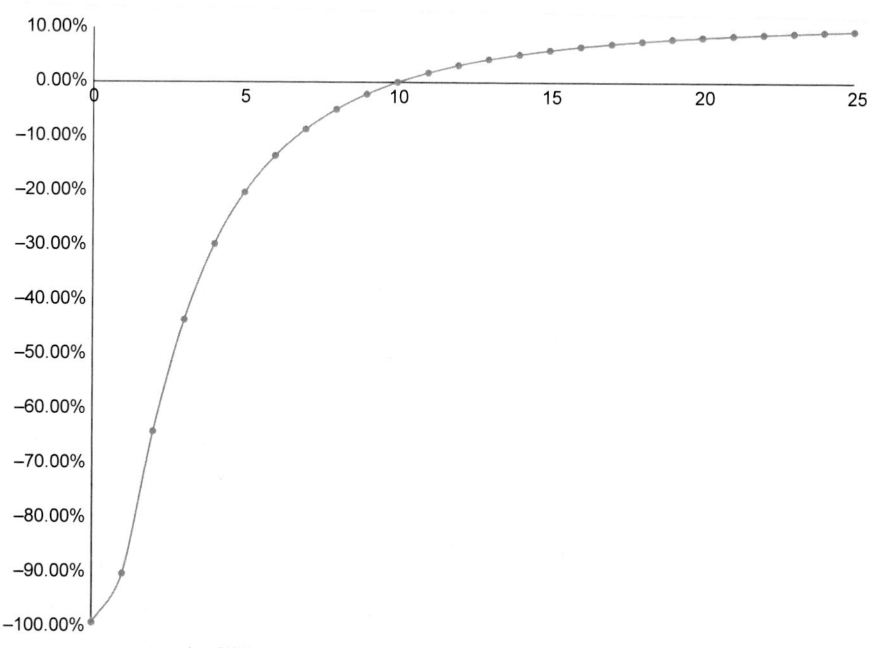

Fig. 4.4 Pre-tax equity IRR.

lien debt has a secondary priority to the collateral, and so on. Furthermore, there may also be senior unsecured debtholders who by definition do not have any direct and explicit security in the project company's assets. Third-lien structure is not common and, in the absence of third-lien debtholders, senior unsecured debtholders, if any, effectively become third-lien debtholders.

These priorities are established in the loan documentation, such as a Collateral Agreement that specifies the priority of the debt holders in the event of a default and eventual restructuring. Since the first-lien debt holders have first priority, they are the first to get paid in the event of liquidation. Any assets left over after satisfying the first-lien debtholders are used to pay out second-lien debt holders. Any assets left over after paying for the second-lien debtholders are then used to pay third-lien debtholders, if any. The senior unsecured lenders are paid off only after secured lenders are paid.

Debtholders ranking at the second level and below are highly exposed to the recovery prospects of the project in case of a restructuring. Second-lien debt is therefore riskier than first-lien debt, and the credit spreads for such lower ranked debt securities are correspondingly higher. The prospects for substantial recovery of the project assets in the event of default may be modest even for second-lien debtholders in renewable projects, depending on the nature of default, the time it takes to restructure debt, and market conditions, but there are cases wherein second-lien debtholders may recover a significant proportion of the principal outstanding. Finally, the aforementioned priority structure is generally respected in the event of a court administered bankruptcy or liquidation. In an out-of-court debt restructuring, the outcome would depend on the negotiations among various debtholders.

Debt-tranching example

The calculations in Table 4.3 provide an extension to the earlier example for debt sizing. In this example, we have assumed that the sponsor decides to add another layer of second-lien debt, which is sized off a DSCR of 1.20x in the P-50 production scenario with an interest rate of 6.5%. There are different ways to prioritize interest and mandatory amortization in the cash flow waterfall (which is discussed further in the next section), but for simplicity, this example assumes that the second-lien debt holders are paid interest and amortization only after first-lien debt interest and amortization is paid.

Table 4.3 calculations also illustrate that the project has second-lien debt capacity of $63.37 million in the base case model, assuming the p-50 production scenario. The debt sizing exercise also illustrates that the introduction of the second-lien debt improves the sponsor's pre-tax equity IRR from 9.63% to 15.05% in this scenario – this, of course, is accompanied by a higher risk associated with the project equity.

Another way to create multiple tranches is to use "**time-tranching**," similar to securitization structures. In this set up, there are multiple tranches of debt within the same class of debt (e.g. first-lien debt), but with different amortization profiles. The simplest arrangement may be to create a number of tranches (e.g. Tranche A, B, C, etc.) within the same class of debt. Then, the principal repayment schedule and tenor of the

Ta le 4.3 Debt-tranching example.

sumptions

		Unit
ant Capacity	200	MW
50 Capacity Factor	40.0%	%
perating Expenses	$40.00	per KW-Year
PA Price (year 1)	$50.00	per MWh
PA Escalation Rate	2.0%	%
flation Factor	2.5%	%
zing DSCR (First-Lien)	1.30x	Ratio
zing DSCR (Second-Lien)	1.20x	Ratio
terest Rate (First-Lien)	4.5%	%
terest Rate (Second-Lien)	6.5%	%
ours in Year	8,760	Hours
onversion to USD 000	1,000	USD
apital Cost	$2,100	per KW
otal Capital Cost	420,000	USD 000

ponsor Returns with Second-lien Debt Added

Pro Forma

Period	0	1	2	3	4	5	6	7	8	9	10	11	12
Electricity Production		700,800	700,800	700,800	700,800	700,800	700,800	700,800	700,800	700,800	700,800	700,800	700,800
PPA Escalation Factor		1.00	1.02	1.04	1.06	1.08	1.10	1.13	1.15	1.17	1.20	1.22	1.24
Inflation Factor		1.00	1.03	1.05	1.08	1.10	1.13	1.16	1.19	1.22	1.25	1.28	1.31
PPA Rate		$50.00	$51.00	$52.02	$53.06	$54.12	$55.20	$56.31	$57.43	$58.58	$59.75	$60.95	$62.17
Electricitiy Revenue		35,040	35,741	36,456	37,185	37,928	38,687	39,461	40,250	41,055	41,876	42,714	43,568
Less: Operating Expense		(8,000)	(8,200)	(8,405)	(8,615)	(8,831)	(9,051)	(9,278)	(9,509)	(9,747)	(9,991)	(10,241)	(10,497)
EBITDA		27,040	27,541	28,051	28,570	29,098	29,636	30,183	30,740	31,308	31,885	32,473	33,071
Cash Flow Available for Debt Service (CFADS)		27,040	27,541	28,051	28,570	29,098	29,636	30,183	30,740	31,308	31,885	32,473	33,071
Mandatory Debt Service (Sizing) - P50		20,800	21,185	21,577	21,977	22,383	22,797	23,218	23,647	24,083	24,527	24,979	25,439
Mandatory Debt Service (Sizing) - P99		18,280	18,606	18,937	19,273	19,616	19,964	20,318	20,678	21,044	21,416	21,794	22,179
Mandatory Debt Service (Sizing) - Minimum		18,280	18,606	18,937	19,273	19,616	19,964	20,318	20,678	21,044	21,416	21,794	22,179
Beginning Principal		324,117	320,422	316,235	311,529	306,275	300,441	293,997	286,909	279,142	270,659	261,423	251,392
Less: Interest Expense		(14,585)	(14,419)	(14,231)	(14,019)	(13,782)	(13,520)	(13,230)	(12,911)	(12,561)	(12,180)	(11,764)	(11,313)
Less: Amortization		(3,695)	(4,187)	(4,706)	(5,255)	(5,833)	(6,444)	(7,088)	(7,767)	(8,483)	(9,236)	(10,030)	(10,867)
Ending Principal	324,117	320,422	316,235	311,529	306,275	300,441	293,997	286,909	279,142	270,659	261,423	251,392	240,526
Total Debt Service (First-Lien)		18,280	18,606	18,937	19,273	19,616	19,964	20,318	20,678	21,044	21,416	21,794	22,179
Actual DSCR (First-Lien)		1.48x	1.48x	1.48x	1.48x	1.48x	1.48x	1.49x	1.49x	1.49x	1.49x	1.49x	1.49x
CFADS after First-Lien		8,760	8,935	9,114	9,296	9,482	9,672	9,865	10,062	10,264	10,469	10,678	10,892
Cash Flow for Second-lien Debt Sizing		4,253	4,345	4,439	4,535	4,632	4,732	4,835	4,939	5,046	5,155	5,266	5,380
Beginning Principal		63,370	63,236	63,001	62,657	62,196	61,606	60,878	60,000	58,961	57,748	56,347	54,743
Less: Interest Expense		(4,119)	(4,110)	(4,095)	(4,073)	(4,043)	(4,004)	(3,957)	(3,900)	(3,832)	(3,754)	(3,663)	(3,558)
Less: Amortization		(134)	(235)	(344)	(462)	(590)	(728)	(878)	(1,039)	(1,213)	(1,401)	(1,604)	(1,822)
Ending Principal	63,370	63,236	63,001	62,657	62,196	61,606	60,878	60,000	58,961	57,748	56,347	54,743	52,921
Total Debt Service (Second-Lien)		4,253	4,345	4,439	4,535	4,632	4,732	4,835	4,939	5,046	5,155	5,266	5,380
Actual DSCR (Second-Lien)		1.20x	1.20x	1.20x	1.20x	1.20x	1.20x	1.20x	1.20x	1.20x	1.20x	1.20x	1.20x
Equity Cash Flows	(32,513)	4,507	4,590	4,675	4,762	4,850	4,939	5,031	5,123	5,218	5,314	5,412	5,512
Cumulative Equity Cash Flows	(32,513)	(28,006)	(23,416)	(18,741)	(13,980)	(9,130)	(4,191)	840	5,963	11,181	16,495	21,908	27,419
Breakeven Year Calculation								6.83					

Period	13	14	15	16	17	18	19	20	21	22	23	24	25
Electricity Production	700,800	700,800	700,800	700,800	700,800	700,800	700,800	700,800	700,800	700,800	700,800	700,800	700,800
PPA Escalation Factor	1.27	1.29	1.32	1.35	1.37	1.40	1.43	1.46	1.49	1.52	1.55	1.58	1.61
Inflation Factor	1.34	1.38	1.41	1.45	1.48	1.52	1.56	1.60	1.64	1.68	1.72	1.76	1.81
PPA Rate	$63.41	$64.68	$65.97	$67.29	$68.64	$70.01	$71.41	$72.84	$74.30	$75.78	$77.30	$78.84	$80.42
Electricitiy Revenue	44,439	45,328	46,235	47,159	48,102	49,064	50,046	51,047	52,068	53,109	54,171	55,255	56,360
Less: Operating Expense	(10,759)	(11,028)	(11,304)	(11,586)	(11,876)	(12,173)	(12,477)	(12,789)	(13,109)	(13,437)	(13,773)	(14,117)	(14,470)
EBITDA	33,680	34,300	34,931	35,573	36,226	36,892	37,568	38,257	38,959	39,672	40,399	41,138	41,890
Cash Flow Available for Debt Service (CFADS)	33,680	34,300	34,931	35,573	36,226	36,892	37,568	38,257	38,959	39,672	40,399	41,138	41,890
Mandatory Debt Service (Sizing) - P50	25,908	26,385	26,870	27,364	27,866	28,378	28,899	29,429	29,968	30,517	31,076	31,644	32,223
Mandatory Debt Service (Sizing) - P99	22,570	22,968	23,372	23,783	24,201	24,625	25,057	25,496	25,942	26,395	26,856	27,324	27,800
Mandatory Debt Service (Sizing) - Minimum	22,570	22,968	23,372	23,783	24,201	24,625	25,057	25,496	25,942	26,395	26,856	27,324	27,800
Beginning Principal	240,526	228,779	216,106	202,459	187,787	172,036	155,152	137,077	117,750	97,107	75,082	51,605	26,603
Less: Interest Expense	(10,824)	(10,295)	(9,725)	(9,111)	(8,450)	(7,742)	(6,982)	(6,168)	(5,299)	(4,370)	(3,379)	(2,322)	(1,197)
Less: Amortization	(11,747)	(12,673)	(13,647)	(14,672)	(15,750)	(16,884)	(18,075)	(19,327)	(20,643)	(22,025)	(23,477)	(25,002)	(26,603)
Ending Principal	228,779	216,106	202,459	187,787	172,036	155,152	137,077	117,750	97,107	75,082	51,605	26,603	0
Total Debt Service (First-Lien)	22,570	22,968	23,372	23,783	24,201	24,625	25,057	25,496	25,942	26,395	26,856	27,324	27,800
Actual DSCR (First-Lien)	1.49x	1.49x	1.49x	1.50x	1.50x	1.50x	1.50x	1.50x	1.50x	1.50x	1.50x	1.51x	1.51x
CFADS after First-Lien	11,110	11,332	11,559	11,790	12,026	12,266	12,511	12,762	13,017	13,277	13,543	13,814	14,090
Cash Flow for Second-lien Debt Sizing	5,496	5,615	5,737	5,861	5,988	6,118	6,250	6,385	6,524	6,665	6,810	6,957	7,108
Beginning Principal	52,921	50,864	48,555	45,975	43,102	39,916	36,393	32,508	28,236	23,547	18,413	12,800	6,674
Less: Interest Expense	(3,440)	(3,306)	(3,156)	(2,988)	(2,802)	(2,595)	(2,366)	(2,113)	(1,835)	(1,531)	(1,197)	(832)	(434)
Less: Amortization	(2,057)	(2,309)	(2,581)	(2,873)	(3,186)	(3,523)	(3,885)	(4,272)	(4,688)	(5,135)	(5,613)	(6,125)	(6,674)
Ending Principal	50,864	48,555	45,975	43,102	39,916	36,393	32,508	28,236	23,547	18,413	12,800	6,674	0
Total Debt Service (Second-Lien)	5,496	5,615	5,737	5,861	5,988	6,118	6,250	6,385	6,524	6,665	6,810	6,957	7,108
Actual DSCR (Second-Lien)	1.20x	1.20x	1.20x	1.20x	1.20x	1.20x	1.20x	1.20x	1.20x	1.20x	1.20x	1.20x	1.20x
Equity Cash Flows	5,613	5,717	5,822	5,929	6,038	6,149	6,261	6,376	6,493	6,612	6,733	6,856	6,982
Cumulative Equity Cash Flows	33,033	38,749	44,571	50,500	56,538	62,686	68,948	75,324	81,817	88,429	95,162	102,019	109,000
Breakeven Year Calculation													

Pre-Tax Equity IRR (with second-lien added)	15.05%	
Equity Breakeven Period	6.83	years
Equity MOIC	4.35x	
Debt WAL	16.76	years

tranches can be created such that Tranche A is amortized fully before Tranche B receives any amortization. Once Tranche B is fully repaid, then debt repayment for Tranche C begins. This type of structure may help improve the liquidity for the debt securities because different types of investors have different investment objectives. For example, banks often prefer shorter maturities whereas insurance companies prefer longer maturities (in order to match the duration of their liabilities). By providing multiple tranches with different tenors, a borrower can increase the number of investors who can participate in debt financing. The three tranches of debt would be priced differently, owing to the differences in tenor and different recovery prospects.

Loan structuring considerations

In this section, we will consider various structural considerations primarily for project finance loans.

Project debt versus holding company loan

As discussed before, most project finance debt is structured at project company level. This implies that the project company is the borrower of record and is responsible for the timely repayment of the debt. The debt is usually secured by substantially all of the assets of the project company, including the equity interests in the project company. Therefore, in the event of a default, the lenders have an ability to foreclose on the collateral to recover the outstanding debt, interest, fees, penalties, etc.

Typically, the equity interests held by the sponsor in the project company are owned through a holding company. Customarily, there is one holding company per project and the holding company can be owned by a portfolio company, which may hold equity interests in multiple holding companies. There may be multiple layers of special purpose companies (SPCs) for various legal, accounting, and tax reasons. The sponsor then owns the portfolio company as a direct or indirect subsidiary.

From a financing perspective, it is possible that a sponsor can raise additional debt at the holding company level and/or the portfolio company level. This leverages the project further and potentially reduces the equity funds that the sponsor must contribute initially at the expense of raising equity risks.

It is important to note, however, that the financing at the holding or portfolio company level does not have the same collateral that the lender at the project company level enjoys. Consequently, the financing is less secure and the prospects for the recovery in the event of default may be limited, and therefore priced more expensively. The holding company debt is generally secured by the sponsor's equity interest in the project company. Therefore, in the event of default, the lender is likely to foreclose on the equity interests and become the *de facto* equity owner.

So how do financial institutions get comfortable with the holding company debt? Generally, the holding company debt is used once the project is commercially operational, at which point the risk of the underlying project performing poorly is significantly reduced. Furthermore, provided that the debt is sized conservatively, the sponsor is likely to continue holding a significant interest in the project and realize free cash flows through the life of the project, which ensures that the interests of the project sponsor and lender are aligned. Finally, the financial institutions may view higher credit spreads as an inducement to justify the higher risks.

Creating holding company debt may not be motivated by the desire for higher project leverage. In the US, tax equity is the predominant method of monetizing the federal subsidies to renewable projects afforded in the form of tax credits. In order to protect their own interests, financial institutions are averse to providing tax equity into a project company that is already levered. In this situation, holding company-level debt (also referred to as "**back-leverage**") is the only way to create financial leverage in the transactions.

Construction loan versus term loan

Project finance loans are often essentially construction loans that automatically convert into term loans once certain conditions are satisfied. The conversion feature allows the parties to structure one loan without having to negotiate the same documents twice. The lenders are more interested in the term loans, which are repaid through the cash flows generated from the project during operating phase. However, the construction loan is the necessary bridge to get the project built.

Since the project does not generate any cash flows during the construction phase, the lenders are willing to accept a smaller credit spread for the construction loan, which is compensated by a higher credit spread in the operational phase. Furthermore, the interest during construction (IDC) typically is capitalized and added to the loan balance. Since the loan is drawn in tranches as the project achieves construction milestones, there is usually a commitment fee to make the construction loan available. This is typically paid based on the construction loan committed but not drawn. Interest payments are calculated based on the construction loan actually drawn. The sum of the commitment fees and interest payment constitute the IDC.

The coupon rate for the term loan is usually linked to LIBOR and the credit spreads have a step-up feature whereby the credit spreads increase every 3—4 years. This may seem counterintuitive because the risk of default often decreases gradually as the project gains operational history. However, the lenders recognize that until the project operations are stabilized, there may be unforeseen capital expenditures or additional operating expenses. Therefore, setting rates with a step-up feature does not encumber the project with a higher interest payment burden during the vulnerable initial phase.

The construction loan converts to term loan upon achievement of several conditions, generally termed as the **conditions precedent** (or "CPs"). The principal condition precedent is the achievement of commercial operation, as specified in the loan documents. If there is a failure to satisfy the CPs and the lender refuses to waive the unsatisfied CPs, the project company is deemed to be in default. This may trigger the acceleration of the loan repayment and recovery process, either through a court-administered bankruptcy process or an out-of-court restructuring.

Fully amortizing loans versus mini-perm loans

Generally, banks are comfortable sizing the loan based on cash flows through the PPA term using the methodology described previously. However, the legal maturity of the loan may be different from the term of the cash flows that determines the debt sizing. Since the financial crisis of 2008, US-based banks have been subject to stringent capital requirements by the Federal Reserve, with longer tenor loans attracting higher capital requirements. As a result, these banks have often limited the legal maturity of project finance loans to 5–10 years. Therefore, these loans may have a defined amortization schedule until the legal maturity of the loan with the remaining principle balance due on the maturity date. The large repayment due on the legal maturity date of the loan is called a "balloon" payment, and the debt repayment profile is termed as "**mini-perm**".

A review of the debt sizing and amortization profile in the previous example in Tables 4.1 and 4.2 can illustrate the mini-perm concept. Compare the same basic example with a fully amortizing debt structure with a mini-perm debt structure with 10-year maturity. The debt size in this example is $324.1 million calculated over the 20-year useful life of the project. Therefore, both fully-amortizing and mini-perm structure would yield the same notional amount of $324.1 million. The debt amortization profile for the first 10 years would also be identical for the two debt structures. However, the mini-perm structure would entail a maturity of 10 years with a balloon or bullet payment at the end of $261.4 million.

In reality, the mini-perm feature is often not truly advantageous to either the sponsor or the lender. This is because the project company has no sources of liquidity other than the project it owns, so the only way the project company can repay the balloon payment is by refinancing the loan. If the market conditions deteriorate by the time the balloon payment is due, the project company may face prospects for default, even if the underlying project works as expected. Of course, the sponsor may avoid this potentially unpleasant situation by refinancing the debt ahead of the legal maturity when market conditions are ripe; however, frequent refinancing may increase transaction expenses, so this option may really depend on the "market timing" skills of the sponsor.

Since mini-perm loan structures create refinancing risks, the stringent capital requirements imposed by the Federal Reserve may render US banks to be less competitive than their foreign counterparts in providing project finance.

Cash flow waterfall

Project finance lending documents define a priority of payments from the cash receipts from the sale of electricity, colloquially referred to as the "cash flow waterfall". The waterfall metaphor treats the cash produced as water which can be captured in a series of containers – only when the cash produced "fills" the first container can any cash flow down to the second container, and so on.

The most important expenses for a project are the operating expenses, regulatory fees, and similar operational costs. These are essential to keep the project running

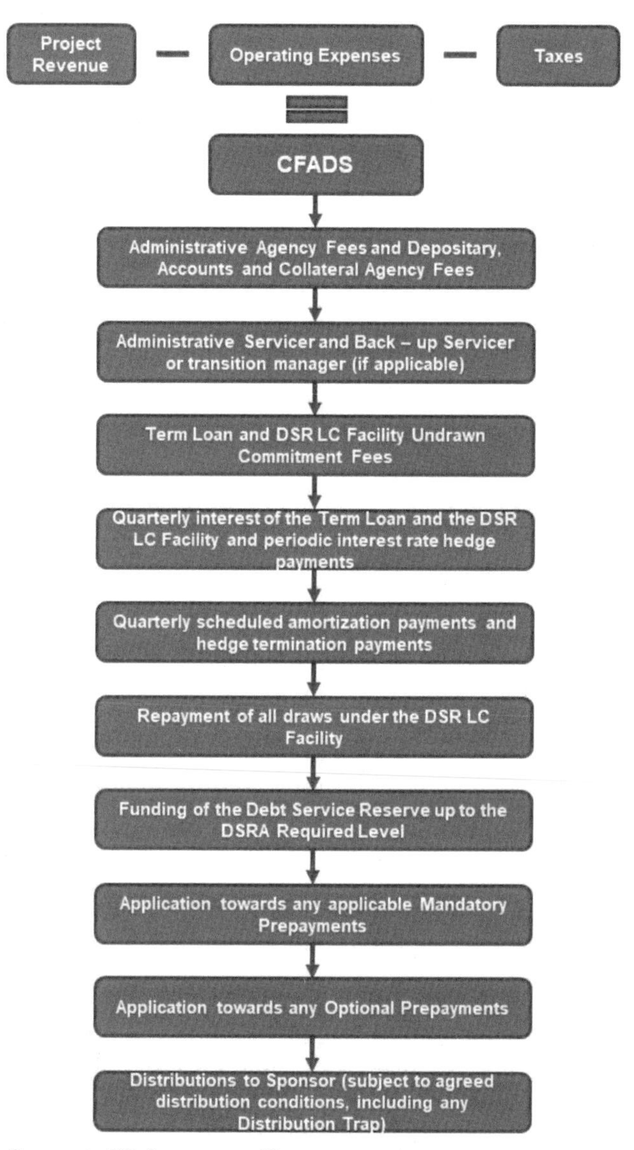

Fig. 4.5 Cash flow waterfall for a renewable energy project.

and produce more cash in the future. These expenses are thus at the top of the waterfall. On the other hand, dividends to the equity sponsor are to be paid last and are at the bottom of the waterfall.

Fig. 4.5 illustrates a simplified priority of payments for a typical renewable energy project.

Letters of credit

The project company is a special purpose company with limited financial resources of its own. Therefore, certain counterparties, especially PPA off-taker counterparties, may require a letter of credit (LC) to be posted to secure the project company's obligations under the relevant contract. A LC is essentially a backstop from a financial institution with sufficient creditworthiness (often defined by a combination of credit rating, typically set at A3/A- or better, and tangible net worth, typically at least $1 billion).

A LC guarantees payment to the counterparty if the project company fails to make the payment required under the contract. The LCs are conditional obligations. Therefore, the provider of the LC only earns a fee. Once the LC is drawn, the amount of the draw converts into a loan, which has its own repayment terms specified in the LC facility documentation, and pays interest at a rate equal to the relevant index rate, such as LIBOR and the applicable credit spread. Sometimes, there is a default rate tacked on to the interest rate for LC loans, typically about 2% per annum.

A well-capitalized project sponsor may have access to revolving credit facilities to provide the necessary LCs on behalf of the project company, or it may issue guarantees in favor of the project company. However, in project finance transactions, it is customary for the banks providing the financing to provide any LCs. Generally, LCs are part of the senior secured loans and are counted toward the overall debt sizing, even though LCs are unfunded obligations at the inception. As a result, the LC facilities enjoy the same security that the senior secured loans have. Consequently, the fees for LCs are often linked to the credit spread applicable for the senior secured loans.

Reserve requirements

Project lenders want to ensure that the project company – which has limited resources of its own – has funds available to meet critical expenses or unforeseen circumstances. For this reason, project finance debt may require several reserves set aside from the debt financing proceeds. For example, funds may be required to be set aside for operating reserves, capital expenditure reserves, maintenance reserves, and debt service reserves.

Each reserve has a different purpose. For example, an operating reserve may be required to ensure that the project has a certain amount of cash set aside in case operating expenses are higher than estimated in certain periods. The capital expenditure reserve ensures that the project company has cash on hand for certain expected capital expenditures planned in the future. Maintenance reserves seek to ensure that the project company always has enough cash to undertake unforeseen maintenance matters. The debt service reserve usually covers six to twelve months of projected mandatory debt service.

The primary concept behind the reserves is that the project is expected to have a long useful life and unforeseen scenarios could push the project company into default due to temporary cash flow shortfalls. Furthermore, if the repairs or capital expenditures are not conducted promptly, an otherwise healthy project may default. Finally, the lenders want to ensure that their interest and mandatory amortization is paid on time to avoid defaults due to temporary cash shortfalls.

If the reserves are drawn upon, the cash flow waterfall is prioritized such that the reserves are promptly replenished from future cash flows. To return to the water metaphor, water flows must be diverted until the reserve "tanks" are refilled before normal repayment priorities resume.

Sponsors generally dislike large reserve accounts since such reserves limit the amount of financing proceeds available to them, increasing their equity investment. Therefore, lenders may in some cases provide the sponsor the flexibility to post LCs drawn from the corporate revolving facility in lieu of maintaining cash reserves. If the sponsor does not have access to appropriate corporate revolving facilities, a LC facility may be carved out from the project financing that can be used to satisfy the cash reserve requirements. Not every bank is comfortable providing such LCs to cover reserve requirements, and the ones that are willing to do so may only do it for a credit spread higher than the underlying loan. The LC provider may also require that the fees for providing such LCs and loan repayments, pursuant to the draws on LCs, are payable before any of the debt service is paid.

DSCR triggers

In order to ensure that the project company always has access to cash flows to run its operations and mitigate unforeseen cash flow shortfalls, the lenders may require that the dividends distributable to the project sponsor are scaled back or trapped at the project company level if certain DSCR triggers are met.

For example, assume that a project finance loan is sized at a DSCR of 1.30x. A lender may then require that if the DSCR falls below 1.20x, only 50% of the cash flows available after the payment of mandatory debt service is released to the project sponsor. Other cash must remain at the project company level. Furthermore, if the project's cash flow continues to deteriorate and the DSCR falls at or below 1.10x, the lender may require that all of the dividends otherwise distributable to the project sponsor be trapped within the project company. These and other cash flow provisions are negotiated upfront and formally documented in the project finance agreements.

Financing merchant power projects

As discussed before, project finance generally depends on stable cash flows. This in turn typically requires that a project have a firm contract (or an equivalent regulatory construct) for its output to fix prices and stabilize expected revenues.

In a merchant power project, the project is exposed to the market price of electricity for its output, and electricity is a commodity with a high degree of price volatility. Given this constraint, merchant power projects, especially in the renewable energy sector, are hard to find.

However, there can be some scope for financing merchant power projects under certain conditions. Usually, loan-to-value for such debt financing is very conservative. The debt term is generally shorter. Moreover, the credit spreads are much wider than those available to contracted power projects. Finally, loans must be structured so that the financing can withstand the ups and downs in cash flows caused by variations in merchant power prices.

As discussed in a prior section, debt financings for contracted power projects have a mandatory debt amortization profile, defining how and when debt amortizes. Since the debt is often sized based on a DSCR of 1.30−1.50x, there is significant cushion should a project undergo temporary underperformance. The structure works when a project has fully contracted sales and cash flows are more predictable. In a merchant power financing, where a substantial portion of the project cash flows are subject to commodity price risks, a cash sweep mechanism is generally used as an alternative structure.

The cash sweep structure

The cash sweep structure essentially reduces the mandatory debt service to the sum of interest accrued in a given period and a minimal amortization, typically 1% of the outstanding principal amount in a given year. Then, in each interest period, an agreed percentage of CFADS remaining after payment of mandatory amortization is "swept" to the lender to repay any outstanding principal balance. By linking the loan repayment amount to the cash flows a project can generate, the structure has a way of cushioning low cash flow years when the merchant power prices are low.

The cash sweep mechanism often further incorporates a deleveraging feature whereby the cash sweep percentage can go higher when the project faces cash flow difficulties due to underperformance or a prolonged cycle of depressed power prices. Since the cash flows remaining after paying for mandatory debt service (comprised of interest payment and a fraction of the outstanding principal balance, as discussed above) are payable to the sponsor as a dividend, the ratcheting of the cash sweep percentage has the effect of reducing or completely shutting off the dividends to the sponsor until the project cash flows are back on track.

Merchant financing example

This sub-section illustrates debt sizing for a merchant power project using the same basic data as used previously. However, in this case there is no PPA, and hence not a fixed or indexed power price for sales. The wind farm instead sells power into the open spot market.

As a basis of comparison, we use the same static power price (before inflation adjustments) in the previous example, but understand this revenue line is simply a forecast and not an agreed contractual price with an off-take counterparty. In a realistic merchant scenario, forecasted power prices are not static and usually derived from a market expert.

As illustrated in Table 4.4, the debt term has been truncated to 10 years (as opposed to 25 years in the previous example) and the interest rate for the loan is set at 6.5% (instead of 4.5% as before). The mandatory amortization is set at 1.0% of the principal outstanding at the beginning of each year. Furthermore, it is assumed that the lender requires that 50% of the excess cash flow ("ECF") available after paying for mandatory debt service is used to repay the debt.

Table 4.4 Debt sizing for a merchant power project.

		Unit
Plant Capacity	200	MW
P-50 Capacity Factor	40.0%	%
Operating Expenses	$40.00	per KW-Year
PPA Price (year 1)	$50.00	per MWh
PPA Escalation Rate	2.0%	%
Inflation Factor	2.5%	%
Mandatory Amortization	1.0%	%
Cash Sweep	50.0%	%
Interest Rate	6.5%	%
Debt Term	10	Years
Hours in Year	8,760	Hours
Conversion to USD 000	1,000	USD
Capital Cost	$2,100	per KW
Total Capital Cost	420,000	USD 000

Sponsor Returns with Term Loan B Cash Sweep

Pro Forma

Period	0	1	2	3	4	5	6	7	8	9	10	11	12
Electricity Production		700,800	700,800	700,800	700,800	700,800	700,800	700,800	700,800	700,800	700,800	700,800	700,800
PPA Escalation Factor		1.00	1.02	1.04	1.06	1.08	1.10	1.13	1.15	1.17	1.20	1.22	1.24
Inflation Factor		1.00	1.03	1.05	1.08	1.10	1.13	1.16	1.19	1.22	1.25	1.28	1.31
PPA Rate		$50.00	$51.00	$52.02	$53.06	$54.12	$55.20	$56.31	$57.43	$58.58	$59.75	$60.95	$62.17
Electricity Revenue		35,040	35,741	36,456	37,185	37,928	38,687	39,461	40,250	41,055	41,876	42,714	43,568
Less: Operating Expense		(8,000)	(8,200)	(8,405)	(8,615)	(8,831)	(9,051)	(9,278)	(9,509)	(9,747)	(9,991)	(10,241)	(10,497)
EBITDA		27,040	27,541	28,051	28,570	29,098	29,636	30,183	30,740	31,308	31,885	32,473	33,071
Cash Flow Available for Debt Service (CFADS)		27,040	27,541	28,051	28,570	29,098	29,636	30,183	30,740	31,308	31,885	32,473	33,071
Beginning Principal		128,652	118,670	108,113	96,958	85,181	72,758	59,661	45,865	31,342	16,064		
Less: Interest Expense		(8,362)	(7,714)	(7,027)	(6,302)	(5,537)	(4,729)	(3,878)	(2,981)	(2,037)	(1,044)		
Less: Mandatory Amortization		(1,287)	(1,287)	(1,287)	(1,287)	(1,287)	(1,287)	(1,287)	(1,287)	(1,287)	(1,287)		
Less: Cash Sweep		(8,696)	(9,270)	(9,868)	(10,490)	(11,137)	(11,810)	(12,509)	(13,236)	(13,992)	(14,777)		
Ending Principal	128,652	118,670	108,113	96,958	85,181	72,758	59,661	45,865	31,342	16,064	(0)		
Sizing	(0)												
Actual DSCR		1.47x	1.51x	1.54x	1.58x	1.62x	1.66x	1.71x	1.76x	1.81x	1.86x		
Equity Cash Flows	(291,348)	8,696	9,270	9,868	10,490	11,137	11,810	12,509	13,236	13,992	14,777	32,473	33,071

Period	13	14	15	16	17	18	19	20	21	22	23	24	25
Electricity Production	700,800	700,800	700,800	700,800	700,800	700,800	700,800	700,800	700,800	700,800	700,800	700,800	700,800
PPA Escalation Factor	1.27	1.29	1.32	1.35	1.37	1.40	1.43	1.46	1.49	1.52	1.55	1.58	1.61
Inflation Factor	1.34	1.38	1.41	1.45	1.48	1.52	1.56	1.60	1.64	1.68	1.72	1.76	1.81
PPA Rate	$63.41	$64.68	$65.97	$67.29	$68.64	$70.01	$71.41	$72.84	$74.30	$75.78	$77.30	$78.84	$80.42
Electricity Revenue	44,439	45,328	46,235	47,159	48,102	49,064	50,046	51,047	52,068	53,109	54,171	55,255	56,360
Less: Operating Expense	(10,759)	(11,028)	(11,304)	(11,586)	(11,876)	(12,173)	(12,477)	(12,789)	(13,109)	(13,437)	(13,773)	(14,117)	(14,470)
EBITDA	33,680	34,300	34,931	35,573	36,226	36,892	37,568	38,257	38,959	39,672	40,399	41,138	41,890
Cash Flow Available for Debt Service (CFADS)	33,680	34,300	34,931	35,573	36,226	36,892	37,568	38,257	38,959	39,672	40,399	41,138	41,890
Equity Cash Flows	33,680	34,300	34,931	35,573	36,226	36,892	37,568	38,257	38,959	39,672	40,399	41,138	41,890

Pre-Tax Equity IRR	5.69%

As shown in the table, for the same set of *pro forma* assumptions, the lender would be willing to extend only $128.7 million, an amount much lower than the leverage offered in the earlier example for the contracted asset. Furthermore, the resultant DSCRs in the base case model are higher – a minimum DSCR of 1.47x and average DSCR of 1.65x – which reflects the lender's required cushion to tolerate the potential for cash flow variability from an asset that bears both price and volume risk. To accommodate the inclusion of a cash sweep mechanism, which, by nature, varies in amount for each year of the *pro forma*, we have to size the debt quantum of the project in a different manner than we have before. Using Excel's Goal Seek tool, this is accomplished by selecting the objective cell, which is the "Ending Principal" balance in Year 10, to yield a value of "0", by changing the "Ending Principal" balance in Year 0.[3]

When a power project needs to be built without any PPAs, quasi-merchant financing is much more palatable to the lenders and more prevalent than a pure merchant power price project. In a quasi-merchant project, the sponsor hedges most of the project output through a physical or financial derivative contract with a financial institution – a forward or swap contract for a renewable energy project or a tolling agreement for a natural gas project – for a term of 5–10 years or as long as the power markets can support. Then, the lenders can put together a two-tranche structure, wherein the first-lien debt is repaid fully through the term of the hedge contract through the contracted cash flows, while most of the second-lien debt remains outstanding when the hedge contract expires. In this structure, the second-lien debt is assuming the refinancing risk and is priced accordingly. The sponsor can mitigate some of the risk for the second-lien debt by extending the hedges mid-way and locking in better prices. Alternatively, the sponsor can refinance the second-lien debt by entering into a new hedge at the expiration of the original hedge contract.

Debt sizing for a quasi-merchant power project is illustrated in Table 4.5 using the same basic data as before. However, assume that the sponsor has been able to get a 10-year PPA or hedge for the wind farm, but after that period, electricity will have to be sold at spot prices for the rest of the project's useful life. As shown in the table, the tenor of the first-loan debt has been truncated to 10 years (as opposed to 25 years in the original example) to coincide with the term of the PPA (or hedge). Since the first-lien debt is expected to be paid off through the PPA/hedge term, the interest rate applicable to the first-lien debt is back to 4.5% as it was originally. The second-lien debt is sized off a merchant refinancing risk at the end of the PPA/hedge term that the lenders are willing to bear. Finally, the second-lien debt is assumed to be interest-only during the 10-year term and carries a coupon of 6.5%. For simplicity, assume only the p-50 production scenario for first-lien debt sizing and a DSCR of 1.30x for the first-lien debt.

[3] Please note that there is no difference between the "Ending Principal" balance in Year 0 and the "Beginning Principal" balance in Year 1.

Table 4.5 Debt sizing for a quasi-merchant project.

Assumptions

	Value	Unit
Plant Capacity	200	MW
P-50 Capacity Factor	40.0%	%
Operating Expenses	$40.00	per KW-Year
PPA Price (year 1)	$50.00	per MWh
PPA Escalation Rate	2.0%	%
Inflation Factor	2.5%	%
Sizing DSCR (First-Lien)	1.30x	Ratio
First-lien Debt Term	10	Years
First-Lien Interest Rate	4.5%	%
Debt Sizing (Second-Lien)	$200.00	$/kW
Interest Rate (First-lien)	4.5%	%
Interest Rate (Second-Lien)	6.5%	%
Mandatory Amortization (Second-Lien)	1.0%	%
Hours in Year	8,760	Hours
Conversion to USD 000	1,000	USD
Capital Cost	$2,100	per KW
Total Capital Cost	420,000	USD 000

Sponsor Returns for Quasi-Merchant Project

Pro Forma

Period	0	1	2	3	4	5	6	7	8	9	10	11	12
Electricity Production		700,800	700,800	700,800	700,800	700,800	700,800	700,800	700,800	700,800	700,800	700,800	700,800
PPA Escalation Factor		1.00	1.02	1.04	1.06	1.08	1.10	1.13	1.15	1.17	1.20	1.22	1.24
Inflation Factor		1.00	1.03	1.05	1.08	1.10	1.13	1.16	1.19	1.22	1.25	1.28	1.31
PPA Rate		$50.00	$51.00	$52.02	$53.06	$54.12	$55.20	$56.31	$57.43	$58.58	$59.75	$60.95	$62.17
Electricity Revenue		35,040	35,741	36,456	37,185	37,928	38,687	39,461	40,250	41,055	41,876	42,714	43,568
Less: Operating Expense		(8,000)	(8,200)	(8,405)	(8,615)	(8,831)	(9,051)	(9,278)	(9,509)	(9,747)	(9,991)	(10,241)	(10,497)
EBITDA		27,040	27,541	28,051	28,570	29,098	29,636	30,183	30,740	31,308	31,885	32,473	33,071
CFADS		27,040	27,541	28,051	28,570	29,098	29,636	30,183	30,740	31,308	31,885	32,473	33,071
CFADS for First-lien Debt Sizing		20,800	21,185	21,577	21,977	22,383	22,797	23,218	23,647	24,083	24,527		
Beginning Principal		177,796	164,997	151,237	136,465	120,629	103,674	85,543	66,175	45,506	23,471	0	-
Less: Interest Expense		(8,001)	(7,425)	(6,806)	(6,141)	(5,428)	(4,665)	(3,849)	(2,978)	(2,048)	(1,056)	-	-
Less: Mandatory Amortization		(12,799)	(13,760)	(14,772)	(15,836)	(16,955)	(18,131)	(19,368)	(20,669)	(22,035)	(23,471)	-	-
Ending Principal	177,796	164,997	151,237	136,465	120,629	103,674	85,543	66,175	45,506	23,471	0	-	-
Total Debt Service (First-Lien)		20,800	21,185	21,577	21,977	22,383	22,797	23,218	23,647	24,083	24,527	-	-
Actual DSCR (First-Lien)		1.30x	1.30x	1.30x	1.30x	1.30x	1.30x	1.30x	1.30x	1.30x	1.30x		
CFADS after First-Lien		6,240	6,356	6,473	6,593	6,715	6,839	6,965	7,094	7,225	7,358	32,473	33,071
Beginning Principal		40,000	39,600	39,200	38,800	38,400	38,000	37,600	37,200	36,800	36,400		
Less: Interest Expense		(2,600)	(2,574)	(2,548)	(2,522)	(2,496)	(2,470)	(2,444)	(2,418)	(2,392)	(2,366)		
Less: Mandatory Amortization		(400)	(400)	(400)	(400)	(400)	(400)	(400)	(400)	(400)	(36,400)		
Ending Principal	40,000	39,600	39,200	38,800	38,400	38,000	37,600	37,200	36,800	36,400	-		
Total Debt Service (Second-Lien)		3,000	2,974	2,948	2,922	2,896	2,870	2,844	2,818	2,792	38,766		
Actual DSCR (Second-Lien)		1.14x	1.14x	1.14x	1.15x	1.15x	1.15x	1.16x	1.16x	1.16x	0.50x		
Equity Cash Flows	(202,204)	3,240	3,382	3,525	3,671	3,819	3,969	4,121	4,276	4,433	(31,408)	32,473	33,071
Cumulative Equity Cash Flows	(202,204)	(198,964)	(195,582)	(192,057)	(188,386)	(184,567)	(180,598)	(176,477)	(172,201)	(167,768)	(199,176)	(166,703)	(133,632)
Breakeven Year Calculation													

Period	13	14	15	16	17	18	19	20	21	22	23	24	25
Electricity Production	700,800	700,800	700,800	700,800	700,800	700,800	700,800	700,800	700,800	700,800	700,800	700,800	700,800
PPA Escalation Factor	1.27	1.29	1.32	1.35	1.37	1.40	1.43	1.46	1.49	1.52	1.55	1.58	1.61
Inflation Factor	1.34	1.38	1.41	1.45	1.48	1.52	1.56	1.60	1.64	1.68	1.72	1.76	1.81
PPA Rate	$63.41	$64.68	$65.97	$67.29	$68.64	$70.01	$71.41	$72.84	$74.30	$75.78	$77.30	$78.84	$80.42
Electricity Revenue	44,439	45,328	46,235	47,159	48,102	49,064	50,046	51,047	52,068	53,109	54,171	55,255	56,360
Less: Operating Expense	(10,759)	(11,028)	(11,304)	(11,586)	(11,876)	(12,173)	(12,477)	(12,789)	(13,109)	(13,437)	(13,773)	(14,117)	(14,470)
EBITDA	33,680	34,300	34,931	35,573	36,226	36,892	37,568	38,257	38,959	39,672	40,399	41,138	41,890
CFADS	33,680	34,300	34,931	35,573	36,226	36,892	37,568	38,257	38,959	39,672	40,399	41,138	41,890
CFADS for First-lien Debt Sizing													
Beginning Principal	-	-	-	-	-	-	-	-	-	-	-	-	-
Less: Interest Expense	-	-	-	-	-	-	-	-	-	-	-	-	-
Less: Mandatory Amortization	-	-	-	-	-	-	-	-	-	-	-	-	-
Ending Principal	-	-	-	-	-	-	-	-	-	-	-	-	-
Total Debt Service (First-Lien)	-	-	-	-	-	-	-	-	-	-	-	-	-
Actual DSCR (First-Lien)													
CFADS after First-Lien	33,680	34,300	34,931	35,573	36,226	36,892	37,568	38,257	38,959	39,672	40,399	41,138	41,890
Beginning Principal													
Less: Interest Expense													
Less: Mandatory Amortization													
Ending Principal													
Total Debt Service (Second-Lien)													
Actual DSCR (Second-Lien)													
Equity Cash Flows	33,680	34,300	34,931	35,573	36,226	36,892	37,568	38,257	38,959	39,672	40,399	41,138	41,890
Cumulative Equity Cash Flows	(99,952)	(65,652)	(30,721)	4,852	41,078	77,969	115,538	153,795	192,754	232,426	272,825	313,963	355,852
Breakeven Year Calculation				15.86									

Pre-Tax Equity IRR	6.09%	
Equity Breakeven Period	15.86	years
Equity MOIC	2.76x	
Debt WAL	6.05	years

As illustrated in Table 4.5, the first-lien debt can be sized at $177.8 million. The second-lien debt tranche is sized based on an initial $200/kW metric, but lenders' underwriting appetite is also determined by the debt quantum that needs to be refinanced at maturity. In this example, only bullet amortization of 1% per annum is applied to the second-lien balance and the resulting balloon at maturity is $36,400,000 or $182/kW. Excluding the repayment of the second-lien principal balance in year 10, the minimum and average DSCRs for the second-lien debt work out to be 1.14x and 1.15x, respectively. As expected, the results indicate that a quasi-merchant project provides significantly better debt capacity and equity IRRs for the sponsor as compared to a fully merchant power project.

Renewable project finance structures and risk allocation

Developing and financing a large-scale renewable energy project such as a wind farm or utility-scale PV facility is complex, and requires a large number of different organizations and entities fulfilling different roles. The legal agreements that define how the financing will work are also highly complicated, and hence most of the financing parties will work with their own legal counsel. In many cases, there are also independent consultants, engineers, and other advisors hired to provide specialist inputs. It is often a slow and complex process to get all of these parties to agree on a solution so the financing can close.

Basic project finance structure

Fig. 5.1 illustrates a simplified project finance structure for a renewable energy project. With some modification, the same diagram can be made applicable to almost any energy or core infrastructure project.

Main parties in a financing

Before delving into the financial and contractual relationships, it is important to have an overview of who does what in the building and financing of a renewable project. Even in this simplified example, there are numerous parties and each of these has a distinct role.

Project sponsor

The project sponsor is the entity that develops the project and is the principal equity holder of the project being financed. The project sponsor is thus the primary architect of the project and the central player in organizing the successful development, construction, operation, and financing of the project. The project sponsor may be for example a renewable energy developer (or subsidiary or affiliate), a private equity firm, or a utility.

Project company

As noted in the previous chapter, project finance operates through special purpose entities (typically a limited liability company). The project company is the single purpose company set up with the sole purpose of owning and operating one or more projects, at times through a series of subsidiaries. It is the counterparty to all contracts that the project needs to operate successfully. The project company is

Renewable Energy Finance. https://doi.org/10.1016/B978-0-12-816441-[illegible]

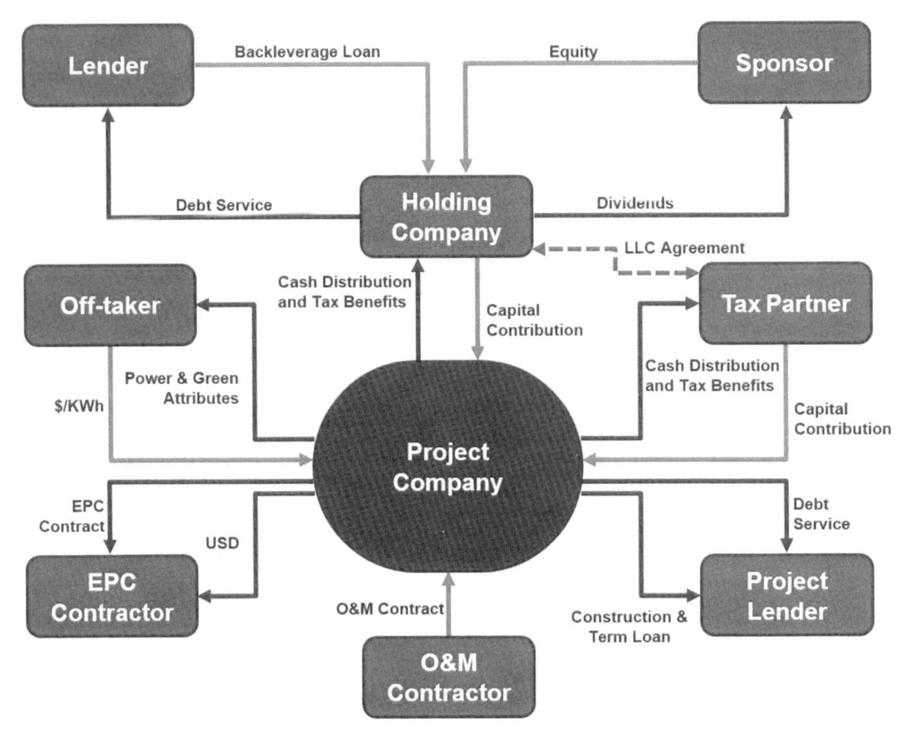

Fig. 5.1 Simplified project finance structure for a renewable energy project.

incorporated in a jurisdiction that allows for the legal ownership and operation of the project. For the purpose of the discussion here, it is assumed that the project company represents the interests of the parent company and all its subsidiaries.

Offtaker

The offtaker is the party that is obligated to purchase the output of the project through a long-term contract (also called an offtake agreement or similar output sales arrangement). For power projects such contracts are often in the form of a Power Purchase Agreement (PPA). For infrastructure projects, this could be a long-term concession signed with a government or quasi-government entity. The offtake agreement is usually the single most important contract for any project. Without a sound offtake agreement with a creditworthy counterparty, most projects will never get off the ground, as the project company needs to sell the project output through a long-term (often take-or-pay) contract at a fixed and determinable price to ensure revenue predictability. This in turn ensures that the project can be financed. Without such long-term contract, the project company will be exposed to merchant price risk (in electricity prices, for example).

When the project company enters into an agreement with an offtaker, the project company is exposed to the credit risk of the offtaker. The concern here is that if the offtaker defaults, the project company will need to enter into a new offtake agreement, which could be at prices lower than the one specified in the original offtake agreement, especially if the power prices have decreased. Therefore, in order to minimize the credit risk to the offtaker, the lenders to the project often require that the offtaker must at least be an **investment grade**-rated company. In the US, many of the buyers for the power output are electric utilities, which are generally rated at investment grade.[1]

EPC contractor

The primary purpose of the EPC ("Engineering, Procurement, and Construction") contract is to design and build the project as per the specifications required by the offtake agreement, federal, state, and local regulations, and the sponsor's specifications. The contract lays out various construction milestones to be achieved, payment schedules, and penalties for missed deadlines or not meeting performance standards.

The EPC contract may be a fully wrapped turnkey contract or could be separated into several subcontracts. For renewable energy projects, the project sponsor will typically procure the primary equipment (e.g. wind turbines for wind farms and solar panels and inverters for solar projects) directly from the manufacturer, through a supply agreement with the vendors. There will also be a separate balance-of-plant (BOP) contract with a reputable construction company. The BOP contract will cover all of the other equipment necessary to construct and commission the project. In such an arrangement, the BOP contractor is responsible only for its obligations under the BOP contract. Banks and financial institutions prefer fully wrapped turnkey contracts, so that there is a single line of command and one party responsible for all equipment and construction obligations. However, such contracts are often more expensive.

The EPC and/or BOP contractors also need to have sufficient financial resources so that the sponsor can claim financial compensation if the project is not completed according to the design specifications and/or in an agreed upon timeframe. Typically, there are provisions for liquidated damages if there are delays in the completion of the project or its construction milestones. The EPC contract may also provide for performance guarantees to ensure that the project meets design specifications. The EPC contractor is generally required to provide a performance bond, LCs and/or parent guarantees to secure its performance for such obligations.

O&M contractor

The O&M contractor provides ongoing operation and maintenance (O&M) services to the project. Similar to the EPC contract, the O&M contract includes performance criteria to be maintained by the contractor. There are often performance bonuses (or penalties) if the actual performance exceeds the criteria (or does not meet them). Also, there are credit support conditions laid out in the contract to ensure that the O&M contractor's obligations under the contract are secured. Sometimes, the project

[1] Edison Electric Institute (EEI), in its 2016 annual report, stated that, according to Standard & Poor's, the average rating for the investor owned utilities (IOUs) comprising the universe of 50 companies was BBB+ and only one IOU was rated below BBB−.

sponsor provides O&M services under a separate contract signed with the project company. Lenders permit such arrangement only if the sponsor has significant experience in operating projects of the same size and type.

In addition to O&M contracts, there are often several other agreements for a power project, such as an Asset Management Agreement (AMA) or Energy Service Agreement (ESA). These contracts cover the provision of management, scheduling, settlement, and similar services to the project. Similar to the O&M contracts, these services are provided by the project sponsor or third-parties for a fee. The AMA covers the miscellaneous administration matters such as procuring and maintaining insurance for the project, providing security services, managing various contracts and financing documents, etc. The ESA defines the management of the project's interface with the local **Independent System Operator (ISO)** or **Regional Transmission Operator (RTO)**, manages the scheduling and delivery of the power produced, and related services.

Project lender

The project lender is the financial institution (or a group of financial institutions) lending money to the project. As discussed in greater detail in later sections, the financing at the project level is usually secured by substantially all of the assets of the company. There may be several classes of debt depending on the type of security and repayment priority.

Tax partner

The tax partner is an equity investor in the project who participates with a view of monetizing the tax benefits. This tax-based structure is relevant only in the United States and will be discussed in Chapter 6.

Back-leverage lender

The project lenders described previously lend directly to the project company, with their loans secured against the assets and income of the project itself. A back-leverage lender provides another potential layer of debt financing, which is not secured by the project company's assets. Instead, the back-leverage lender has a security interest in the sponsor's equity interests in the project. Therefore, in the event of a default under the back-leverage, the back-leverage lender can foreclose on the sponsor's equity interest to effectively become a project owner and guide the destiny of the failed project. Given the limited security associated with the back-leverage and potentially limited avenues for improving recovery prospects in the event of default, this back-leverage debt is priced more expensively relative to the debt at the project level.

As we will discuss in detail in Chapter 6, back-leverage is the preferred debt instrument if the project needs to access tax equity financing. If tax equity financing is not the primary objective, another application of back-leverage is to lever the project further, thereby reducing the sponsor's equity investment.

Risk allocation

As has been seen, project development is a long and continuous process, with various contracts signed along the way. The process may take several years and, in certain cases, longer than 10 years. Situations may change over the long gestation period – laws, permit requirements and/or market conditions may change, for example. Consequently, when the project is ready for financing, while all the contracts and permits may be in place, there may be gaps or deficiencies that render financing difficult. As a result, all the parties who are contractually bound to the project at the time of financing often need to work collaboratively through the negotiation process to ensure successful closing.

A simple analogy for considering the risk allocation process is to imagine all of the parties sitting around a large table. Each party has its interests and differing levels of influence about the final structuring of the transaction. A fundamental principle of project finance is that while risks can be reallocated, not all risks can be removed. The final deal structure – in which everyone benefits if an agreement can be reached – has to provide a reasonable risk/reward structure for each of the key parties since each party has the right to walk away from the table.

Reallocation of risk example

A simple example helps illustrate how risk sharing and allocation may work in real life project finance negotiations. Assume the project sponsor has a tight construction timeline due to delays in the site-permitting process. In this situation, a lender may be concerned that further construction delays may cause the project to miss the deadline for commercial operation (after any applicable grace period) agreed upon in the PPA, which could allow the offtaker to exercise its right to terminate the PPA. Undermining the certainty of the offtake contract could easily undermine the basic economic structure of the project, making financing impossible.

One way to fix the situation may be to go back to the offtaker and revise the deadlines. This is often easier said than done, as the offtaker could be a regulated utility that needs regulatory approval for such a change. Consequently, a sponsor may be forced to seek other alternatives. The sponsor can potentially go to the EPC contractor and renegotiate the contract with more stringent completion deadlines. The EPC contractor might perhaps agree to a tighter deadline by asking for a higher price for the EPC contract. The EPC contractor may in turn go to its subcontractors and vendors to get a clearer view on the equipment delivery schedule and so on down the line. By negotiating with the relevant parties involved, the sponsor may diffuse the risk across the multiple parties in a way that allows the project lenders to become sufficiently comfortable with the construction risk so that debt underwriting can close successfully.

Assessment of project risks

As discussed before, the financial institution is exposed to risks specific to the project it is financing by lending to a single project through a single-purpose entity. While lenders can lend to multiple projects and try to gain the diversification benefits, each project must be evaluated individually so that the risks are well understood and can be mitigated as a part of the underwriting process. As will be discussed in this section, analysis of project risks crosses various disciplines, such as engineering, environmental, accounting, regulatory, legal, real estate, etc. To help understand and address these myriad risks, financial institutions often have teams of engineers and outside advisors to help with project finance due diligence.

The remainder of this section provides a high-level overview of some typical renewable energy project risks and how they are often addressed in a projcct financing structure. Many of these items are discussed in more detail later in this book.

Technology choice

Technology choice is an important consideration for financial institutions under project financing. The project finance community is inherently risk averse because the lenders are lending into a single-purpose entity with no recourse to the project sponsor. Few lenders wish to be the first to lend to a project with a new and unproven technology, or even for a new design of an old technology, such as wind turbines.

Therefore, unless the sponsor can demonstrate significant track record for the technology being used in the project, and there are back-stops with performance guarantees or other protections from the EPC contractor and/or key equipment manufacturer(s), lenders are unlikely to underwrite the performance risk. As a result, newer technologies often have a hard time getting financed.

Construction

As will be discussed later, project finance loans typically consist of a construction loan (for building the project) that automatically converts into term loan (for the operational life of the project). Therefore, banks providing the construction loans are exposed to construction risks. Unfortunately, construction is also the most vulnerable phase for any project because every project (especially energy projects) is made up of many elements, and often all must work before the project can produce any sellable energy.

As a result of the inherent risks during construction phase, rating agencies have an unfavorable view of construction loans, and, in the absence of a fully wrapped turnkey contract with a proven technology, rating agencies are unlikely to rate a project financing highly, even for an otherwise strong project with an attractive offtake agreement.

Banks can mitigate construction risks in multiple ways. First, the banks require that the equity sponsor puts its equity in the project first before drawing upon any of the loans. Second, the banks may require a contingency reserve to be funded upfront, so that any construction cost overruns can be covered without having to chase the project sponsor for an additional equity contribution. Third, in addition to extensive due diligence by outside advisors such as lawyers and engineers, the banks may require that the construction loan be drawn in multiple tranches upon successful achievement of defined construction milestones. Fourth, the banks may require that the project sponsor and/or EPC contractor provide monthly construction supplemental information to be verified by an Independent Engineer to ensure the construction is proceeding as planned. Finally, banks typically have a lower loan-to-value ratio during the construction phase than in the operating phase to limit their exposure to construction risk.

Despite these safeguards, projects do fail or are completed with actual performance below expectations, relative to the design specifications. Rating agencies routinely publish data confirming that the default risk for project finance loans is highest during the construction phase.

The construction phase is one area in which wind and solar projects typically pose the least construction risks among all power generation technologies. The wind turbines and solar panels are often modular in nature and can be installed in a plug-and-play fashion. Their construction timeline is also relatively short, making project planning and execution more predictable than many other technologies.

Operational risks

Once a project achieves commercial operation, the risks usually subside materially. However, the project still faces operational risks, such as poor equipment availability. These risks can be mitigated significantly by having an experienced operator provide O&M services. Banks can further mitigate the risks by requiring the sponsor to provide periodic operating reports to ensure that the project continues operating as expected.

Commodities market and fuel supply

As discussed before, for any energy project, the revenues are dependent on commodities prices. To shield the project's cash flows from commodities price volatility and ensure cost effective financing, a long-term offtake agreement is essential.

Traditional power projects and biomass projects need feedstock, whose prices are also subject to market volatility. Therefore, having a long-term fixed price fuel contract can help cash flow predictability, improving the financing of the projects and reducing related costs.

Offtake credit quality

As discussed before, the offtake agreement is the most important contract in a project financing, as it helps define the project revenues necessary to pay for all of the capital and operating costs incurred. The offtake agreement can therefore make or break commercial viability of a project. As such, underwriters scrutinize the offtake agreement and the offtaker's credit extensively. Financial institutions generally require that the offtaker be rated investment grade. The concern here is that if the offtaker defaults, the project company will need to enter into a new offtake agreement, which could be at prices lower than the one specified in the original offtake agreement. Banks cannot underwrite such contingent commodities risk, hence the typical insistence on an investment grade rating for the offtaker.

The investment grade rating is only the starting point since the banks have internal credit review processes of their own. Additionally, each offtaker and counterparty that is crucial for the successful operation of the project must be analyzed thoroughly before the underwriting can proceed.

Transmission interconnection

A power project cannot connect to a grid unless the project has an **interconnection agreement** with the grid operator. In the United States, most transmission interconnection agreements must be approved by the Federal Energy Regulatory Commission (FERC). Many other countries also require some form of regulatory approval for transmission interconnections that allow generation projects to generate and transmit energy through the power grid.

The location where the project delivers electricity to the offtaker is also important. If the delivery point is away from the interconnection point to the transmission grid, the project company may be exposed to basis risk. We will discuss transmission congestion and basis in Chapters 8 and 9. However, transmission basis risk, if not managed properly, could potentially imperil the profitability of the project.

Sometimes, the project company is required to transmit the power through multiple transmission regions. In this case, the project company may need to procure long-term transmission rights for delivery of electricity to the delivery point agreed upon in the off-take agreement.

Resource

Renewable energy projects are often inherently intermittent sources of electricity. Neither the project sponsor nor the lenders can control how much the wind blows or how much the sun shines. Therefore, project sponsors typically engage an independent resource expert to assess the renewable energy generation potential and its likely variability. If the electricity generation turns out to be systemically lower than what the resource expert projected, the project company could face a cash flow shortfall that jeopardizes debt service. Assessing future generation variability – and the metrics used to assess it - was discussed in Chapter 4.

Regulatory

While specific laws and regulations may differ country to country, the energy industry in most countries is heavily regulated. In the US, the FERC is the agency responsible for regulating the wholesale power market in most of the country. For renewable energy projects, FERC jurisdiction primarily comes into play for anti-trust and interstate transmission matters. The underwriters need to ascertain that the project complies with all applicable ISO and FERC regulations and policies.

Environmental and permitting matters

Each project needs to comply with the applicable environmental laws. In the US, the project sponsor may need to get environmental permits from federal, state, and local authorities, which can make the permitting process difficult and cumbersome. The sponsor also needs to conduct an environmental site assessment to evaluate environmental risks (e.g. chemical spills in the past) associated with the site.

Renewable energy projects must comply with regulations and permitting requirements concerning birds and other wildlife. The presence of various species of endangered birds, bats and other species poses environmental regulation risks that could prevent permitting or limit project operations (for a wind farm, for example).

Natural risks

Depending on the location, projects can be exposed to natural perils such as floods, windstorms, earthquakes, etc. In order to protect the investors, the project company is usually required to procure insurance against such perils. Lenders would naturally prefer that project sponsors insure the projects for the replacement cost of the projects. However, since renewable energy projects, especially wind farms, are often spread across a large area, project sponsors have often been able to convince the financing parties to determine the insurance coverage based on **probabilistic maximum loss** (PML) studies. PML studies are conducted with the help of an insurance consultant and determine the potential worst-case loss, due to a hurricane or windstorm for example.

Real estate matters

The project sponsors must secure the land for the site in the early stages of project development process. Usually, the project company has long-term leases for the land where the project is situated. Renewable energy projects, especially wind projects, need a large amount of land - often stretching across thousands of acres. Consequently, the project company may have to deal with numerous landlords unless the project is located on government-owned lands. The real estate due diligence for a project typically involves ensuring all the lease agreements are in place and have the necessary usage rights and rights-of-way for the generation project itself (e.g. wind turbines or solar panels), as well as the transmission line connecting the project to the transmission grid.

Other project finance characteristics

There are other broad aspects of a project which play a major role in deciding if and how project financing is viable. Several of these are illustrated in Fig. 5.2, then discussed further. Of course, every project is different - many of these risk factors may not be relevant in all cases, while others not listed could be critical. A successful project finance structuring requires a holistic overview of project risks and how these are allocated legally to be successful.

Domicile

Domicile plays an important role not only because one has to consider the laws and regulations applicable to the project, but also where the company owning the project is domiciled. For projects domiciled in emerging markets, especially the ones in high-risk countries, expropriation risks are often critical. A host government might have an incentive to confiscate a profitable project owned by a foreign company if economic conditions decline or if the government changes.

Another consideration for successful project financing is the choice of governing law. The choice of governing law may dictate the recovery prospects if the project

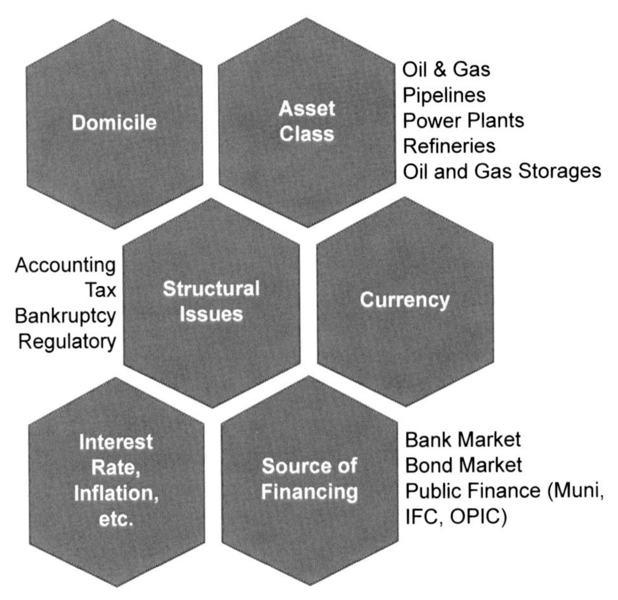

Fig. 5.2 Aspects of project finance.

economics do not work out as expected or if contractual disputes arise. For international project finance transactions, it is customary to submit to New York or English law, and the choice between the two depends upon the parties involved and the nature of financing. For completely domestic transactions, local law is often used.

Asset class

Asset class, of course, poses its own idiosyncrasies. There are unique construction and operations risks to be considered for each asset class, which may govern the financing structure. Financing a power project is much different than financing a refinery. Depending on the nature of the project, the sponsor needs to not only secure a long-term contract for the project output, but also ensure there is a secure supply for the feedstock. Since wind or solar projects do not need any fuel input, financing such projects relieves this constraint.

Currency

Currency risks are relevant for international project financing. Currency risks may arise to the extent revenues, operating expenses, and fuel costs are denominated in a currency different from the one for financing. Typically, debt financings for large international projects are denominated in major currencies such as euros, US dollars, and Japanese yen. Therefore, a **cross-currency swap** may be necessary in order to hedge the currency risks to the extent any currency mismatch exists.

The tenor for debt financing is linked to the term of the revenue contract, which may extend to the useful life of the project, especially in projects with a useful life that extends 20 years or beyond. Certain currencies, especially in emerging markets, do not have the depth in the commercial swaps market to absorb the size of the deals for the tenor-matching financing. Therefore, sponsors may need to access cross-currency swap facilities available from multilateral institutions.

Interest rate and inflation

Similar to the currency risks, the projects may face inflation risks because costs or revenues may be affected by inflation. Inflation risks typically cannot be hedged perfectly. Therefore, the financial institutions may need to study these risks and get comfortable as a part of the underwriting process.

Bank loans are typically floating rate instruments, often linked to the LIBOR. The index rate differs based on the currency. In any case, an **interest rate swap** is necessary so that the project is not exposed to interest rates rising through the term of the financing. Often the banks providing the loans are willing to provide such swaps through a separate derivatives contract. However, similar to cross-currency swaps, interest rate swaps may be difficult to secure due to liquidity constraints or other additional structural challenges.

Source of financing

The project sponsors can tap into bond markets or bank loan markets for financing (discussed in detail in the next section). The project sponsors may also be faced with the task of bringing along other equity investors, especially for very large projects. The task is further complicated because the equity investment can take several forms (e.g. joint venture, straight equity, mezzanine, junior or subordinated loans, etc.) depending on the parties involved and their financial objectives. In the US, there may be the possibility of tapping into public finance in the form of municipal financing or quasi-government agency financing, which may provide certain tax benefits because the investors in the municipal debt are exempt from paying federal income tax on the interest income. While such tax-efficient financing, if available, has the benefit of reducing the financing costs, it adds another layer of complexity.

Structural issues

When structuring a project finance transaction, all the parties need to consider their respective accounting, tax, and regulatory issues. Large-scale projects by definition go through a multitude of regulatory and permitting processes, which can be time consuming. There may be international regulatory matters to consider for cross-border projects or projects located in international waters. Accounting matters can complicate project finance structure, especially if there is a desire on the part of the sponsor not to consolidate the project company on its balance sheet. Furthermore, the financing needs to be structured in a tax efficient manner so that the project being financed is not overly burdened by tax liabilities. For cross-border transactions, there may be opportunities to prevent double taxation or minimize the tax burden. In the US, renewable energy transactions usually need tax equity financing in order to monetize the PTCs and ITCs available. Finally, legal considerations such as bankruptcy may constrain the structuring alternatives. As discussed before, the choice of governing law has a significant bearing on the success of project financing.

These considerations for structuring a financing transaction may conflict with each other, and there are often multiple parties involved, each with conflicting objectives. Therefore, structuring a project financing transaction and closing successfully can be an arduous task. The project sponsor is usually pivotal in bringing the parties with their conflicting issues together and structuring a transaction that makes the financing work.

Tax structures for financing renewable energy projects in the U.S.

The United States has a long history of creating business incentives for public policy goals through the use of federal corporate tax incentives. Rather than creating direct subsidies, US legislators have often chosen to subsidize certain activities by creating tax credits for companies that make qualifying investments. Although the economic effects are similar, the general view is that tax credits are less controversial than directly funding subsidies from government revenues.

Since these tax break mechanisms cannot be used by all investors equally, the effect of these policies has often been the need to create sophisticated tax-oriented financial structures to maximize the value of the incentives offered. Historically, this has been especially true in the US renewable energy sector, where at least a large focus of project finance has been the development of deal structures to accommodate "tax equity" investors. As the rules around these tax incentives have been modified numerous times over the years, renewable energy investors have needed to constantly follow the changing federal renewable energy policy landscape in creating project structures which are consistent with the latest federal tax laws.

Overview of renewable energy tax incentives in the US

The primary federal tax incentives for renewable energy are in the form of tax credits. Tax credits are different from tax deductions in that they can be used directly to offset income tax liability. Tax deductions, on the other hand, can offset taxable income. In this regard, tax credits are more valuable per dollar amount for companies, as they directly offset tax liability on a dollar for dollar basis rather than just reducing taxable income.

There are three primary tax credits available to the renewable energy industry in addition to certain depreciation benefits. Each is discussed in the following sections.

Production tax credit

The Production Tax Credit (PTC) was established as a part of the 1992 *Energy Policy Act*. This tax credit is a production-based incentive whereby the owner of a power project that employs certain renewable energy technologies (e.g. open-loop biomass, small irrigation power, landfill gas, waste-to-energy, hydropower, and marine and hydrokinetic facilities) can claim a tax credit based on the annual electricity generation. The credit is calculated by multiplying the government-mandated PTC rate by the amount of electricity generated in a given tax year. The 1992 *Energy Policy Act* set the PTC

Renewable Energy Finance. https://doi.org/10.1016/B978-0-12-816441-9.00006-[illegible]

rate at $15/MWh and the rate is indexed to the inflation rate. The US Internal Revenue Service (IRS) publishes the PTC rate annually based on its inflation indexing formula.

For the year 2018, the PTC rate was $24/MWh for wind projects and was $12/MWh (50% of the applicable PTC rate for wind projects) for other qualifying renewable energy technologies. The PTCs are available for a term of 10-years from the date a project is placed in service. This term is defined by the IRS, which essentially coincides with the date that the wind project (or other PTC-eligible project) achieves substantial completion.

According to the 1992 *Energy Policy Act*, the PTC was scheduled to be terminated in July 1999. However, the US government extended the PTCs seven times. Consequently, the wind energy industry has been through multiple peaks and troughs that coincided with the expected expiration and/or extension of the PTCs. Recently, the *Protecting Americans from Tax Hikes Act of 2015* ("PATH Act") extended the termination of the PTC eligibility dates to January 1, 2020. However, the PATH Act also required that PTCs be phased out over time based on when construction of the PTC eligible project begins, starting in 2016. Accordingly, for wind projects, projects beginning construction in year 2016 qualify for 100% of the PTC rate, projects beginning construction in year 2017 qualify for 80% of the PTC rate, projects beginning construction in year 2018 qualify for 60% of the PTC rate, and projects beginning construction in year 2019 qualify for 40% of the PTC rate. Projects beginning construction on or after January 1, 2020 will not qualify for the PTC.

The PTC qualification criteria envisaged in the 1992 *Energy Policy Act* was predicated on the notion of projects seeking PTC qualification being placed in service by the eligibility termination date. However, with the *American Taxpayer Relief Act of 2012,* the US government changed the PTC qualification eligibility, stipulating that projects must start construction by the termination date to qualify, rather than being placed in service. Consequently, the IRS has issued several notices providing guidance as to what it means to "begin construction," ultimately stating that a developer must have incurred at least 5% of the project cost by the end of such year or have begun physical construction of significant nature. The latest such IRS notice issued in 2017 provides developers four years from the year construction began to complete an eligible project in order to qualify for the PTC rate applicable for the year in which such project begins construction. If it takes longer than four years to complete an eligible project, it may still qualify for the PTC rate applicable for the year in which it began construction, as long as the developer can demonstrate continuous construction or development until the project is placed in service.

Investment tax credit

The Investment Tax Credit (ITC) is a tax credit available for solar, fuel cell, small wind, and geothermal power projects, among others. The tax credit amounts to 30% of the **tax basis** of the eligible property for solar, fuel cell, and small wind projects, and 10% for geothermal and other qualifying projects. The tax basis for the eligible property qualifying for ITC is calculated based on the cost of the power project after making allowances for the project components that do not qualify for ITC (primarily real estate, transmission equipment, buildings, and intangible assets). Broadly speaking, about 90–95% of the cost of a typical solar power project qualifies for the ITC. To the extent a project claims an ITC, the depreciable basis (i.e., the initial cost) of the ITC eligible property is reduced by 50% of the ITC claimed.

The ITC is subject to a recapture period of 5 years, so that the ITC claimed for eligible property may be recaptured (i.e., required to be paid back to the Treasury) if the subject property is sold or ceases to operate within the five-year recapture period. The recapture liability is subject to a step-down through the recapture period. Specifically, if an eligible property claiming an ITC suffers a recapture event in the first year of operation, the recapture liability is assessed at 100% of the ITC claimed. The recapture liability steps down by 20% for each subsequent year of operation such that after 5 years of recapture period, the recapture liability is reduced to zero.

The US *Energy Policy Act of 2005* created a 30% tax credit for residential and commercial solar energy systems placed in service during the calendar years 2006 and 2007. The credits were extended for one year pursuant to the *Tax Relief and Health Care Act of 2006*. The *Emergency Economic Stabilization Act* signed in 2008 extended the ITC for eight years and eliminated the monetary cap on residential solar installations. The PATH Act of 2015 extended the ITC until 2023, subject to a phase down described next, and the *Bipartisan Budget Act of 2018* changed the eligibility requirement from a "placed-in-service" standard to a "beginning-of-construction" standard similar to PTCs. The IRS issued guidance similar to what was issued under the PTC for purposes of determining what it means to "begin construction." According to the latest law, the ITC is expected to be phased down for solar projects as follows:

- 30% ITC for solar projects that begin construction on or before December 31, 2019
- 26% ITC for solar projects that begin construction in calendar year 2020
- 22% ITC for solar projects that begin construction in 2021
- 10% ITC for solar projects that begin construction on or after (i) December 31, 2021 or (ii) before January 1, 2022 but not placed in service before January 1, 2024

PTC in lieu of ITC

The *American Recovery and Reinvestment Act of 2009* (ARRA) provided an irrevocable election for certain PTC-eligible properties placed in service after December 31, 2008 to claim the ITC in lieu of PTCs. The PATH Act extended the credits for projects seeking ITC in lieu of the PTCs subject to the phase out according to the following schedule (which is consistent with the phase-out for PTCs):

- 30% ITC for projects beginning construction on or before December 31, 2016
- 24% ITC for projects beginning construction in calendar year 2017
- 18% ITC for projects beginning construction in calendar year 2018
- 12% ITC for projects beginning construction in calendar year 2019

The IRS beginning-of-construction guidance for PTC eligible properties also applies to properties claiming ITC in lieu of PTCs.

ITC in lieu of PTCs is valuable for projects that have high capital costs and/or low expected generation profile. Potential examples could be wind projects located in the northeastern parts of the US (where the wind resource is less favorable resulting in lower annual generation) or offshore wind projects (where capital costs are likely to be high).

Eligibility for tax credits

The renewable energy tax credits are available only if for-profit entities that pay corporate income taxes are the owners of the projects claiming the credits. If tax-exempt or non-profit organizations, such as endowments and pension funds, own the projects directly or indirectly, renewable energy tax credits may be disallowed. Certain tax-exempts or non-profits may be eligible for renewable energy tax credits if they invest through a taxable **blocker corporation** (provided the tax-exempt or non-profit do not own more than 50% of the value of the stock) or make an election to cause income from the project to be subject to tax.

In many project finance structures, the project company which directly owns the project will often be a special-purpose entity that may not have taxable income for many years, and thus, under US law, would not directly be eligible for the valuable PTC or ITC tax credits. Therefore, it is common for tax equity investors that invest in the special-purpose entity to take advantage of the ITC or PTC generated by the project owned by the entity. These structures are discussed in detail later in this chapter.

Depreciation benefits

In addition to the tax credits outlined above, renewable energy projects qualify for certain depreciation benefits. Since renewable energy projects are generally quite capital intensive, these depreciation benefits may be valuable to both tax equity investors and sponsors.

Most assets comprising a renewable energy project will qualify for depreciation over five years using the Modified Accelerated Cost Recovery System (MACRS). Some assets with longer useful lives may depreciate over 7-, 15-, 25-, or 39- years using MACRS. The following table illustrates the five-year depreciation schedule using MACRS. As illustrated in Table 6.1, MACRS essentially "front-loads" the depreciation providing valuable tax deductions to offset taxable income during the initial 5–6 years of operation.

If the entity owning a qualified renewable energy projects elects for ITC, the depreciable basis of the project is reduced by 50% of the ITC claimed. Therefore, with a 30% of ITC for an eligible basis of the project, the depreciable basis is reduced to 85% (calculated as 100–50% * 30%).

Table 6.1 MACRS depreciation schedule.

Year of operation	Depreciation
1	20.00%
2	32.00%
3	19.20%
4	11.52%
5	11.52%
6	5.76%
Total	**100%**

Bonus depreciation

In the past, Congress has allowed for bonus depreciation in the amount of 50–100% of the capital cost for qualified energy projects. The federal *Tax Cuts and Jobs Act of 2017* (TCJA) modified the bonus depreciation provision by providing for:

(i) 100% bonus depreciation for qualified property placed in service after September 27, 2017 and before January 1, 2023
(ii) 80% bonus depreciation for qualified property placed in service in 2023
(iii) 60% bonus depreciation for qualified property placed in service in 2024
(iv) 40% bonus depreciation for qualified property placed in service in 2025
(v) 20% bonus depreciation for qualified property placed in service in 2026

As a result, if placed in service before 2023, a renewable energy project can "expense" (or write off) 21% of the project value for tax purposes in the first year of operation using the corporate tax rate of 21% enacted by the TCJA. It should be noted that the depreciation benefit of 21% essentially is repatriated to the tax authorities because 100% of the tax shield afforded by depreciation is depleted in the first year of operation, which causes the taxable income (and the subsequent tax bill) in the following years to be higher (relative to the alternative depreciation method such as MACRS or straight-line). With an upward sloping interest rate curve, this still generates substantial savings for a taxpayer, especially given the long useful life of renewable energy projects.

Tax equity financing structures

As noted above, while renewable energy projects can qualify for tax credits and depreciation benefits, the project companies, which are formed as single-purpose LLCs to own the projects, do not have any means of utilizing the tax benefits due to limited tax capacity of their own. Although renewable energy projects have a long useful life, the project can directly utilize the tax benefits only to the extent of the pre-tax positive income generated by the project, which often is limited. Therefore, absent any financial engineering, there is no way a project company can monetize the tax benefits, as shown in the following extended example.

Illustration of the tax absorption problem for a wind project

The following stylized example illustrates the tax benefit absorption problem:

Assume a project consists of 88 2.3 MW wind turbines with a total capacity of 202.4 MW. The project has a capital cost of $1,575 per MW or approximately $318.78 million total. The project has a 25-year PPA matching the useful life for $55/MWh, escalating at 2.5% per annum. The project has operating expenses that vary from $9.73 million in the first year to $17.59 million in the 25th year of operation. Accordingly, the project has a healthy EBITDA margin of 70.31%. For simplicity, assume that the project qualifies for depreciation benefits alone and the project costs can be allocated according to the schedule in Table 6.2.

Based on cash flow projections for the project, we can assess the project has a debt capacity of $271.87 million, assuming a debt term of 18 years and an interest rate of 4.0% if the debt is sized at a DSCR of 1.30x. Furthermore, after debt sizing, the project has a levered pre-tax IRR of 15.7%. These calculations are summarized briefly in Table 6.3.

Table 6.3 shows the pre-tax benefits to the owner of this sample project. The after-tax IRR to the sponsor depends on how the tax benefits are analyzed. Hypothetically, at one extreme, the project could face a "use-it-or-lose-it" tax regime: if a depreciation benefit cannot be used in that year, its value is lost forever. As shown in Table 6.4, this tax regime assumption provides a levered after-tax IRR of 12.3%. Much of the value of the depreciation benefits is lost as the project company does not have the taxable income each year to absorb the deprecation benefits.

To look at another extreme case, assume that the value of tax benefits can be absorbed efficiently in each year they are generated. This (highly stylized) example provides a levered after-tax IRR of 27.1%, as illustrated in Table 6.5.

Finally, consider a more realistic scenario in which depreciation tax benefits are utilized in the year they are generated, to the extent of any inherent tax liability of

Table 6.2 Project example depreciation assumptions.

Year of operation	Depreciation
5-year MACRS	90%
15-year MACRS	7%
20-year straight-line	2%
39-year straight-line	1%
Total	**100%**

Table 6.3 Example project pre-tax IRR.

Example Model
Assumptions

Assumptions

Project		
# of WTGs	88	#
WTG capacity	2.3	MW
Project capacity	202.4	MW

Capital cost	$/kW	$k	%
WTG	1,000	202,400	63.5%
BOP	300	60,720	19.0%
Contingency	100	20,240	6.3%
Grid connection	100	20,240	6.3%
Development	75	15,180	4.8%
Financing	0	0	0.0%
IDC	0	0	0.0%
Total	1,575	318,780	100.0%

Operating Expenses		
Royalties	3%	% of revenues
Turbine	6.25	$/MWh
BOP	2.00	$/MWh
Utilities	0.40	$/MWh
Project Mgmt	1.25	$/MWh
Insurance	1.00	$/MWh
Property Taxes	2.00	$/MWh
Other Services	1.00	$/MWh
Contingency	5%	% of Operating Expenses
Escalation factor	2.5%	%
WTG warranty	5	years

Production Scenario	NCF
P50	33.6%
P75	32.2%
P90	30.4%
P95	28.6%
P99	27.3%

Offtake	
PPA rate ($/MWh)	55
PPA term (years)	25

Term Loan	
Debt term (years)	18
Debt sizing	1.30x

Backleverage Loan	
Debt term (years)	10
Debt sizing	1.50x

Financing		
Debt	271,866	85.3%
Equity	46,914	14.7%
Total	318,780	100.0%

Other	
Tax Rate	21%
Treasury grant	0%

Continued

Table 6.3 Example project pre-tax IRR.—cont'd

Project-Level Cash Flows													
Period		**1**	**2**	**3**	**4**	**5**	**6**	**7**	**8**	**9**	**10**	**11**	**12**
Revenue													
Production (MWh)		595,736	595,736	595,736	595,736	595,736	595,736	595,736	595,736	595,736	595,736	595,736	595,736
PPA rate ($/MWh)		55.00	56.38	57.78	59.23	60.71	62.23	63.78	65.38	67.01	68.69	70.40	72.16
Revenue		**32,765**	**33,585**	**34,424**	**35,285**	**36,167**	**37,071**	**37,998**	**38,948**	**39,922**	**40,920**	**41,943**	**42,991**
Operating Expense													
Royalties		(983)	(1,008)	(1,033)	(1,059)	(1,085)	(1,112)	(1,140)	(1,168)	(1,198)	(1,228)	(1,258)	(1,290)
Turbine		(3,723)	(3,816)	(3,912)	(4,010)	(4,110)	(4,213)	(4,318)	(4,426)	(4,537)	(4,650)	(4,766)	(4,885)
BOP		(1,191)	(1,221)	(1,252)	(1,283)	(1,315)	(1,348)	(1,382)	(1,416)	(1,452)	(1,488)	(1,525)	(1,563)
Utilities		(238)	(244)	(250)	(257)	(263)	(270)	(276)	(283)	(290)	(298)	(305)	(313)
Project Mgmt		(745)	(763)	(782)	(802)	(822)	(843)	(864)	(885)	(907)	(930)	(953)	(977)
Insurance		(596)	(611)	(626)	(642)	(658)	(674)	(691)	(708)	(726)	(744)	(763)	(782)
Property Taxes		(1,191)	(1,221)	(1,252)	(1,283)	(1,315)	(1,348)	(1,382)	(1,416)	(1,452)	(1,488)	(1,525)	(1,563)
Other Services		(596)	(611)	(626)	(642)	(658)	(674)	(691)	(708)	(726)	(744)	(763)	(782)
Contingency		(463)	(475)	(487)	(499)	(511)	(524)	(537)	(551)	(564)	(578)	(593)	(608)
Total OpEx		**(9,727)**	**(9,970)**	**(10,219)**	**(10,475)**	**(10,737)**	**(11,005)**	**(11,280)**	**(11,562)**	**(11,851)**	**(12,148)**	**(12,451)**	**(12,763)**
EBITDA		**23,039**	**23,615**	**24,205**	**24,810**	**25,430**	**26,066**	**26,718**	**27,386**	**28,070**	**28,772**	**29,491**	**30,229**
Unlevered Cash Flows													
Unlevered After-Tax Cash Flows (Full tax benefit absorption)													
Income tax calculation													
Depreciation (tax)		(63,031)	(100,951)	(61,423)	(37,639)	(37,455)	(19,602)	(1,838)	(1,838)	(1,840)	(1,838)	(1,840)	(1,838)
EBIT		(39,993)	(77,336)	(37,218)	(12,829)	(12,025)	6,464	24,880	25,548	26,230	26,934	27,651	28,391
Income tax benefit/(expense)		8,398	16,241	7,816	2,694	2,525	(1,357)	(5,225)	(5,365)	(5,508)	(5,656)	(5,807)	(5,962)
Unlevered CF to project		**31,437**	**39,855**	**32,021**	**27,504**	**27,955**	**24,709**	**21,493**	**22,021**	**22,562**	**23,116**	**23,685**	**24,267**
Capital costs	(318,780)												
Treasury grant	0												
Unlevered cash flows	(318,780)	31,437	39,855	32,021	27,504	27,955	24,709	21,493	22,021	22,562	23,116	23,685	24,267
IRR	**7.2%**												
Levered Cash Flows													
Levered Pre-Tax Cash Flows													
EBITDA		23,039	23,615	24,205	24,810	25,430	26,066	26,718	27,386	28,070	28,772	29,491	30,229
Less: Debt service		(17,722)	(18,165)	(18,619)	(19,085)	(19,562)	(20,051)	(20,552)	(21,066)	(21,593)	(22,132)	(22,686)	(23,253)
Levered CF to project		5,317	5,450	5,586	5,725	5,869	6,015	6,166	6,320	6,478	6,640	6,806	6,976
Capital costs	(46,914)												
Treasury grant	0												
Levered cash flows	(46,914)	5,317	5,450	5,586	5,725	5,869	6,015	6,166	6,320	6,478	6,640	6,806	6,976
IRR	**15.7%**												

Project-Level Cash Flows													
Period	**13**	**14**	**15**	**16**	**17**	**18**	**19**	**20**	**21**	**22**	**23**	**24**	**25**
Revenue													
Production (MWh)	595,736	595,736	595,736	595,736	595,736	595,736	595,736	595,736	595,736	595,736	595,736	595,736	595,736
PPA rate ($/MWh)	73.97	75.82	77.71	79.66	81.65	83.69	85.78	87.93	90.12	92.38	94.69	97.05	99.48
Revenue	**44,066**	**45,168**	**46,297**	**47,454**	**48,641**	**49,857**	**51,103**	**52,381**	**53,690**	**55,032**	**56,408**	**57,818**	**59,264**
Operating Expense													
Royalties	(1,322)	(1,355)	(1,389)	(1,424)	(1,459)	(1,496)	(1,533)	(1,571)	(1,611)	(1,651)	(1,692)	(1,735)	(1,778)
Turbine	(5,007)	(5,133)	(5,261)	(5,393)	(5,527)	(5,666)	(5,807)	(5,952)	(6,101)	(6,254)	(6,410)	(6,570)	(6,735)
BOP	(1,602)	(1,642)	(1,684)	(1,726)	(1,769)	(1,813)	(1,858)	(1,905)	(1,952)	(2,001)	(2,051)	(2,102)	(2,155)
Utilities	(320)	(328)	(337)	(345)	(354)	(363)	(372)	(381)	(390)	(400)	(410)	(420)	(431)
Project Mgmt	(1,001)	(1,027)	(1,052)	(1,079)	(1,105)	(1,133)	(1,161)	(1,190)	(1,220)	(1,251)	(1,282)	(1,314)	(1,347)
Insurance	(801)	(821)	(842)	(863)	(884)	(906)	(929)	(952)	(976)	(1,001)	(1,026)	(1,051)	(1,078)
Property Taxes	(1,602)	(1,642)	(1,684)	(1,726)	(1,739)	(1,813)	(1,858)	(1,905)	(1,952)	(2,001)	(2,051)	(2,102)	(2,155)
Other Services	(801)	(821)	(842)	(863)	(884)	(906)	(929)	(952)	(976)	(1,001)	(1,026)	(1,051)	(1,078)
Contingency	(623)	(639)	(654)	(671)	(688)	(705)	(722)	(740)	(759)	(778)	(797)	(817)	(838)
Total OpEx	**(13,082)**	**(13,409)**	**(13,744)**	**(14,087)**	**(14,440)**	**(14,801)**	**(15,171)**	**(15,550)**	**(15,939)**	**(16,337)**	**(16,746)**	**(17,164)**	**(17,593)**
EBITDA	**30,984**	**31,759**	**32,553**	**33,367**	**34,201**	**35,056**	**35,932**	**36,831**	**37,751**	**38,695**	**39,663**	**40,654**	**41,671**
Unlevered Cash Flows													
Unlevered After-Tax Cash Flows (Full tax benefit absorption)													
Income tax calculation													
Depreciation (tax)	(1,840)	(1,838)	(1,840)	(1,133)	(429)	(429)	(429)	(429)	(87)	(87)	(87)	(87)	(87)
EBIT	29,144	29,921	30,713	32,234	33,772	34,627	35,504	36,402	37,664	38,608	39,575	40,567	41,583
Income tax benefit/(expense)	(6,120)	(6,283)	(6,450)	(6,769)	(7,092)	(7,272)	(7,456)	(7,644)	(7,909)	(8,108)	(8,311)	(8,519)	(8,732)
Unlevered CF to project	**24,864**	**25,475**	**26,103**	**26,598**	**27,109**	**27,784**	**28,477**	**29,186**	**29,842**	**30,588**	**31,352**	**32,135**	**32,938**
Capital costs													
Treasury grant													
Unlevered cash flows	24,864	25,475	26,103	26,598	27,109	27,784	28,477	29,186	29,842	30,588	31,352	32,135	32,938
Levered Cash Flows													
Levered Pre-Tax Cash Flows													
EBITDA	30,984	31,759	32,553	33,367	34,201	35,056	35,932	36,831	37,751	38,695	39,663	40,654	41,671
Less: Debt service	(23,834)	(24,430)	(25,041)	(25,667)	(26,308)	(26,966)	0	0	0	0	0	0	0
Levered CF to project	7,150	7,329	7,512	7,700	7,893	8,090	35,932	36,831	37,751	38,695	39,663	40,654	41,671
Capital costs													
Treasury grant													
Levered cash flows	7,150	7,329	7,512	7,700	7,893	8,090	35,932	36,831	37,751	38,695	39,663	40,654	41,671

Table 6.4 After-tax IRR under a "use-it-or lose-it" tax regime.

Levered After-tax Cash Flows (No Loss Carry-forward)														
Income tax calculation														
EBITDA		23,039	23,615	24,205	24,810	25,430	26,066	26,718	27,386	28,070	28,772	29,491	30,229	30,984
Interest expense		(10,875)	(10,601)	(10,298)	(9,965)	(9,601)	(9,202)	(8,768)	(8,297)	(7,786)	(7,234)	(6,638)	(5,996)	(5,306)
Depreciation (tax		(63,031)	(100,951)	(61,423)	(37,639)	(37,455)	(19,602)	(1,838)	(1,838)	(1,840)	(1,838)	(1,840)	(1,838)	(1,840)
Income tax expense		0	0	0	0	0	0	(3,384)	(3,623)	(3,873)	(4,137)	(4,413)	(4,703)	(5,006)
EBITDA		23,039	23,615	24,205	24,810	25,430	26,066	26,718	27,386	28,070	28,772	29,491	30,229	30,984
Less: Income tax expense		0	0	0	0	0	0	(3,384)	(3,623)	(3,873)	(4,137)	(4,413)	(4,703)	(5,006)
Less: Debt service		(17,722)	(18,165)	(18,619)	(19,085)	(19,562)	(20,051)	(20,552)	(21,066)	(21,593)	(22,132)	(22,686)	(23,253)	(23,834)
Levered CF		5,317	5,450	5,586	5,725	5,869	6,015	2,782	2,697	2,604	2,503	2,393	2,273	2,144
Capital costs	(46,914)													
Treasury grant	0													
Levered cash flows	(46,914)	5,317	5,450	5,586	5,725	5,869	6,015	2,782	2,697	2,604	2,503	2,393	2,273	2,144
IRR	12.3%													

Levered After-tax Cash Flows (No Loss Carry-forward) (continued)												
Income tax calculation												
EBITDA	31,759	32,553	33,367	34,201	35,056	35,932	36,831	37,751	38,695	39,663	40,654	41,671
Interest expense	(4,564)	(3,770)	(2,919)	(2,009)	(1,037)	0	0	0	0	0	0	0
Depreciation (tax	(1,838)	(1,840)	(1,133)	(429)	(429)	(429)	(429)	(87)	(87)	(87)	(87)	(87)
Income tax expense	(5,325)	(5,658)	(6,156)	(6,670)	(7,054)	(7,456)	(7,644)	(7,909)	(8,108)	(8,311)	(8,519)	(8,732)
EBITDA	31,759	32,553	33,367	34,201	35,056	35,932	36,831	37,751	38,695	39,663	40,654	41,671
Less: Income tax expense	(5,325)	(5,658)	(6,156)	(6,670)	(7,054)	(7,456)	(7,644)	(7,909)	(8,108)	(8,311)	(8,519)	(8,732)
Less: Debt service	(24,430)	(25,041)	(25,667)	(26,308)	(26,966)	0	0	0	0	0	0	0
Levered CF	2,004	1,854	1,544	1,222	1,036	28,477	29,186	29,842	30,588	31,352	32,135	32,938
Levered cash flows	2,004	1,854	1,544	1,222	1,036	28,477	29,186	29,842	30,588	31,352	32,135	32,938

GAAP Depreciation

Captial cost 318,780

	% of project	Total	1	2	3	4	5	6	7	8	9	10	11	12	13	14	15	16	17	18	19	20
5-year MACRS	90%	100.00%	20.0%	32.0%	19.2%	11.5%	11.5%	5.8%														
15-year MACRS	7%	100.00%	5.0%	9.5%	8.6%	7.7%	6.9%	6.2%	5.9%	5.9%	5.9%	5.9%	5.9%	5.9%	5.9%	5.9%	5.9%	3.0%				
20-year SL	2%	100.00%	5.0%	5.0%	5.0%	5.0%	5.0%	5.0%	5.0%	5.0%	5.0%	5.0%	5.0%	5.0%	5.0%	5.0%	5.0%	5.0%	5.0%	5.0%	5.0%	5.0%
39-year SL	1%	100.00%	2.6%	2.6%	2.6%	2.6%	2.6%	2.6%	2.6%	2.6%	2.6%	2.6%	2.6%	2.6%	2.6%	2.6%	2.6%	2.6%	2.6%	2.6%	2.6%	2.6%
Total	**100.0%**	**100.0%**	**18.5%**	**29.6%**	**18.0%**	**11.0%**	**11.0%**	**5.7%**	**0.5%**	**0.5%**	**0.5%**	**0.5%**	**0.5%**	**0.5%**	**0.5%**	**0.5%**	**0.5%**	**0.3%**	**0.1%**	**0.1%**	**0.1%**	**0.1%**
Tax depreciation		318,780	58,897	94,329	57,394	35,170	34,998	18,316	1,717	1,717	1,719	1,717	1,719	1,717	1,719	1,717	1,719	1,059	401	401	401	401

	21	22	23	24	25	26	27	28	29	30	31	32	33	34	35	36	37	38	39
5-year MACRS																			
15-year MACRS																			
20-year SL																			
39-year SL	2.6%	2.6%	2.6%	2.6%	2.6%	2.6%	2.6%	2.6%	2.6%	2.6%	2.6%	2.6%	2.6%	2.6%	2.6%	2.6%	2.6%	2.6%	2.6%
Total	**0.0%**	**0.0%**	**0.0%**	**0.0%**	**0.0%**	**0.0%**	**0.0%**	**0.0%**	**0.0%**	**0.0%**	**0.0%**	**0.0%**	**0.0%**	**0.0%**	**0.0%**	**0.0%**	**0.0%**	**0.0%**	**0.0%**
Tax depreciation	82	82	82	82	82	82	82	82	82	82	82	82	82	82	82	82	82	82	82

Tax Book Depreciation Schedule (FMV)

Captial cost	318,780
Development Fee Mark-up (15%)	47,817
Total Fair Market Value	366,597

	% of project	Total	1	2	3	4	5	6	7	8	9	10	11	12	13	14	15	16	17	18	19	20
5-year MACRS	90%	100.00%	20.0%	32.0%	19.2%	11.5%	11.5%	5.8%														
15-year MACRS	7%	100.00%	5.0%	9.5%	8.6%	7.7%	6.9%	6.2%	5.9%	5.9%	5.9%	5.9%	5.9%	5.9%	5.9%	5.9%	5.9%	3.0%				
20-year SL	2%	100.00%	5.0%	5.0%	5.0%	5.0%	5.0%	5.0%	5.0%	5.0%	5.0%	5.0%	5.0%	5.0%	5.0%	5.0%	5.0%	5.0%	5.0%	5.0%	5.0%	5.0%
39-year SL	1%	100.00%	2.6%	2.6%	2.6%	2.6%	2.6%	2.6%	2.6%	2.6%	2.6%	2.6%	2.6%	2.6%	2.6%	2.6%	2.6%	2.6%	2.6%	2.6%	2.6%	2.6%
Total	**100.0%**	**100.0%**	**18.5%**	**29.6%**	**18.0%**	**11.0%**	**11.0%**	**5.7%**	**0.5%**	**0.5%**	**0.5%**	**0.5%**	**0.5%**	**0.5%**	**0.5%**	**0.5%**	**0.5%**	**0.3%**	**0.1%**	**0.1%**	**0.1%**	**0.1%**
Book depreciation		366,597	67,731	108,478	66,003	40,445	40,248	21,064	1,975	1,975	1,977	1,975	1,977	1,975	1,977	1,975	1,977	1,218	461	461	461	461

	21	22	23	24	25	26	27	28	29	30	31	32	33	34	35	36	37	38	39
5-year MACRS																			
15-year MACRS																			
20-year SL																			
39-year SL	2.6%	2.6%	2.6%	2.6%	2.6%	2.6%	2.6%	2.6%	2.6%	2.6%	2.6%	2.6%	2.6%	2.6%	2.6%	2.6%	2.6%	2.6%	2.6%
Total	**0.0%**	**0.0%**	**0.0%**	**0.0%**	**0.0%**	**0.0%**	**0.0%**	**0.0%**	**0.0%**	**0.0%**	**0.0%**	**0.0%**	**0.0%**	**0.0%**	**0.0%**	**0.0%**	**0.0%**	**0.0%**	**0.0%**
Book depreciation	94	94	94	94	94	94	94	94	94	94	94	94	94	94	94	94	94	94	94

Table 6.5 After-tax IRR with all tax benefits used in the year that they are generated.

Levered After-tax Cash Flows (Full tax benefit absorption)																										
EBITDA		23,039	23,615	24,205	24,810	25,430	26,066	26,718	27,386	28,070	28,772	29,491	30,229	30,984	31,759	32,553	33,367	34,201	35,056	35,932	36,831	37,751	38,695	39,663	40,654	41,671
Income tax benefit/(expense)		10,682	18,467	9,978	4,787	4,541	575	(3,384)	(3,623)	(3,873)	(4,137)	(4,413)	(4,703)	(5,006)	(5,325)	(5,658)	(6,156)	(6,670)	(7,054)	(7,456)	(7,644)	(7,909)	(8,108)	(8,311)	(8,519)	(8,732)
Debt Service		(17,722)	(18,165)	(18,619)	(19,085)	(19,562)	(20,051)	(20,552)	(21,066)	(21,593)	(22,132)	(22,686)	(23,253)	(23,834)	(24,430)	(25,041)	(25,667)	(26,308)	(26,966)	0	0	0	0	0	0	0
AT Levered CF		15,999	23,916	15,564	10,512	10,410	6,590	2,782	2,697	2,604	2,503	2,393	2,273	2,144	2,004	1,854	1,544	1,222	1,036	28,477	29,186	29,842	30,588	31,352	32,135	32,938
Capital costs	(46,914)																									
Treasury grant	0																									
Levered cash flows	(46,914)	15,999	23,916	15,564	10,512	10,410	6,590	2,782	2,697	2,604	2,503	2,393	2,273	2,144	2,004	1,854	1,544	1,222	1,036	28,477	29,186	29,842	30,588	31,352	32,135	32,938
IRR	27.1%																									

the project, while any leftover tax benefits are deferred until offset by tax liability generated in future years. The results of using this assumption are illustrated in Table 6.6. Under this assumption, the levered after-tax IRR settles at a figure of 14.9%.[1]

These examples demonstrate that tax benefits (in this case, accelerated depreciation) can greatly affect the after-tax project returns that are critical to investors. However, unless these tax benefits can be monetized through appropriate financial structures, they are likely to be lost in an ordinary project finance structure.

Fortunately, IRS rules allow for a way of monetizing the tax benefits through a variety of structured transactions that have collectively come to be known as "tax equity" transactions. The three most common tax equity structures are discussed in the remainder of this chapter.

The partnership flip structure

US partnership tax rules allow for disproportionate allocations of economic and tax benefits among the partnership's partners. Using these rules, a sponsor developing a renewable energy project can find a tax equity investor who has tax liabilities of its own and form a partnership with an objective of monetizing the renewable energy project's tax benefits.

In its simplest form, the tax equity investor and the sponsor form a partnership with the tax equity investor funding approximately 35–75% of the capital cost of the project (depending on the project's details) and the sponsor funding the remainder (Fig. 6.1). The tax equity investor is allocated 99% taxable income, gain, loss, deduction, and credit from the partnership until the "Flip Date", while the Sponsor takes the remaining 1%. The Flip Date is the date on which the tax equity investor realizes its pre-defined return target (often called a "Target IRR" or "Flip Yield"). After the Flip Date, the allocation of taxable income, gain, loss, deduction, and credit to the tax equity investor drops to 5%, and goes up to 95% for the sponsor. The share of cash distributed to the tax equity investor and sponsor depends on the economic motivations of the sponsor and the tax equity investors, but there is a general consensus in the tax equity community that the tax equity investor shall take at least 5% of the cash generated by the renewable energy project through the life of its investment.

Most tax equity investors size cash distributions so that the pre-tax cash-on-cash IRR over the useful life of the project (usually 25 years) is at least 2.0%. For the purpose of calculating a pre-tax IRR, both cash distributions and tax credits are considered and treated as cash. There is no specific guidance from the IRS in terms of IRR requirements for these investments, and hence the sponsor and tax equity investor can engineer a mutually satisfactory outcome in order to get to the 2% pre-tax IRR. Finally, most tax equity investors require that the after-tax IRR over the useful life of the

[1] The example here assumes 100% of the losses otherwise not utilized in a given year can be deferred to the following year. This has generally been the case in the United States. However, the recently enacted Tax Cuts and Jobs Act limits the amount of losses than can be carried forward in a subsequent year to 80%.

Table 6.6 After-tax IRR with deferred tax benefits.

Levered After-tax Cash Flows (Loss Carry-forward)													
Income tax calculation													
Yearly Tax Payable		10,682	18,467	9,978	4,787	4,541	575	(3,384)	(3,623)	(3,873)	(4,137)	(4,413)	(4,703)
Tax Loss Carryforward	0	10,682	29,149	39,127	43,914	48,455	49,030	45,647	42,024	38,151	34,014	29,601	24,898
Actual Tax Paid		0	0	0	0	0	0	0	0	0	0	0	0
EBITDA		23,039	23,615	24,205	24,810	25,430	26,066	26,718	27,386	28,070	28,772	29,491	30,229
Less: Income tax expense		0	0	0	0	0	0	0	0	0	0	0	0
Less: Debt service		(17,722)	(18,165)	(18,619)	(19,085)	(19,562)	(20,051)	(20,552)	(21,066)	(21,593)	(22,132)	(22,686)	(23,253)
AT Levered CF		5,317	5,450	5,586	5,725	5,869	6,015	6,166	6,320	6,478	6,640	6,806	6,976
Capital costs	(46,914)												
Treasury grant	0												
Levered cash flows	(46,914)	5,317	5,450	5,586	5,725	5,869	6,015	6,166	6,320	6,478	6,640	6,806	6,976
IRR	14.9%												

Levered After-tax Cash Flows (Loss Carry-forward) (cont.)													
Income tax calculation													
Yearly Tax Payable	(5,006)	(5,325)	(5,658)	(6,156)	(6,670)	(7,054)	(7,456)	(7,644)	(7,909)	(8,108)	(8,311)	(8,519)	(8,732)
Tax Loss Carryforward	19,892	14,567	8,909	2,753	0	0	0	0	0	0	0	0	0
Actual Tax Paid	0	0	0	0	(3,918)	(7,054)	(7,456)	(7,644)	(7,909)	(8,108)	(8,311)	(8,519)	(8,732)
EBITDA	30,984	31,759	32,553	33,367	34,201	35,056	35,932	36,831	37,751	38,695	39,663	40,654	41,671
Less: Income tax expense	0	0	0	0	(3,918)	(7,054)	(7,456)	(7,644)	(7,909)	(8,108)	(8,311)	(8,519)	(8,732)
Less: Debt service	(23,834)	(24,430)	(25,041)	(25,667)	(26,308)	(26,966)	0	0	0	0	0	0	0
AT Levered CF	7,150	7,329	7,512	7,700	3,975	1,036	28,477	29,186	29,842	30,588	31,352	32,135	32,938
Capital costs													
Treasury grant													
Levered cash flows	7,150	7,329	7,512	7,700	3,975	1,036	28,477	29,186	29,842	30,588	31,352	32,135	32,938

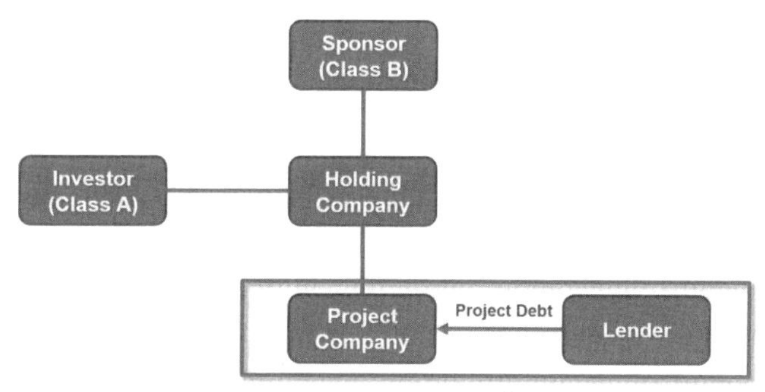

Fig. 6.1 Simplified diagram of a partnership flip structure[2].

project (usually 25 years) should be at least 50 bps higher than the Target IRR. Again, there is no specific IRS guidance on such matters, but the tax equity market has acted fairly consistently in requiring these metrics. The 50 bps bump in the after-tax IRR is achieved by adjusting the cash distributions to the tax equity investors allocable after the Flip Date higher or lower with a proviso that the tax equity investor receives at least 5% of the cash distributions after the Flip Date. Consequently, the sponsor desires to keep the cash distributions allocable to the tax equity investor as close to 5% as possible because the upfront investment from the tax equity investor typically does not depend on the cash distributions the investor receives after the Flip Date. In other words, the tax equity investor receives all cash distributions after the Flip Date for "free."

The transactions are structured so that, at the inception of the tax equity partnership, the tax equity investor is expected to achieve the Target IRR by a certain date ("Target Flip Date"). If the tax equity investor does not achieve the Target IRR by the Target Flip Date, the partnership documents require that the tax equity investor receives up to 100% of the partnership cash until the tax equity investor achieves the Target IRR. This is an important consideration from a financing perspective because the cash sweep provision limits sponsor back-leverage (discussed in later sections) and may create other inter-creditor complications.

The tax equity investor's IRR for the investment is calculated using the highest marginal federal tax rate without consideration for state income tax rate. Therefore, if there is a change in tax rate (similar to what was enacted in 2017 pursuant to the TCJA), the tax equity investor is insulated from the lower tax benefits resulting from a lower tax rate. Specifically, if the change in tax rate reduces the *net* depreciation benefits allocable to the tax equity investor, the Flip Date would be delayed. As a result, up to 100% of the cash available from the project would be swept to the tax equity investor, which will generate its return through a correspondingly higher cash flow distribution. This return would accrue through cash sweeps after the Target Flip Date because the tax equity investor may have to stay in the deal longer in order to make up for the lost tax benefits. Please note, the timing of the future tax rate change is

[2] With project-level debt, the partnership structure is termed as a levered partnership. Without project level debt, the structure is termed as an unlevered partnership.

important. If the tax rate change occurs in the later part of the tax equity financing when most of the depreciation benefits have already been claimed, the flip date may accelerate and occur prior to the Target Flip Date.

Purchase options in a partnership flip structure

Typically, the sponsor has an option to purchase the tax equity investor's equity interest in the project at "fair market value" (as certified by an independent appraiser) at pre-determined time intervals after the Flip Date. In another variation, the fair market value governing the purchase option is fixed at the inception of the transaction as a good faith estimate provided by an independent appraiser. Usually, tax equity investors set the purchase option exercise price at 102–105% of such fixed purchase price estimates for an added margin of comfort, so as to avoid additional scrutiny from the IRS which might claim that the purchase option gives the tax equity investor certainty of exit from the partnership.[3]

Variations in the partnership flip structure[4]

In a variation of the partnership flip structures, the Flip Date is fixed at the start, leading to a structure known as a "Fixed Flip" structure. In this structure, the tax equity investor receives cash distributions equal to 2% of its original investment on a preferred basis and some proportion of the remaining cash on a *pro rata* basis for a fixed time period ("Fixed Flip Term"). Once the Fixed Flip Term is over, the Sponsor may exercise the purchase option at a predetermined time interval. However, if the sponsor chooses not to do so, the tax equity investor typically has a withdrawal option whereby for a period of 2 years, the investor receives a predetermined level of cash from the partnership, termed as withdrawal cash. However, if the full withdrawal price is not recovered from the cash available from the project within the first two years after the Fixed Flip Term, the tax equity investor can take over the project.

Renewable energy developers generally prefer the Fixed Flip structure because the structure allows them to retain most of the cash, against which they can raise additional debt (in the form of "back-leverage," discussed in later sections). The structure also has several debt-like features, which ensures that the tax equity investor exits the partnership at the end of the fixed flip term.

In another variation of partnership structure, the tax equity investor makes investments in two parts – an initial upfront investment and deferred pay-as-you-go (called "PAYGO") contributions that are linked to the PTCs received by the project. If the generation from the project is lower than expected at the time of financing, the PAYGO rate automatically adjusts downward according to a predetermined scale.

[3] The fixed purchase options are allowed by IRS pursuant to the IRS Notice 2009-69 (https://www.irs.gov/pub/irs-drop/a-09-69.pdf).

[4] Keith Martin (2017, April 6). "Partnership Flips", Norton Rose Fulbright. Retrieved from https://www.nortonrosefulbright.com/en-us/knowledge/publications/bde454cb/partnership-flips.

The tax equity investment is sized such that the upfront investment is at least 75% of the sum of upfront investment and PAYGO contributions in order to comply with the IRS Revenue Procedure 2007-65 (discussed further in subsequent sections). This structure with deferred PAYGO contributions enables the tax equity investor to mitigate some of the wind generation risks, as the amount it contributes are linked to PTCs received by the project, which are directly related to the level of wind production. This structure also alleviates potential underperformance due to transmission curtailments from the perspective of the tax equity investor.

Partnership structure requirements

In a partnership flip structure, the tax equity investor receives three streams of benefits: a portion of the partnership cash distributions, tax credits, and tax benefits or losses that can be used against taxable income. The tax equity investor makes an upfront investment based on its view of the present value of the benefit streams, net of any potential tax liabilities. A typical partnership flip structure raises 35–50% of the project capital for solar ITC transactions and 50–80% of total project capital for wind PTC transactions.[5] The partnership flip structures are typically documented in an LLC Agreement (a partnership formation document), which identifies the role and responsibility of each partner, income allocations, future cash distributions, etc. The document also designates the sponsor as the Managing Member who retains the day-to-day control of the project. The tax equity investor serves as a passive investor so as not to have to consolidate the investment on its financial statements. The investor does have certain consent rights to ensure that the project is run and operated prudently. The tax equity investor also receives monthly operational reports and periodic financial statements. If the tax equity investor has reason to believe that the project is being run in contravention of the governing documents or if the sponsor has committed willful misconduct or gross negligence, they typically have the right to remove the sponsor as the Managing Member of the LLC.

Partnership flip structures have been around for a long time and were used for early transactions in the wind sector. The market for tax equity remained fragmented, however, as there was limited consensus as to how the transactions were to be structured. Several sponsors approached the IRS to get a private letter ruling to validate the structure for the tax equity financing they raised. Consequently, in 2007, the IRS published Revenue Procedure RP2007-65 to establish a safe harbor for partnership structures for wind projects.[6] In 2009, IRS Notice N2009-69 further clarified the Revenue Procedure 2007-65's safe harbor guidelines.[7] The guidelines provide that the IRS will generally not scrutinize a wind project company as a partnership or investors as partners in the partnership if the following requirements are satisfied:

[5] As the PTCs and ITC step down pursuant to the PATH act, the capitalization rate are expected to change.
[6] IRS Revenue Procedure 2007-65 (https://www.irs.gov/pub/irs-drop/rp-07-65.pdf).
[7] IRS Notice 2009–69 (https://www.irs.gov/pub/irs-drop/n-09-69.pdf).

- The sponsor has a minimum 1% interest in each material item of partnership income, gain, loss, deduction, and credit in all items during the existence of the project company; each tax equity investor has a minimum 5% interest in each material item of partnership income or gain for each taxable year.
- On or before the later of the dates that the wind farm is placed in service or the date the investor acquires its interest in the project company, the investor must make a minimum unconditional investment in the project company. The investor minimum investment must be equal to at least 20% of the sum of the fixed capital contributions plus reasonably anticipated contingent capital contributions required to be made by the investor under the partnership agreement. The investor must not be protected against loss of any portion of the investor-minimum investment through any agreement directly or indirectly with the sponsors, any other investor, the turbine supplier, the power purchaser, or any affiliates thereof.
- At least 75% of the sum of an investor's fixed capital contribution, plus reasonably anticipated contingent capital contributions (i.e., PAYGO contributions) to be contributed by an investor with respect to the interest in the project company, must be fixed and determinable obligations that are not contingent in amount or certainty of payment.[8]
- Neither the sponsor, the tax equity investors, nor any related parties may have a contractual right to purchase at any time the wind farm, any property included in the wind farm, or an interest in the project company at a price less than its fair market value determined at the time of exercise of the contractual right to purchase, and further that the sponsor may not have a contractual right to purchase the wind farm or an interest in the project company earlier than 5 years after the wind farm is placed in service.[9]
- The project company may not have a contractual right to cause any party to purchase the wind farm or any property included in the wind farm from the project company. An investor may not have a contractual right to cause any party to purchase its partnership interest in the project company.
- No person may guarantee or otherwise ensure the tax equity investor the right to allocation of the PTC. The project company must bear the risk that available wind resource is not as great as anticipated. The sponsor, the turbine supplier, or any power purchaser may not provide guarantee that the wind resource will be available at a certain level. A guarantee for the wind resource availability may be provided by a third party not related to the sponsor, the turbine supplier, any power purchaser, or any other project participant if the project company or the tax equity investor pays the cost of or a premium for such a guarantee.[10] A long-term power purchase agreement between the project company and any unrelated party may not constitute a guarantee. However, a take-or-pay contract between related parties would constitute a guarantee and is not permissible. The sponsor may not lend any tax equity investor the funds to acquire any part of the tax equity investor's interest in the project company or guarantee any indebtedness incurred in relation to the tax equity investor's interest in the project company.

[8] The IRS safe harbor specified in Revenue Procedure 2007-65 belies the PAYGO structure described earlier.

[9] The IRS notice 2009-69 clarified that the fair market value governing the buyout option may be fixed at the inception of the transaction if it is a good faith estimate provided by an independent appraiser.

[10] For example, a weather derivative contract between the project company and an insurance company is an acceptable guarantee.

In a 2015 memo, the IRS clarified that the safe harbor guidelines listed above do not apply for solar transactions or other renewable energy projects claiming ITC. The memo directed that the general partnership principles should apply to test as to whether the tax equity investor is really a partner in a partnership. The IRS has not issued any guidelines for partnership flip structure for ITC transactions. However, the tax equity investor community has grown comfortable adopting most of the structuring elements from PTC to ITC partnership flip structures.

Partnership flip structures are the most common structure for tax credit monetization. Almost 80% of the solar ITC transactions and almost 100% of the wind PTC transactions used the partnership flip structure in recent years.

Structural considerations governing the partnership flip structure[11]

The partnership flip model places constraints on how financing transactions may be structured, with respect to transaction timing, accounting, additional leverage (back-leverage), and other factors.

Transaction timing

For wind project PTC transactions, tax equity investors can commit to an investment as soon as the project is ready for construction. Such commitment, when made early on, gives the sponsor confidence that the project will get tax equity funding when the project reaches commercial operation. Moreover, once a tax equity commitment is in place, lenders feel comfortable extending incremental construction loans that bridge a portion of the tax equity commitment (usually 95%). The legal documents governing such a commitment specify the conditions that need to be satisfied before the tax equity investor is legally obligated to fund the investment (the "Conditions Precedent"). Such conditions are similar to the ones lenders would typically ask for in conversion of construction loan into term loan. Some of the important conditions are:

(i) the project reaches commercial operation
(ii) the project company has all material documents in full force and effect
(iii) the project company is not in default or has not commenced bankruptcy proceeding
(iv) the project qualifies for and has met the requirements for tax credits and depreciation benefits, etc.

Once the conditions are satisfied, the sponsor can notify the tax equity investors certifying that all Conditions Precedent have been satisfied, at which time the tax equity investor can fund the investment. Tax equity transactions for solar ITC projects work generally the same way, with one important exception. Tax equity investors typically fund wind project investments upon commercial operation, and there is no timing

[11] Please note that for the purpose of this section, the terms "book", "equity", etc. refer to "tax book" and "equity" from a tax perspective. The terms shouldn't be confused with the similar terms used from a financial accounting perspective.

constraint as to when a tax equity investor can fund the investment. However, for solar ITC transactions, the partnership needs to be formed and the tax equity investor needs to be a partner in the partnership before the project is placed in service.[12] There is no strict rule as to exactly when the tax equity investor needs to make the investment for projects claiming ITC. However, for the sake of conservatism, most tax equity investors make an investment at or before the project achieves mechanical completion.[13] Of course, a tax equity investor does not wish to fully fund the investment at mechanical completion and take the risk, albeit fairly small, that the project never achieves commercial operation. Therefore, tax equity investors typically fund investments in two installments: the first installment occurs at or immediately before mechanical completion (in the amount of at least 20% of the overall tax equity investment) and the second installment at or after commercial operation.[14]

Contribution versus purchase model

There are two ways to create a partnership flip structure. The tax equity investor may initially contribute its share of capital in exchange for an interest in the project company (i.e., the tax equity partnership). Alternatively, the tax equity investor may "purchase" an interest in the project company directly from the sponsor. The choice between the two is dictated by the intended use of the proceeds, taxation, and step-up in basis. In the contribution model, the proceeds from the investor's contribution can be distributed to the sponsor - but it makes more sense if the project company can use the funds to pay for any outstanding construction costs. In this model, the sponsor may avoid having to pay taxes on the tax equity investor's contributions provided that the transaction abides by the safe harbor for "disguised sale", which essentially states that the sponsor can treat the distribution of tax equity proceeds as the reimbursement of capital expenditures made on the project during the preceding two years. In order to qualify for the safe harbor and avoid taxation, the project cannot be worth more than 120% of the tax basis of the project at the time of the formation of the partnership with the tax equity investor. This has the effect of limiting the step-up in basis for the project, which may affect the ITC and depreciation benefits that the project company can claim.

[12] As discussed before, "placed in service" is a technical term under the tax law which generally fixes the time at which the relevant asset is eligible for tax credits and depreciation. In determining if an asset is placed in service, there is a five-factor test, which specifies that a project is deemed to be placed in service based on whether the following five conditions are satisfied: (1) approval of required licenses and permits; (2) passage of control of the facility to taxpayer; (3) completion of critical tests; (4) commencement of daily or regular operations; and, (5) synchronization into a power grid for generating electricity to produce income. See IRS Private Letter Ruling PLR 144,688-12 (June 28, 2013); See also Moran, Gambino, Chase, & Ludwig, Renewable Power Facilities: Placed-in-Service Issues, Tax Notes (May 23, 2016).

[13] Mechanical completion is yet another milestone that is defined in EPC contracts. The milestone typically occurs once all mechanical equipment is completely installed but before the project substation is energized.

[14] Most of the tax equity investors rely upon the safeharbor established by the IRS Revenue Procedure 2007-65 when sizing the first installment at 20%. However, there are some tax equity investors, albeit in minority, who believe that the first installment can be sized as small as 5%.

In the purchase model, the proceeds from the tax equity investor are remitted directly to the sponsor who is treated as selling a share of project assets to the tax equity investor. Consequently, the sponsor is required to pay taxes on the gains that result from the share of the project assets deemed sold. However, the purchase model may be beneficial in that the higher basis resulting from a step-up allows for a higher ITC and depreciation benefits, albeit at the expense of immediate tax liability resulting from a proportionately higher gain for the sponsor. The potential for higher ITC and depreciation benefits make the purchase model preferable for projects claiming ITC. The model is less relevant for wind projects claiming PTCs because the incremental benefit arising from depreciation is less meaningful.

It should be noted that the tax equity investor ultimately bears the risk of any step-up in basis for ITC transactions. This risk arises as the IRS may dispute the ITC basis through periodic tax audits. There have been several lawsuits pending wherein the IRS disputed that because the ITC basis was not justified, the project company should have received a smaller tax credit.[15]

There is guidance from the IRS that suggests that a development fee mark-up of approximately 15–20% is acceptable. As a result, tax equity investors typically limit the step-up in basis to 15–20% of the project costs if justified by the appraisal of the fair market value of the project. As an additional protection, tax equity investors typically require that the sponsor indemnify the tax equity investors for any reduction in ITC pursuant to a successful IRS dispute of the ITC basis. The sponsor is often required to support the indemnity through a creditworthy parental guarantee. If the parent is not sufficiently creditworthy and/or as an additional protection, the tax equity investor may require a cash sweep to the extent any of the indemnity claims remain unfulfilled. This provision allows tax equity investors to claim additional distributions until any deficiency in ITC is fully repaid (with an appropriate adjustment for taxes and time value of money). As will be seen in later sections, the cash sweep provision often creates frictions if the sponsor employs back-leverage secured by its own interests in the project company.

Capital accounts and deficit restoration obligations

In partnership transactions, the tax rules require that the partners track the capital account and **outside tax basis** for their investments. A capital account tracks each partner's share of equity in a partnership. At the inception of the partnership, each partner's capital account starts with the equity contributed to the partnership. Thereafter, the capital account balance for each partner is increased by adding book income and

[15] One such example of the dispute is the 2018 case Alta Wind I Owner-Lessor C et al. v. United States, 897 F.3d 13,651,365 argued in front of the US Federal Court of Appeals. See Moran, Broome, Gambino, Odell, & Chase, Federal Circuit Reverses and Remands Alta Wind, Holding that a Portion of the Asserted Section 1603 Basis May Be Allocated Entirely to Intangibles, Energy Law Report, Vol. 19-3 (march, 2019). Another case California Ridge Wind Energy, LLC & Invenergy Wind, LLC v. United States was also decided in favor of the government[, but at the time of this publication, the opinion had not yet been released.] See, California Ridge Wind Energy, LLC & Invenergy Wind, LLC v. United States, No. 14–250 C (Fed. Cl. filed Jan. 7, 2019).

capital contributions allocated to the respective partners, whereas the capital account balance for each partner is reduced by deducting book losses and cash distributions allocated to the respective partners. If any of the partner's capital account balance reaches zero, any further book losses shift to the other partner. Given the structure of the tax equity transactions, the tax equity investor's capital account balance typically reaches zero first. Following that, any book losses and tax credits are reallocated to the sponsor. Since the primary objective of the tax equity transactions is to utilize tax benefits efficiently, reallocating book losses to the sponsor is not ideal.

However, if the tax equity investor agrees to a Deficit Restoration Obligation (DRO), then its capital account balance can go negative in an amount up to the DRO and the investor can continue to receive the losses. The DRO obligates the tax equity investor to contribute capital in the amount of DRO in the event the partnership liquidates. The DRO is a contingent liability and is crystallized only in the rare event of partnership liquidation.[16] The deficits can be higher if the partnership claims bonus depreciation or 100% expensing (as per TCJA) because the book losses are front-loaded into the transaction with these claims. Similarly, the PAYGO transactions explained earlier can lead to higher deficits. In this structure, the tax equity investor contributes up to 25% of the capital on a deferred basis, which reduces the up-front capital account balance attributable to the tax equity investor. Given the potential for higher deficits, the tax equity investor may need to sign up for higher DROs.

Tax equity investors are usually leery about signing up for high DROs. Typical market transactions cap the DRO liability to 40% of the capital contributed upfront. The DROs are relatively more bothersome for PTC transactions because once the DRO cap is reached, unless it is raised, both depreciation benefits and PTCs are reallocated to the sponsor, which often impairs the economics for the sponsor if the sponsor has limited tax capacity. For the ITC transactions, it is customary to allow for reallocations once the ITC is claimed, which effectively brings down the income allocation to the tax equity investor from 99% to a lower number. In order to avoid recapture of the ITC, the tax rules require that income allocations to the tax equity investor should not drop below 67% until the expiration of the 5-year recapture period. There is also a general consensus that the tax equity investor should keep its share of income allocations at 99% for a meaningful period following the date the project is placed in service, owing to the concern that the IRS may argue that 99% of income allocation was meant only to capture the ITC when available and the income allocation changed when profits were received. Sometimes, tax equity transactions are structured such that the income allocations to the tax equity investors are reduced to 67% for a 4-year period following the first anniversary of the placed-in-service date in order to manage the DROs to a lower level. All of these considerations differ from institution to institution, which can make structuring tax equity transactions very challenging.

[16] Commercial Aspects of Deficit Restoration Obligations in LLC and Partnership Transactions - Akin Gump Publication by David Burton (https://www.akingump.com/images/content/4/1/v2/41603/DRO-Commercial-Aspects-Project-Perspectives-Article.pdf).

If the tax equity investor has a negative capital amount balance (or a DRO) on the Flip Date, tax equity investors typically require that they receive the maximum permissible income allocation. At such time, the partnership is likely to be outside of the tax loss situation and should be generating taxable income. Accordingly, any additional income allocation (beyond the minimum post-flip 5% allocation agreed upon in the LLC Agreement), result in reducing the DRO (or reducing the absolute value of the capital account balance). However, any such allocation to "cure" the DRO after the Flip Date triggers a tax liability (in an amount equal to the additional income allocation multiplied by the prevailing federal tax rate) for the tax equity investor. Therefore, the tax equity investor is often allocated additional cash to cover the incremental tax liability.

It is possible that the sponsor will end up with a negative capital account balance. In such cases, the tax equity investor may require the sponsor to sign up for a DRO. Such situations are rare and can be dealt with in the LLC Agreement.

As a part of the up-front structuring, one way to "manage" the DRO is to introduce a "cash holiday" whereby the tax equity investor takes no cash during the first 5–7 years as contractually agreed upon initially. By doing so, the tax equity investor can manage the stress on the capital accounts since any cash distributions are deducted from the capital account balance. Once the cash holiday is over, the tax equity investor takes a much higher share of cash distributions (50–100%) until the Target Flip Date.

Another way to manage the DRO is to add non-recourse debt at the project level. Such levered tax equity transactions are still infrequent as tax equity investors are highly averse to the potential for any tax credit recapture caused by foreclosure of the project assets in the event of a default. So far, our discussion has been limited to unlevered tax equity transactions, in which there is no non-recourse debt at the project level. Levered tax equity transactions are not discussed here for the sake of simplicity. However, these transactions may become more prevalent in the future, especially for solar projects, once tax credits start stepping down pursuant to the PATH Act.

Explanation of inside and outside tax basis

As with capital accounts, partners are also required to track their "inside" and "outside" tax basis. The initial outside basis of a partner is typically equal to its initial capital account balance; however, there are certain situations beyond the scope of this book that may cause outside basis to differ from its capital account balance. Inside basis is generally the partner's share of the tax basis of partnership assets (i.e., the cost of such assets reduced by depreciation).

Once the outside tax basis of a partner reaches zero, any tax losses allocated to said partner are suspended. Such suspended losses can be offset only by the future income allocated to said partner. The partner with the suspended losses may continue receiving "excess cash distributions" because they are cash distributions in excess of the partner's outside tax basis; these must be reported as capital gains, and taxed accordingly.

Due to the loss limitations caused by outside tax basis, there are limitations on how much tax losses a tax equity investor can absorb despite agreeing to a higher DRO. It should be noted that since suspended losses do not have any economic value for

Table 6.7 A Simplified tax equity financing model using partnership flip structure.

Example Model
Tax Equity and Partnership

Project capacity	202.4
Production selected	P50
Escalation rate	2.5%
Tax rate	21%

Target Flip Date (Years)	10
CPI Inflation Rate	2%
2018 PTC Rate ($/MWh)	24
Tax Equity Target IRR	6.50%
Tax Equity Full-Term After Tax IRR	7.19%
Backelverage Term	10
Backleverage Sizing DSCR	1.50x <<— *For simplicity, backleverage is sized at a DSCR of 1.30x for P50 production level*

Factor	NCF	MWh
P50	33.6%	595,736
P75	32.2%	570,914
P90	30.4%	538,999
P95	28.6%	507,085
P99	27.3%	484,036

	Cash Allocations		Income Allocations	
Tax Equity	20%	5%	99%	5%
Sponsor	80%	95%	1%	95%

	Income Reallocation Limits	
Tax Equity	Max. Allocation	99%
Tax Equity	Min. Allocation	5%
Sponsor	Effective Allocation	95%

	Deficit Restoration Obligation		
Tax Equity	% of Contribution	35%	60,000
Sponsor	% of Contribution	15%	30,000

Partnership Cash Flows

Period	1	2	3	4	5	6	7	8	9	10	11	12	13
Revenue													
Production (MWh)	595,736	595,736	595,736	595,736	595,736	595,736	595,736	595,736	595,736	595,736	595,736	595,736	595,736
PPA rate	55.00	56.38	57.78	59.23	60.71	62.23	63.78	65.38	67.01	68.69	70.40	72.16	73.97
Revenues	**32,765**	**33,585**	**34,424**	**35,285**	**36,167**	**37,071**	**37,998**	**38,948**	**39,922**	**40,920**	**41,943**	**42,991**	**44,066**
Operating Expense													
Royalties	(983)	(1,008)	(1,033)	(1,059)	(1,085)	(1,112)	(1,140)	(1,168)	(1,198)	(1,228)	(1,258)	(1,290)	(1,322)
Turbine	(3,723)	(3,816)	(3,912)	(4,010)	(4,110)	(4,213)	(4,318)	(4,426)	(4,537)	(4,650)	(4,766)	(4,885)	(5,007)
BOP	(1,191)	(1,221)	(1,252)	(1,283)	(1,315)	(1,348)	(1,382)	(1,416)	(1,452)	(1,488)	(1,525)	(1,563)	(1,602)
Utilities	(238)	(244)	(250)	(257)	(263)	(270)	(276)	(283)	(290)	(298)	(305)	(313)	(320)
Project Mgmt	(745)	(763)	(782)	(802)	(822)	(843)	(864)	(885)	(907)	(930)	(953)	(977)	(1,001)
Insurance	(596)	(611)	(626)	(642)	(658)	(674)	(691)	(708)	(726)	(744)	(763)	(782)	(801)
Property Taxes	(1,191)	(1,221)	(1,252)	(1,283)	(1,315)	(1,348)	(1,382)	(1,416)	(1,452)	(1,488)	(1,525)	(1,563)	(1,602)
Other Services	(596)	(611)	(626)	(642)	(658)	(674)	(691)	(708)	(726)	(744)	(763)	(782)	(801)
Contingency	(463)	(475)	(487)	(499)	(511)	(524)	(537)	(551)	(564)	(578)	(593)	(608)	(623)
Total OpEx	**(9,727)**	**(9,970)**	**(10,219)**	**(10,475)**	**(10,737)**	**(11,005)**	**(11,280)**	**(11,562)**	**(11,851)**	**(12,148)**	**(12,451)**	**(12,763)**	**(13,082)**
EBITDA	**23,039**	**23,615**	**24,205**	**24,810**	**25,430**	**26,066**	**26,718**	**27,386**	**28,070**	**28,772**	**29,491**	**30,229**	**30,984**
Depreciation (tax)	(67,731)	(108,478)	(66,003)	(40,445)	(40,248)	(21,064)	(1,975)	(1,975)	(1,977)	(1,975)	(1,977)	(1,975)	(1,977)
Partnership Taxable Income	**(44,693)**	**(84,864)**	**(41,798)**	**(15,635)**	**(14,817)**	**5,002**	**24,743**	**25,411**	**26,093**	**26,797**	**27,514**	**28,254**	**29,007**
PTC Rate ($/MWh)	24	25	25	26	26	27	28	28	29	29	30	30	31
PTCs Earned by the Partnership	14,298	14,893	14,893	15,489	15,489	16,085	16,681	16,681	17,276	17,276	0	0	0
Income Allocation to Tax Equity	99%	99%	99%	99%	99%	99%	99%	99%	99%	99%	5%	5%	5%
Income Allocation to the Sponsor	1%	1%	1%	1%	1%	1%	1%	1%	1%	1%	95%	95%	95%
Cash Allocation to Tax Equity	20%	20%	20%	20%	20%	20%	20%	20%	20%	20%	5%	5%	5%
Cash Allocation to the Sponsor	80%	80%	80%	80%	80%	80%	80%	80%	80%	80%	95%	95%	95%
Taxable Income to Tax Equity	(44,246)	(84,015)	(41,380)	(15,479)	(14,669)	4,952	24,496	25,157	25,832	26,529	1,376	1,413	1,450
Taxable Income to Sponsor	(447)	(849)	(418)	(156)	(148)	50	247	254	261	268	26,138	26,841	27,557

Period	14	15	16	17	18	19	20	21	22	23	24	25
Revenue												
Production (MWh)	595,736	595,736	595,736	595,736	595,736	595,736	595,736	595,736	595,736	595,736	595,736	595,736
PPA rate	75.82	77.71	79.66	81.65	83.69	85.78	87.93	90.12	92.38	94.69	97.05	99.48
Revenues	**45,168**	**46,297**	**47,454**	**48,641**	**49,857**	**51,103**	**52,381**	**53,690**	**55,032**	**56,408**	**57,818**	**59,264**
Operating Expense												
Royalties	(1,355)	(1,389)	(1,424)	(1,459)	(1,496)	(1,533)	(1,571)	(1,611)	(1,651)	(1,692)	(1,735)	(1,778)
Turbine	(5,133)	(5,261)	(5,393)	(5,527)	(5,666)	(5,807)	(5,952)	(6,101)	(6,254)	(6,410)	(6,570)	(6,735)
BOP	(1,642)	(1,684)	(1,726)	(1,769)	(1,813)	(1,858)	(1,905)	(1,952)	(2,001)	(2,051)	(2,102)	(2,155)
Utilities	(328)	(337)	(345)	(354)	(363)	(372)	(381)	(390)	(400)	(410)	(420)	(431)
Project Mgmt	(1,027)	(1,052)	(1,079)	(1,105)	(1,133)	(1,161)	(1,190)	(1,220)	(1,251)	(1,282)	(1,314)	(1,347)
Insurance	(821)	(842)	(863)	(884)	(906)	(929)	(952)	(976)	(1,001)	(1,026)	(1,051)	(1,078)
Property Taxes	(1,642)	(1,684)	(1,726)	(1,769)	(1,813)	(1,858)	(1,905)	(1,952)	(2,001)	(2,051)	(2,102)	(2,155)
Other Services	(821)	(842)	(863)	(884)	(906)	(929)	(952)	(976)	(1,001)	(1,026)	(1,051)	(1,078)
Contingency	(639)	(654)	(671)	(688)	(705)	(722)	(740)	(759)	(778)	(797)	(817)	(838)
Total OpEx	**(13,409)**	**(13,744)**	**(14,087)**	**(14,440)**	**(14,801)**	**(15,171)**	**(15,550)**	**(15,939)**	**(16,337)**	**(16,746)**	**(17,164)**	**(17,593)**
EBITDA	**31,759**	**32,553**	**33,367**	**34,201**	**35,056**	**35,932**	**36,831**	**37,751**	**38,695**	**39,663**	**40,654**	**41,671**
Depreciation (tax)	(1,975)	(1,977)	(1,218)	(461)	(461)	(461)	(461)	(94)	(94)	(94)	(94)	(94)
Partnership Taxable Income	**29,784**	**30,576**	**32,149**	**33,740**	**34,595**	**35,472**	**36,370**	**37,657**	**38,601**	**39,569**	**40,560**	**41,577**
PTC Rate ($/MWh)	32	32	33	34	34	35	36	36	37	38	39	39
PTCs Earned by the Partnership	0	0	0	0	0	0	0	0	0	0	0	0
Income Allocation to Tax Equity	5%	5%	5%	5%	5%	5%	5%	5%	5%	5%	5%	5%
Income Allocation to the Sponsor	95%	95%	95%	95%	95%	95%	95%	95%	95%	95%	95%	95%
Cash Allocation to Tax Equity	5%	5%	5%	5%	5%	5%	5%	5%	5%	5%	5%	5%
Cash Allocation to the Sponsor	95%	95%	95%	95%	95%	95%	95%	95%	95%	95%	95%	95%
Taxable Income to Tax Equity	1,489	1,529	1,607	1,687	1,730	1,774	1,819	1,883	1,930	1,978	2,028	2,079
Taxable Income to Sponsor	28,295	29,047	30,542	32,053	32,866	33,698	34,552	35,775	36,671	37,590	38,532	39,498

		1	2	3	4	5	6	7	8	9	10	11	12	13	14	15	16	17	18	19	20	21	22	23	24	25
Benefits Attributable to Tax Equity																										
Cash Distributions		4,608	4,723	4,841	4,962	5,086	5,213	5,344	5,477	5,614	5,754	1,475	1,511	1,549	1,588	1,628	1,668	1,710	1,753	1,797	1,842	1,888	1,935	1,983	2,033	2,084
Tax Benefits/(losses)		9,292	17,643	8,690	3,251	3,081	(1,040)	(5,144)	(5,283)	(5,425)	(5,571)	(289)	(297)	(305)	(313)	(321)	(338)	(354)	(363)	(372)	(382)	(395)	(405)	(415)	(426)	(437)
PTCs		14,155	14,744	14,744	15,334	15,334	15,924	16,514	16,514	17,104	17,104	0	0	0	0	0	0	0	0	0	0	0	0	0	0	0
Total Benefits	(171,569)	28,054	37,111	28,275	23,547	23,501	20,097	16,713	16,708	17,293	17,287	1,186	1,215	1,245	1,275	1,307	1,331	1,356	1,390	1,424	1,460	1,492	1,529	1,568	1,607	1,647
Tax Equity Contribution	171,569																									
Benefits Attributable to Sponsor																										
Cash Distributions		18,431	18,892	19,364	19,848	20,344	20,853	21,374	21,909	22,456	23,018	28,017	28,717	29,435	30,171	30,925	31,698	32,491	33,303	34,136	34,989	35,864	36,760	37,679	38,621	39,587
Tax Benefits/(losses)		94	178	88	33	31	(11)	(52)	(53)	(55)	(56)	(5,489)	(5,637)	(5,787)	(5,942)	(6,100)	(6,414)	(6,731)	(6,902)	(7,077)	(7,256)	(7,513)	(7,701)	(7,894)	(8,092)	(8,295)
PTCs		143	149	149	155	155	161	167	167	173	173	0	0	0	0	0	0	0	0	0	0	0	0	0	0	0
Total Benefits		18,668	19,219	19,601	20,036	20,530	21,003	21,489	22,022	22,574	23,134	22,528	23,081	23,648	24,229	24,825	25,285	25,760	26,401	27,059	27,733	28,351	29,060	29,786	30,530	31,292
Sponsor Contribution	195,028																									
Sponsor Backleverage																										
Cash Flow Available for Backleverage		18,431	18,892	19,364	19,848	20,344	20,853	21,374	21,909	22,456	23,018	28,017	28,717	29,435	30,171	30,925	31,698	32,491	33,303	34,136	34,989	35,864	36,760	37,679	38,621	39,587
CF Available for Debt Service		12,287	12,594	12,909	13,232	13,563	13,902	14,249	14,606	14,971	15,345	0	0	0	0	0	0	0	0	0	0	0	0	0	0	0
Backleverage Amount	107,955																									
Sponsor Levered Cash Flows	(87,074)	6,144	6,297	6,455	6,616	6,781	6,951	7,125	7,303	7,485	7,673	28,017	28,717	29,435	30,171	30,925	31,698	32,491	33,303	34,136	34,989	35,864	36,760	37,679	38,621	39,587
704(b) Capital Accounts and Outside Basis																										
Period		**1**	**2**	**3**	**4**	**5**	**6**	**7**	**8**	**9**	**10**	**11**	**12**	**13**	**14**	**15**	**16**	**17**	**18**	**19**	**20**	**21**	**22**	**23**	**24**	**25**
Partnership 704(b) Capital Account																										
Note: 734(b) Step-ups ignored																										
Beginning Balance	0	366,597	298,866	190,387	124,385	83,939	43,692	22,628	20,653	18,679	16,702	14,727	12,750	10,775	8,798	6,823	4,846	3,628	3,168	2,707	2,247	1,786	1,692	1,598	1,504	1,410
Plus: Basis Contribution	366,597	0	0	0	0	0	0	0	0	0	0	0	0	0	0	0	0	0	0	0	0	0	0	0	0	0
Plus: Allocation of Income / (Loss)	0	(44,693)	(84,864)	(41,798)	(15,635)	(14,817)	5,002	24,743	25,411	26,093	26,797	27,514	28,254	29,007	29,784	30,576	32,149	33,740	34,595	35,472	36,370	37,657	38,601	39,569	40,560	41,577
Less: Cash Distribution	0	(23,039)	(23,615)	(24,205)	(24,810)	(25,430)	(26,066)	(26,718)	(27,386)	(28,070)	(28,772)	(29,491)	(30,229)	(30,984)	(31,759)	(32,553)	(33,367)	(34,201)	(35,056)	(35,932)	(36,831)	(37,751)	(38,695)	(39,663)	(40,654)	(41,671)
Ending Balance	366,597	298,866	190,387	124,385	83,939	43,692	22,628	20,653	18,679	16,702	14,727	12,750	10,775	8,798	6,823	4,846	3,628	3,168	2,707	2,247	1,786	1,692	1,598	1,504	1,410	1,316
Tax Equity 704(b) Capital Account																										
Note: 734(b) Step-up is ignored																										
Beginning Balance	0	171,569	122,715	33,977	(12,243)	(32,684)	(52,440)	(52,701)	(33,548)	(13,869)	6,349	27,124	27,012	26,899	26,786	26,672	26,558	26,481	26,441	26,401	26,360	26,319	26,295	26,271	26,247	26,222
Plus: Basis Contribution	171,569	0	0	0	0	0	0	0	0	0	0	0	0	0	0	0	0	0	0	0	0	0	0	0	0	0
Less: Cash Distribution	0	(4,608)	(4,723)	(4,841)	(4,962)	(5,086)	(5,213)	(5,344)	(5,477)	(5,614)	(5,754)	(1,475)	(1,511)	(1,549)	(1,588)	(1,628)	(1,668)	(1,710)	(1,753)	(1,797)	(1,842)	(1,888)	(1,935)	(1,983)	(2,033)	(2,084)
Plus: Initial Allocation of Income / (Loss)	0	(44,246)	(84,015)	(41,380)	(15,479)	(14,669)	4,952	24,496	25,157	25,832	26,529	1,376	1,413	1,450	1,489	1,529	1,607	1,687	1,730	1,774	1,819	1,883	1,930	1,978	2,028	2,079
Interim Balance #1	171,569	122,715	33,977	(12,243)	(32,684)	(52,440)	(52,701)	(33,548)	(13,869)	6,349	27,124	27,026	26,913	26,800	26,687	26,573	26,497	26,458	26,418	26,378	26,337	26,314	26,290	26,266	26,242	26,217
Reallocations from Tax Equity to Sponsor	0	0	0	0	0	0	0	0	0	0	0	0	0	0	0	0	0	0	0	0	0	0	0	0	0	0
Reallocations from Sponsor to Tax Equity	0	0	0	0	0	0	0	0	0	0	0	(14)	(14)	(15)	(15)	(15)	(16)	(17)	(17)	(18)	(18)	(19)	(19)	(20)	(20)	(21)
DRO Special Income Allocation	0	0	0	0	0	0	0	0	0	0	0	0	0	0	0	0	0	0	0	0	0	0	0	0	0	0
Income / (Loss) Allocation	0	(44,246)	(84,015)	(41,380)	(15,479)	(14,669)	4,952	24,496	25,157	25,832	26,529	1,362	1,399	1,436	1,474	1,513	1,591	1,670	1,712	1,756	1,800	1,864	1,911	1,959	2,008	2,058
Interim Balance #2	171,569	122,715	33,977	(12,243)	(32,684)	(52,440)	(52,701)	(33,548)	(13,869)	6,349	27,124	27,012	26,899	26,786	26,672	26,558	26,481	26,441	26,401	26,360	26,319	26,295	26,271	26,247	26,222	26,196
DRO Contribution / Distribution	0	0	0	0	0	0	0	0	0	0	0	0	0	0	0	0	0	0	0	0	0	0	0	0	0	(26,196)
Ending Balance	171,569	122,715	33,977	(12,243)	(32,684)	(52,440)	(52,701)	(33,548)	(13,869)	6,349	27,124	27,012	26,899	26,786	26,672	26,558	26,481	26,441	26,401	26,360	26,319	26,295	26,271	26,247	26,222	0
Tax Equity Deficit Restoration Obligation																										
Tax Equity DRO Cap	60,000	60,000	60,000	60,000	60,000	60,000	60,000	60,000	60,000	60,000	60,000	0	0	0	0	0	0	0	0	0	0	0	0	0	0	0
DRO Cap %	35.0%																									
Balance		0	0	12,243	32,684	52,440	52,701	33,548	13,869	0	0	0	0	0	0	0	0	0	0	0	0	0	0	0	0	0
Max Balance	52,701																									
Max Balance %	30.7%																									
Sponsor 704(b) Capital Account																										
Note: 734(b) Step-ups ignored																										
Beginning Balance	0	195,028	176,150	156,410	136,628	116,624	96,131	75,329	54,202	32,547	10,352	(12,398)	(14,262)	(16,124)	(17,988)	(19,849)	(21,712)	(22,853)	(23,273)	(23,694)	(24,113)	(24,533)	(24,603)	(24,673)	(24,743)	(24,812)
Plus: Basis Contribution	195,028	0	0	0	0	0	0	0	0	0	0	0	0	0	0	0	0	0	0	0	0	0	0	0	0	0
Less: Cash Distribution	0	(18,431)	(18,892)	(19,364)	(19,848)	(20,344)	(20,853)	(21,374)	(21,909)	(22,456)	(23,018)	(28,017)	(28,717)	(29,435)	(30,171)	(30,925)	(31,698)	(32,491)	(33,303)	(34,136)	(34,989)	(35,864)	(36,760)	(37,679)	(38,621)	(39,587)
Plus: Initial Allocation of Income / (Loss)	0	(447)	(849)	(418)	(156)	(148)	50	247	254	261	268	26,138	26,841	27,557	28,295	29,047	30,542	32,053	32,866	33,698	34,552	35,775	36,671	37,590	38,532	39,498
Interim Balance #1	195,028	176,150	156,410	136,628	116,624	96,131	75,329	54,202	32,547	10,352	(12,398)	(14,276)	(16,138)	(18,002)	(19,864)	(21,727)	(22,869)	(23,290)	(23,711)	(24,131)	(24,551)	(24,622)	(24,692)	(24,762)	(24,832)	(24,901)
Reallocations from Sponsor to Tax Equity	0	0	0	0	0	0	0	0	0	0	0	14	14	15	15	15	16	17	17	18	18	19	19	20	20	21
Reallocations from Tax Equity to Sponsor	0	0	0	0	0	0	0	0	0	0	0	0	0	0	0	0	0	0	0	0	0	0	0	0	0	0
DRO Special Income Allocation	0	0	0	0	0	0	0	0	0	0	0	0	0	0	0	0	0	0	0	0	0	0	0	0	0	0
Income / (Loss) Allocation	0	(447)	(849)	(418)	(156)	(148)	50	247	254	261	268	26,152	26,855	27,571	28,310	29,062	30,558	32,070	32,883	33,716	34,570	35,793	36,690	37,610	38,552	39,518
Interim Balance #2	195,028	176,150	156,410	136,628	116,624	96,131	75,329	54,202	32,547	10,352	(12,398)	(14,262)	(16,124)	(17,988)	(19,849)	(21,712)	(22,853)	(23,273)	(23,694)	(24,113)	(24,533)	(24,603)	(24,673)	(24,743)	(24,812)	(24,880)
DRO Contribution / Distribution	0	0	0	0	0	0	0	0	0	0	0	0	0	0	0	0	0	0	0	0	0	0	0	0	0	24,880
Ending Balance	195,028	176,150	156,410	136,628	116,624	96,131	75,329	54,202	32,547	10,352	(12,398)	(14,262)	(16,124)	(17,988)	(19,849)	(21,712)	(22,853)	(23,273)	(23,694)	(24,113)	(24,533)	(24,603)	(24,673)	(24,743)	(24,812)	0
Sponsor Deficit Restoration Obligation																										
Sponsor DRO Cap	30,000	30,000	30,000	30,000	30,000	30,000	30,000	30,000	30,000	30,000	30,000	0	0	0	0	0	0	0	0	0	0	0	0	0	0	0
DRO Cap %	15.4%																									
Balance		0	0	0	0	0	0	0	0	0	12,398	14,262	16,124	17,988	19,849	21,712	22,853	23,273	23,694	24,113	24,533	24,603	24,673	24,743	24,812	0
Max Balance	24,812																									
Max Balance %	12.7%																									

tax equity investors, when modeling tax equity transactions, the suspended losses should not be counted toward calculating the tax equity investor's IRR. Furthermore, the IRR calculation should consider any capital gain taxes payable on the excess capital distributions.

Table 6.7 provides a tax equity sizing model using the same parameters we used in Table 6.3. However, the example has been modified to include PTCs to make it more relevant to the current market. The tax equity sizing model incorporates the capital account and inside/outside tax basis concepts we discussed in prior sections. The spreadsheet example illustrates a tax equity sized at $171.60 million for the sizing parameters listed in Table 6.7. The tax equity is sized with a Target IRR of 6.50% and the investment yields a full-term IRR of 7.19%. The tax equity investor has a maximum DRO of $60 million (approximately 35% of the up-front investment). The sponsor has a DRO of $30 million (approximately 15% of the up-front investment).

Back-leverage considerations

As discussed above, tax equity investors are averse to project-level debt due to the concern that in the event of a default, the lenders would foreclose on the project assets, which in turn could trigger a recapture of the tax credits. The risk is more pronounced for ITC transactions, which are subject to recapture during the first five years of operation upon a change of control. Sometimes, in order to avoid the potential for recapture, lenders to the project-level debt may agree to forbearance during the recapture period. However, project-level debt still remains a rare feature of these transactions because, with tax equity capital in short supply, tax equity investors are able to maintain a strong position in setting transaction terms, which keeps unlevered partnerships as the primary structure to monetize tax credits.

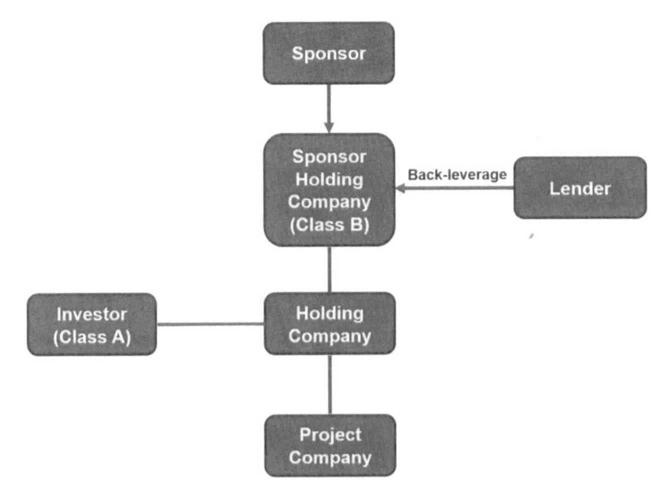

Fig. 6.2 Back leverage in a project finance structure.

As a result, back-leverage transactions have become more prevalent when the sponsor needs additional capital. Back-leverage transactions are debt-financing transactions secured by the sponsor's equity interest in the project company, as illustrated in Fig. 6.2.

In a back-leverage loan, the lender is one step away from the project company's assets. Specifically, if the back-leverage loan defaults, the lender can foreclose upon the sponsor's equity in the project company and not on the project itself. Consequently, back-leverage debt financings are thus perceived to be less secure and more expensive than project level debt.

Other risk allocation issues

Tax equity transactions may create additional complications from a structuring and risk allocation perspective. First, as discussed before, tax equity investors need access to cash otherwise distributable to the sponsor in a partnership flip transaction if there is a delay in achieving Target IRR beyond the Target Flip Date, or if the Sponsor is unable to fulfill its indemnity obligations (for example, if the ITC is reduced pursuant to a successful IRS audit). If a cash sweep is triggered, there is less cash to pay for the debt service, which may impair debt service. Second, if there is a default under back-leverage, the lender may foreclose on the sponsor's equity interest. Tax equity investors usually impose significant operational experience requirements on the parties to whom the sponsor's equity interests can be transferred in order to manage operational risks. Both of these issues may lead to protracted tripartite negotiations among the sponsor, tax equity investor, and back-leverage lenders before financing can close.

Finally, as discussed before, the transaction timing is complicated for projects qualified for ITC because tax equity investors typically wish to stage their investments in two installments. Usually, the first installment, 20% of the upfront tax equity investment, is not sufficient to repay the construction loan. As a result, the construction lenders are unlikely to give up their security in the project assets at this stage with a paydown of the construction loan for a mere 20% of the tax equity commitment. Subsequently, until the project achieves commercial operation, the tax equity investors are effectively invested into a levered tax equity transaction. Moreover, the tax equity investors are unlikely to recover any meaningful portion of their first installment if the project fails to achieve commercial operation. In such circumstances, the tax equity investors are faced with a dilemma to either take the risk that the first installment may never be recovered, or structure mitigating recovery measures for the transaction within the auspices of the IRS guidelines.

Consensus has emerged to resolve these thorny inter-creditor issues so that financing can close in a reasonable time frame. For example, tax equity investors may limit the cash sweep to an amount that leaves enough cash for debt service. In terms of change of control in the event of a back-leverage debt default, tax equity investors may require provisions whereby a back-leverage lender can fulfill the operational experience requirement by hiring a third-party operator that possesses the minimum qualification requirements. Insurance companies have also stepped up to provide an ITC loss insurance wherein the insured gets a payment equal to the loss of ITC pursuant to an IRS audit

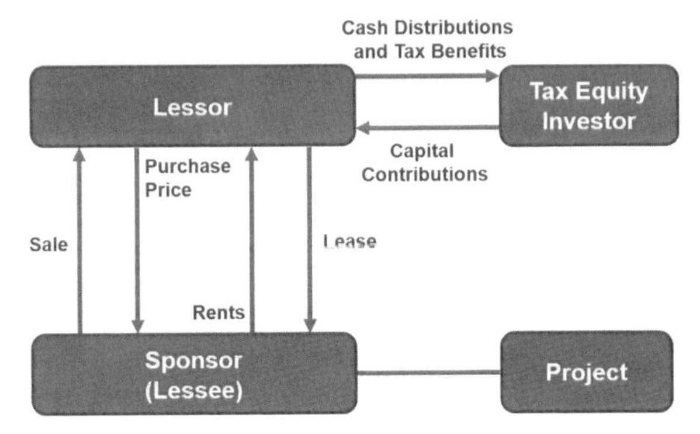

Fig. 6.3 Sale/leaseback basic structure.

(subject to an agreed-upon deductible). The insurance cost may range from 2 to 5% of the maximum payable amount under the insurance policy. A well-structured insurance policy issued in the name of the tax equity investor may convince the investor to agree to a lower cap on the cash sweep for any ITC losses.

The sale/leaseback structure

Sale/leaseback structures have been used in all sorts of financings for decades. The basic sale/leaseback structure as it applies to renewable energy projects is simple. The sponsor develops and constructs a project and, once the project reaches commercial operation, sells the project to a tax equity investor. The tax equity investor immediately leases the project back to the sponsor in consideration for periodic lease payment. This structure is illustrated in Fig. 6.3.

The lease has a finite term that is no more than 80% of the useful life of the project, but the developer receives periodic options to repurchase the project from the tax equity investor at fair market value, determined through an independent appraisal. If the transaction is well-structured as a "true lease" for tax purposes, the tax equity investor is the true owner of the project and receives the tax credits and depreciation benefits generated by the project, which are therefore factored into the lease financing rate for the tax equity investor. The sponsor incurs capital gain tax liability to the extent the sale price to the tax equity investor is higher than the costs incurred to build the project.

It is customary for the sponsor to prepay a certain amount of rent – usually 15–20% of the sale price - which then reduces the upfront consideration to the sponsor. The rent prepayment must comply with Section 467 of the Internal Revenue Code, which treats the prepayment as an implicit loan from the lessee to the lessor and assesses "Section 467 Interest" as part of the income.

In addition to Section 467 considerations, the tax equity financing may need to be optimized for another constraint called "Uneven Rent Test". The Test essentially constrains the annual pattern of rental allocations (which is independent of the cash flow pattern for the underlying project). A lessor may use non-recourse debt financing for the purpose of acquiring a project before leasing to the lessee. Such transactions are called "levered leased" transactions. The IRS has issued Revenue Procedure 2001-28, which supersedes Revenue Procedure 75-21, dealing with leveraged leases. This Revenue Procedure serves as safe harbor, stating that the IRS will consider the lessor in a sale/leaseback transaction to be the owner of the property and subject to a lease if the following conditions are met:[17]

- The lessor must have made a minimum unconditional "at risk" investment in the property from the inception to the end of the lease term in an amount equal to 20% of the cost of the property.
- The lessee doesn't have a conditional right to purchase the property from the lessor at a price less than its fair market value at the time the right is exercised.
- No part of the cost of the property, improved modifications, or addition to the property may be furnished by the lease. However, the costs of general repair, maintenance, and/or insurance may be borne by the lessee.
- The lessee may not lend to the lessor any of the funds necessary to acquire the property, or guarantee any indebtedness created in connection with the acquisition of the property by the lessor.
- The lessor must represent and demonstrate that it expects to receive a profit from the lease transaction apart from the value of the tax benefits resulting from such transaction.
- Overall profit - The aggregate amount paid by the lessee over the lease term plus the value of the residual must exceed an amount equal to the sum of the aggregate disbursements required to be paid to the lessor in connection with the ownership of the property and the lessor's equity investment in the property, including any direct costs to finance the equity investment.
- Positive cash flow – The aggregate amount required to be paid to the lessor over the lease term exceeds by a reasonable amount the aggregate disbursements required to be paid by the lessor in connection with the ownership of the property.

The lessor needs to represent and demonstrate certain facts relating to the fair market value and estimated remaining useful life of the property at the end of the lease term. This provision is intended, in part, to ensure that the lessor has not transferred the use of the property to the lessee for substantially its entire useful life.

Unlike the partnership flip transactions, the Revenue Procedure 2001-28 does not permit fixed purchase options, which are estimates of the fair market value determined by an independent appraiser at the inception of the lease agreement (similar to the fixed purchase options discussed earlier in the Partnership Flip Structures). However, there is case law wherein the Supreme Court has blessed lease transactions with fixed purchase options, which have rendered strict compliance with the Revenue Procedure impractical.[18]

Advantages and disadvantages of the sale/leaseback structure

The sale/leaseback structure has several advantages over the partnership flip structure. First, partnership flip structures are inherently complex. As a result, structuring a

[17] IRS Revenue Procedure 2001-28 (https://www.irs.gov/pub/irs-irbs/irb01-28.pdf).
[18] "Foundation: Sale Leasebacks" by David Burton (https://www.akingump.com/images/content/3/5/v2/35209/Sale-Leasbacks-Feb-2015-D-Burton.pdf).

partnership flip transaction is time-consuming and expensive. Sale/leaseback transactions have been around for decades and are simpler to execute.

Second, unlike partnership flip structures, a sale/leaseback can commence up to 90 days after a project is placed in service and still qualify for ITC. This is an important benefit because the sponsor is not stuck in sourcing tax equity before the project achieves commercial operation. The flexibility in terms of when tax equity can come into a transaction also simplifies the complex structuring around tax equity funding in installments.

Third, sale/leaseback structure allows for a more efficient monetization of tax benefits. Since there are no structural considerations such as capital accounts, outside tax basis, etc., accelerated depreciation benefits (such as 100% expensing or bonus depreciation) can be absorbed more efficiently.

Fourth, sale/leaseback transactions yield a much better advance rate for the Sponsors. Sale/leaseback transactions may, in theory, raise up to 100% of the fair market value. Even if the Internal Revenue Code Section 467 prepaid rent provisions are considered, sale/leaseback transactions can raise substantial capital for the sponsors, reducing or eliminating the need for back-leverage or other debt financing. As a result, sponsors can recycle capital more efficiently.

Fifth, financial institutions prefer the sale/leaseback structure due to attractive capital requirement provisions. As a result, financial institutions can provide more competitive financing rates for sale/leaseback transactions.[19]

Despite all these benefits, sale/leaseback transactions are less prevalent than partnership flip transactions. As mentioned earlier, almost all wind PTC transactions and more than 80% of the solar sector's transactions are financed through a partnership flip. Sale/leaseback transactions are confined mostly to small commercial/industrial installations. There are two reasons why partnership flip is a preferred transaction structure.

First, while sale/leaseback structure allows for a better advance rate than partnership flip structure, in this structure the sponsors are essentially selling the project to tax equity investors. There are early buyout options, but such options are very expensive owing to the "at least 20% residual value" rule stipulated by the IRS. The residual interest in partnership flip transactions is close to 5%. Therefore, once the Flip Date is achieved, a sponsor can buy out the tax equity investor's interest and recapitalize the project company with more efficient (and cheaper) debt financing. This is an especially attractive recapitalization strategy for solar projects given the shorter flip terms.

Second, tax equity investors are invested for a longer period of time in a sale/leaseback transaction, and have more capital tied up while receiving essentially the same amount of tax benefits as a partnership flip structure. As tax equity capacity is limited, this is unattractive economically.

[19] These benefits can be even more powerful if a leveraged lease is used. A leveraged lease, however, brings with it similar risks as we see in back-leverage situations. See Moran, Chase. Renewable Structures: Choices and Challenges, Pratt's Energy Law Report, Vol. 15, No. 7, July 2015.

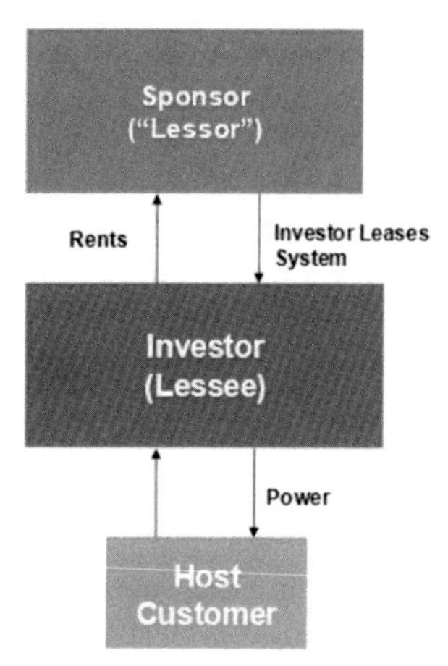

Fig. 6.4 Inverted lease structure.

The inverted lease structure[20]

Popular for residential solar installations, the inverted lease structure works only for ITC-eligible projects. It differs from sale/leaseback structure in that the roles for the sponsor and tax equity investor are reversed. Here, a sponsor leases the project to a tax equity investor through a master lease and the tax equity investor claims the ITC. The ITC is calculated based on the fair market value of the project without regard to the project cost. The sponsor retains the depreciation benefits. The tax equity investor can then sell the electricity output through a PPA or sub-lease the project to a customer. Once the lease term is over, the project reverts back to the sponsor (Fig. 6.4).

Sponsors often find the inverted lease structure more attractive compared to the sale/leaseback structure because they get the project for no additional consideration at the end of the lease term (instead of having to buy at fair market value in a sale/leaseback transaction). Furthermore, while the ITC is based on the fair market value of the project, there is no tax to be paid on the capital gain realized over the project cost. However, the structure is predicated upon the sponsor's ability to absorb depreciation benefits, which can be a limiting factor since not all sponsors have tax capacity.

There are no IRS safe harbor guidelines for inverted lease structures. However, in Notice 2014-12, the IRS acknowledged that inverted lease structures for ITC projects are similar to those common in the historic tax credit market. Applying the guidelines for equivalent historic tax credit transactions, a tax equity investor needs to demonstrate that it has both upside and downside risks in order for the lease to be treated as a true lease (the "merchant tail"). One way to interpret this requirement is that the master-lease (to which the sponsor is counterparty) is at least 20% longer than

[20] Keith Martin (2017, June) "Inverted Leases", Norton Rose Fulbright. Retrieved from https://www.nortonrosefulbright.com/en-us/knowledge/publications/f93752a0/inverted-leases.

the customer agreement or customer sub-lease. This can be problematic because standard PPA contracts often run up to 25 years. Alternatively, some tax equity investors require that the lessee prepay at least 20% of the rent to comply with the merchant tail requirement.

Consideration should also be shown to how the lease payments are structured. The most conservative way to structure inverted leases is to have the tax equity investor make fixed rent payments, which are firm "hell or high water" obligations of the investor. At the other extreme, rent payments can be made contingent upon the output or cash flow generated by the project. However, such creative structuring may expose the tax equity investor to IRS scrutiny.

The structuring challenge for inverted leases is how the capital moves from the lessee (i.e. tax equity investor) to the lessor (i.e. sponsor). The most conservative way to structure an inverted lease is to characterize such payment as prepaid rent. However, certain tax equity investors are comfortable making the capital contribution for 49% equity interest in the LLC that serves as the lessor with the sponsor owning the remaining 51% of the LLC. The structure has an additional benefit that the tax equity investor can claim 49% of the depreciation benefit by virtue of its ownership in the LLC. However, such a structure may add to the tax risks assumed by the tax equity investor due to its potential for IRS scrutiny.

The inverted lease is the least prevalent of the three tax equity structures. Because it is without an explicit safe harbor, it is difficult to structure transactions without incremental tax risks. Furthermore, the tax equity market has not evolved to the point of standardizing a structure. As previously noted, there are several variations in financial structuring that may be adopted, but they either introduce additional tax risks or commercial risks depending on the structure. Tax equity investors often refuse to accept these additional risks easily, given the limited tax equity capital available to the renewable energy market.

A critical assessment of tax credits as an incentive mechanism

As discussed in Chapter 2, tax credits are more palatable as an incentive mechanism than direct subsidies. However, as the history of renewable energy in the US suggests, using tax credits as an incentive mechanism is not ideal from a financing perspective.

First, as discussed earlier, the tax credits have been through multiple cycles of extensions, expiries, and reenactments, which has created significant uncertainties for the market participants. As a result, the industry has witnessed a boom-bust cycle, especially in the wind sector, that coincided with the tax law changes applicable to the tax credits.

Second, financing large energy projects using project finance structure is time consuming and complex as is. The tax equity structures make the exercise even more challenging. While the IRS safe harbor and subsequent guidance issued by the IRS have provided an element of certainty, both tax equity investor and sponsor are unsure as to how these transactions would fare in the event of an audit. As a result, each transaction needs to be structured based on its own merits, which increases the execution timeframe and transaction expenses.

Third, the tax equity investor base is very small and limited to a select few financial institutions, which typically have a consistent tax base *and* the project finance expertise to make these investments. In 2012, the Obama administration invited Fortune 100 companies to the White House to encourage them to invest into tax equity products. Since then, however, the response has been lukewarm and sporadic. Only a limited number of companies like Intel, Starbucks, and Facebook have been able to enter the tax equity market with any scale. The tepid response is primarily due to the complexities of the tax equity structure. Even some of the regional banks or insurance companies who are eager to enter the market have been hobbled by the complexities of the internal product approval. This is because tax equity is an equity product (as opposed to a loan product) and the tax equity structure needs to be vetted essentially by every corner of the bank – tax, accounting, regulatory, treasury, credit, risk management, to name a few. The implementation of the Dodd Frank and Volcker rules has introduced further complexity.[21]

Fourth, the IRS directive that the tax equity investor shall be a true equity owner have caused the cost of financing to be expensive (relative to a project finance loan). Consequently, the tax equity investments are essentially structured as preferred equity instruments and do not enjoy the same foreclosure right that a lender may have in a typical non-recourse loan. Therefore, the pricing for the tax equity instrument needs to reflect the limited prospects for the recovery in the event the project faces defaults.[22] The incremental risk along with a limited investor base makes the tax equity investments incrementally expensive. Another related issue is that the tax equity

[21] The Volcker Rule prohibits a banking entity from acquiring an ownership interest in a covered fund. "Covered fund" is defined in the Volcker Rule as an issuer that would be an investment company as defined in the Investment Company Act of 1940 (the "1940 Act") but for sections 3(c) (1) and 3(c) (7) of the 1940 Act. However, the banking industry has come to its own conclusion that, for purposes of the analysis of tax equity investments under the Volcker Rule, the LLCs in which the banks invest do not qualify as investment companies. Because the issuers are not investment companies, they do not need to rely on the exceptions set forth 3(c) (1) and 3(c) (7) of the 1940 Act, and, therefore, do not qualify as covered funds. Separately, regulatory authority for energy credit monetization transactions similar to the ones discussed in this chapter has been premised on an OCC ruling, issued in 1994, which found that a national bank can provide financing for tax credit monetization in the form of purchasing energy producing properties. In a letter addressed to Union Bank of California dated February 2006 regarding its investment in a wind farm producing §45 tax credits, the OCC confirmed its findings in the 1994 letter. See, also, OCC Interpretive Letter 1139 (November 2013) regarding solar projects producing §48 tax credits.

[22] The perception in the industry that the tax equity investments are very expensive is somewhat misplaced. As discussed in this chapter, the tax equity investments are structured with a pre-tax IRR of as little as 2%. This is essentially the actual cost that sponsor pays for monetizing the tax credits and cash flow distributions. While the after-tax IRR that the investor earns does have some sting (due to the cash sweep provisions if the Target IRR is not met by the Target Flip Date), the difference between the pre-tax IRR and after-tax IRR is made up by depreciation benefits, which, in the absence of a tax equity investment the sponsor would not have been able to monetize anyway.

investments are not truly non-recourse compared to project finance loans. The sponsor needs to indemnify the tax equity investors for the project related risks (that a typical acquirer may require from a seller) and additional tax indemnities. Furthermore, such indemnities need to be guaranteed by a creditworthy entity. The indemnification and the creditworthiness criteria make tax equity investments challenging to asset management companies such as private equity firms and hedge funds.

The tax credit as an incentive mechanism does have certain benefits from the public policy perspective.

First, the government can limit its role in formulating the policy, designing the incentive mechanism, and providing the oversight of the subsidy program through audits. Consequently, the private investors decide which projects "deserve" the subsidies. Such an approach is more capitalistic and as it saves the government vast amounts of resources in due diligence and oversight and avoids the perception of choosing the winners and losers.

Second, since the subsidies are passed on in the form of tax credits, which can be offset only by the tax liabilities payable in the same country, the subsidies and the direct/indirect benefits of the subsidies stay in the same country. The direct subsidies and other mechanisms (such as feed-in-tariff mechanisms) implemented in Europe have attracted negative publicity because any sponsor – domestic or foreign – could qualify for the benefits; as a result, these subsidies and the benefits could pass outside the national boundaries for the projects sponsored by foreign owners.

Financing distributed generation projects

7

Introduction to distributed generation

The design of the modern electric power system relies on a vast network of transmission and distribution (T&D) lines to connect large electricity generators to loads. Transmission lines deliver electricity across long distances at high voltages while distribution lines deliver electricity directly to load centers (i.e. residences, businesses, etc.) at lower voltages. Conventional electric generation usually generates electricity at 34.5 or 69 KV. The voltage is amplified through step-up transformers and fed into the transmission lines, which operate at higher voltages. As the power reaches the load centers through distribution lines, the voltage is reduced through step-down transformers, because most electric equipment utilized in residential, commercial, and industrial applications is designed to operate at low voltage. As the electric power moves through the transmission lines, some of the active power is lost through thermal and reactive losses, collectively referred to as transmission losses herein. The EIA estimates that T&D losses account for approximately 5% of the power generated by the electric power system in the United States.

Distributed Generation (DG) turns the power system design on its head by generating power onsite (i.e. at the load) at distribution voltage levels. Conventional power plants are centralized and need to be large enough to take advantage of economies of scale. However, wind and solar power, which are inherently modular in design, make it economically feasible to erect renewable energy projects, especially rooftop solar, in urban areas to generate power for local use. A potential benefit of distributed generation is the reduction of T&D losses.

While the concept of distributed generation has been around for decades, implementation in most countries was limited until about 10 years ago. DG took off due primarily to the rapid decline in solar component costs. While the decrease in solar installation costs made DG solar more cost competitive, the technical implementation of DG is not as straightforward. Modern distribution systems were designed with relays and circuit breakers (collectively termed as "switchgear"), which safeguard the system in the event of an extreme situation, like an electric fault. The switchgear has traditionally been designed under the premise that electric power moves in one direction from high voltage systems to individual distribution and utilization points. With DG, however, power can move in both directions; this results from the fact that the amount of DG power produced may never equal the power being consumed at a load onsite. If the power generated exceeds the load consumption, the excess power must find its way over the distribution network to serve other loads. Accordingly, the switchgear for the distribution network must be reconfigured to allow the reverse flow of power but avoid tripping an otherwise healthy system. In addition to

Renewable Energy Finance. https://doi.org/10.1016/B978-0-12-816441-9.00007-6

this switchgear redesign issue, DG poses other technical challenges such as voltage control, harmonics, distribution grid stability, and reliability caused by the intermittent nature of the generation and the solid-state electronics involved.

The redesign of the distribution network to accommodate distributed generation can be significant. As T&D networks are cost regulated, the costs of redesigning grids to accommodate DG may fall on a wide range of customers — raising distributional issues among different groups of customers.

Net metering

In addition to tackling the technical challenges of DG, successful implementation must address an important policy challenge: how to facilitate the absorption of excess electricity generated by a DG installation. In many countries, regulators have adopted net metering for this purpose. In simple terms, under a net metering mechanism, consumers can connect a DG source (i.e. a wind turbine or rooftop solar installation) and inject the excess generation from that source into the distribution grid, subject to certain limits. The distribution company is then required to compensate the owner of the DG resource at a price approved by regulators. There are no specific benchmarks or market standards with respect to pricing for excess generation — the power can be purchased at a certain minimal rate, the average price applicable to the total consumption for the subject customer, or a marginal price specified in a regulator-approved retail tariff. The net metering programs may put certain eligibility conditions to ensure the integrity and reliability of the distribution networks.

For a home with a solar installation and net metering, the electric meter installed to measure actual electricity consumption may turn backwards, allowing the consumer to receive credits for the excess generation that will offset their on-site consumption. Consequently, the customer is charged for the "net" energy consumption at his or her home. According to the Solar Energy Industries (SEIA), on average, 20%–40% of a solar energy system's output ever goes onto the grid, and this exported solar electricity serves nearby customers' loads.

Virtual net metering

In order to qualify for net metering, regulatory policies require co-location of a DG resource and the end user of the electricity. For example, a house or commercial building that uses the electricity generated by a solar installation on the roof of the home or building would qualify for net metering benefits. Such co-location is not always feasible due to space or other constraints. Because of this, in some jurisdictions policymakers have established rules to allow the DG resource to be sited at a different location than the customer. There may be other restrictions — for example, the DG resource may need to be in the same utility area. Apart from the DG resource location rules, virtual net metering works in the same general way as net metering.

Rooftop solar has been the main beneficiary of net metering rules since it lends itself well to offsetting on-site electric consumption. Regulators also often created a positive environment for rooftop solar through specific rules that cater to rooftop solar. Because of this, we will devote rest of this chapter to rooftop solar installations.

Tax credits for distributed generation in the US

In the US, owners of rooftop solar installations can qualify for an ITC and the tax credit is available for individual taxpayers as well as corporations who may pool multiple residential solar installations for commercial purposes. The rules for claiming the ITC for households are slightly different and had a different sunset schedule but the PATH Act brought the ITC available for individual households more in line with those that can be claimed by the corporations.

Business models for distributed generation and financing structures

In the US, owners of retail electricity installations can qualify for an ITC; they can claim the ITC against their tax liability for the year that the rooftop solar is installed. However, in order to benefit from the tax credit subsidy, the system owner needs a clear view of their projected tax liability for the year. The credit can be deferred if there is insufficient tax liability in a given year, but if such deferral is elected, the economic value of the credit significantly decreases due to the time-value of money.

Third-party ownership model

The initial wave of rooftop solar projects installed in the US was mainly through private financing from individual electricity consumers who wanted solar on their home. Individual consumers purchased the necessary equipment directly, or through a distributor, and were responsible for ongoing maintenance of the system. The effort and informational costs associated with the procurement, construction, and operation of the installation, in addition to the personal financing, were burdensome and this self-financing option did not initially gain wide traction.

The residential solar sector in the US experienced a breakthrough when the Third-Party Ownership (TPO) business model was introduced. In the TPO model, a solar developer creates a large pool of retail customers across a large geographic area who may be interested in a residential solar installation. Customers are offered attractive electricity rates through long-term contracts extending up to 20 years. The contracts typically provide retail customers with an up front discount relative to their current retail electricity rate and an escalation rate that can be proven to be lower than the historical escalation rate for their current retail electricity. The solar developer then arranges third-party financing for construction, back-leverage, and tax equity

The TPO business model was an instant hit for residential solar and allowed companies like SolarCity to grow quickly. Fig. 7.1 below illustrates SolarCity's growth over the period 2010–15. Before its acquisition by Tesla, SolarCity catered to both the residential and Commercial & Industrial (C&I) sectors. However, most of the company's growth came from the residential solar sector.

The TPO business model became popular for several reasons. First, customers were not required to do any of the work to procure or install the solar system; the solar developer was responsible for this work in addition to the ongoing O&M of the system through the useful life of the systems. Second, customers were able to lock in 15%–20% of upfront savings, relative to their then-current rates, with a demonstrably low and fixed escalation rate. Third, solar developers were able to achieve economies of scale, from the sheer volume of installations that resulted from their large lead generation, and they passed the associated benefits on to customers. Fourth, developers were able to aggregate the installations into portfolios, allowing them to negotiate favorable financing terms and employ innovative financing techniques, like securitization. Finally, developers were able to extract significant tax benefits from the aggregated installation portfolios and pass some of the benefits on to customers through an up-front retail rate discount. The tax benefits were generated in some cases as the developers claimed ITC based on a fair market value of the aggregated systems, rather than the installation costs, as often used for individual residential solar systems. Furthermore, since the pools are owned by special purpose LLCs, these LLCs were able to claim MACRS depreciation benefits that are otherwise not available to individual taxpayers.

Fig. 7.2 illustrates the structure for financing a pool of residential solar systems, which follows the principles of non-recourse project finance with a couple of important differences. First, the pool of retail electricity customers serves as the offtaker, rather than a typical investment grade-rated utility offtaker or a corporation; each retail customer has its own net metering arrangement with the local utility, and continues to be served by the utility for any electricity needs that are not satisfied by the solar installation. Second, the financing parties are relying on the diversification benefits of a large pool of retail customers to satisfy its credit analysis, rather than the credit rating of a traditional offtaker. The pooled offtaker structure lends itself well to a securitization style of underwriting, rather than the traditional underwriting process for project financing.

Although securitization can be effectively utilized to facilitate the financing of residential solar at scale, the market has been slow to develop in the United States. Rating agencies were slow to react to the burgeoning market, but more importantly, tax equity proved to be a significant challenge. As discussed in Chapter 6, an important provision in a partnership flip structure is the requirement that sponsors indemnify the tax equity investor against the loss of the ITC in the event of a successful challenge of the ITC basis by the IRS. As was previously stated, the ITC basis for a pool of residential solar installations is driven by the fair market value of the bundle, and that

[1] SolarCity Corp – Form 10 K for the period ending on December 31, 2015.

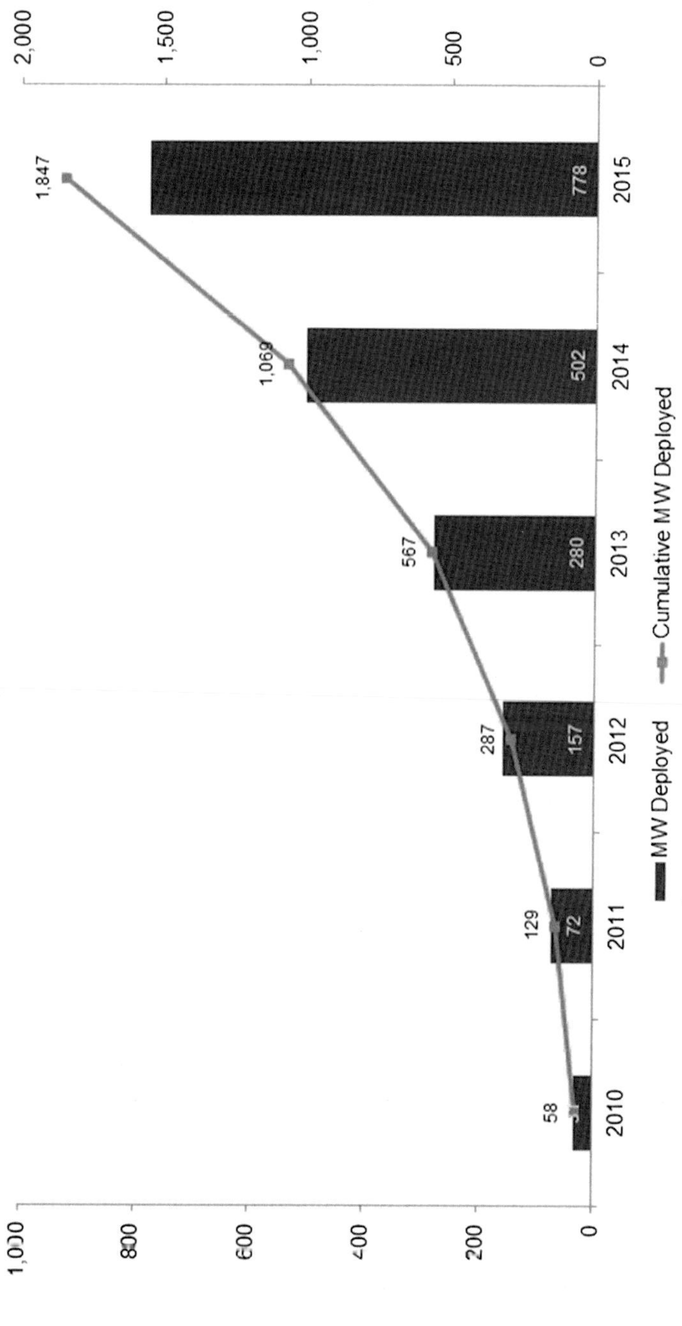

Fig. 7.1 SolarCity installation volume growth.[1]

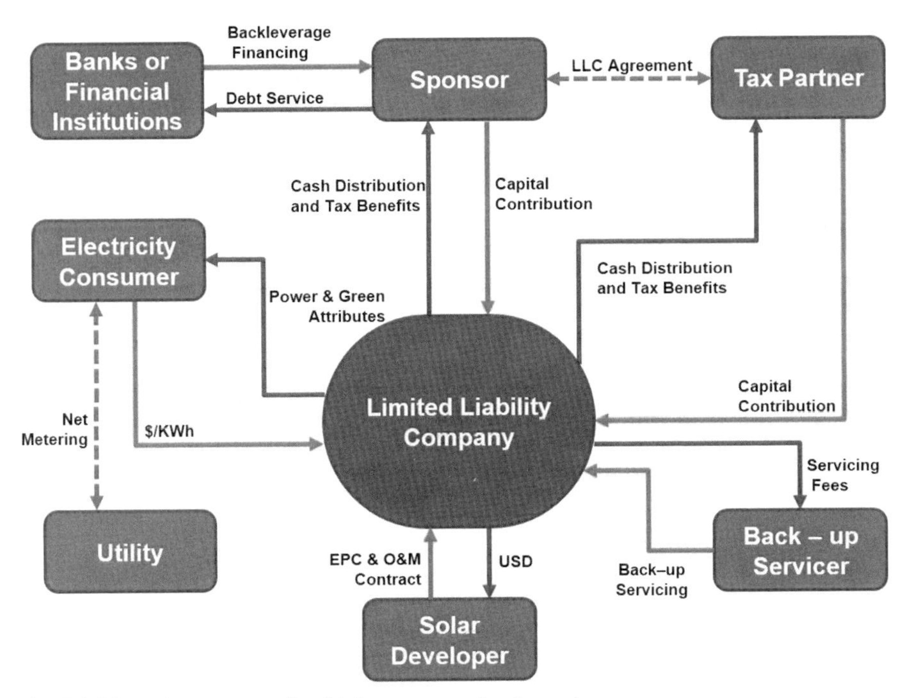

Fig. 7.2 Financing structure for third-party owned solar systems.

value can be significantly higher than the actual installed costs of the systems. The higher the difference between the two values, the greater the risk that the IRS could challenge and audit the ITC basis associated with the pool of systems. Through the indemnification mechanism, the tax equity investor is protected from the economic losses suffered from the potential disallowance of the portion of the ITC basis. If the sponsor does not or cannot fulfill this obligation, the tax equity investor has a right to "sweep" – or take - the cash that would otherwise be distributed to the sponsor. This cash sweep can pose structural challenges should the sponsor raise back-leverage that is secured by the distributable cash flows that would be used to make the tax equity investor whole, especially if the back-leverage is sourced through securitization. In order to combat this problem, certain insurance companies created an ITC insurance product that insures the loss of the ITC pursuant to a successful IRS audit challenge. The insurance product has its own complexities, but market participants now view it favorably for providing adequate protection and paving the way for securitization. The securitization market has grown from $54.4 million in 2013, the year of the first SolarCity deal, to more than $2.2 billion in 2018.[2] Most of the issuances are aimed at

[2] Mendelsohn, M. (November 16, 2018). *Raising capital in very large chunks: The rise of solar securitization*. Retrieved from https://pv-magazine-usa.com/2018/11/16/raising-capital-in-very-large-chunks-the-rise-of-solar-securitization/.

qualified institutional buyers, pursuant to the Securities Act Rule 144A offerings. As more investors get comfortable with this financing structure, the securitization market is expected to continue to grow.

The tax equity financing process for TPO modeled residential solar is similar to the process for utility-scale projects. The only difference is how the tax equity investor interacts with the shorter construction timeline for residential installations. As discussed in Chapter 6, a tax equity investor investing through a partnership flip structure needs to be a financing partner before a solar project is placed in service; in other words, the tax equity investor needs to make a commitment for a programmatic investment (colloquially termed as a "Tax Equity Fund"). Once the Tax Equity Fund commitment is in place, the sponsor can draw the equity commitment in multiple tranches as new residential customers are signed up and progress through development and construction. However, this structure requires additional time and resource commitment from the developer and tax equity investor because both parties need to agree upon the terms of the portfolio up-front. Since tax capacity is a scarce resource for most of the tax equity investors, they need protections to ensure that the sponsor deploys the committed capital and completes the solar installations, as contemplated at the closing of the equity commitment.

Solar loan financing model

Despite the success of the TPO business model, the individual homeownership model has gained traction in the recent years with the advent of the loan structure. In this structure, a third-party installer or developer installs and maintains the solar system but the homeowner owns it. The homeowner claims the ITC, similar to the self-financed residential solar ownership model, and accesses long-term loan financing through the developer for a term of up to 30 years. The ITC cannot be claimed until individual homeowners file their tax return so the developer will advance the expected amount of the tax credit. In its initial form, the loan structure obligated the homeowner to pay down the loan in the amount of the tax credit. However, as described below, the recent versions of the loan financing business model provide flexibility to the homeowner in terms of loan repayment.

Typically, the loan has a repayment term of up to 30 years. The repayment of the loan starts once the local utility issues a Permission-to-Operate (PTO) Letter. The homeowner has 18 months to claim the ITC on his or her personal tax return and make prepayment for the corresponding amount. The homeowner may choose to prepay a higher or lower amount than the ITC claimed. Accordingly, the monthly payment schedule could be reset lower or higher. There is no event of default under the solar loan if the homeowner chooses not to prepay the loan for the amount of ITC claimed. Furthermore, the homeowners can often repay a part or all of the loan any time during the loan term. Similar to the TPO business model, systems are installed and maintained by the solar developer.

Developers usually tap securitization markets for financing the installation of residential solar systems backed by a portfolio of loans. Since homeowners are claiming the ITC, the frictions associated with the tax equity financing in the TPO business model can be avoided.

One important drawback of the loan financing product to the homeowners is that it may affect the homeowner's credit score. However, there are several reasons why the individual homeownership model is gaining popularity with the help of loan financing products:

- Homeowners are increasingly opting for owning a solar system outright instead of going through a third-party-ownership channel. The decreased cost of systems has made outright purchases affordable. Moreover, the ITC for homeowners has come down with the reduction in system prices making tax planning for the homeowners easier
- Consumer perception that loans are more cost effective than third party ownership
- Increased transparency regarding differences among the financing options, which affords the homeowners more choices and comfort around their financial decisions
- The homeowners are getting more comfortable with solar energy systems, their operation and performance, which makes them willing to take the risk with loan products
- The developers find loan products attractive because the product doesn't rely on otherwise constrained tax equity market. Furthermore, the loan products rely on a service-based business model, thereby providing immediate cash for the systems sold to the homeowners since most of the loans originated can be securitized relatively easily

PACE financing

Households in the US can also tap into another form of loan financing called Property Assessed Clean Energy financing (PACE). This loan-financing product is available through a property tax assessment for certain qualified clean energy projects, like energy efficiency and residential solar. The loans are repaid through a property tax assessment. Therefore, the interest on the PACE loan is not subject to income tax at the state or federal level. That said, the *Tax Cuts and Jobs Act of 2017* limited the exemption allowed for State and Local Taxes (SALT), thus diminishing the benefit of the interest tax shield created by PACE financing.

Another homeowner benefit of PACE financing is that it does not affect credit scores due to the non-recourse nature of the financing. However, the flip side of the non-recourse financing is that if the homeowner elects to sell the home, the loan is passed to the new owner through the property tax assessment. Anecdotal evidence suggests that this pass-through may cloud the sale of the property via a requirement from the potential buyer to either fully repay the PACE loan prior to sale or lower the asking price.[3] The Federal Housing Finance Agency (FHFA), which governs the Federal National Mortgage Association ("Fannie Mae") and the Federal Home Loan Mortgage Corporation ("Freddie Mac"), has taken an adversarial position toward PACE

[3] Dolan, P. (October 23, 2018). *PACE securitizations: Norton Rose Fulbright*. Retrieved from https://www.projectfinance.law/publications/2018/october/pace-securitizations/.

financing, objecting to the priority lien position of a PACE loan over mortgage financing, which worsens the position of the existing mortgage lenders. Consequently, the FHFA has instructed correspondent banks to reject any mortgage loans for properties with PACE financing attached. For a typical household, the value of the PACE financing is much lower than the mortgage financing, so at least some of the correspondent banks have not followed the FHFA guidance.[4] Oftentimes, mortgage lenders are not even notified at the time of the PACE financing despite the priority of the PACE financing over the mortgage loan, given the amount of the financing involved, which may prove troublesome.

PACE financing is typically more expensive than traditional sources of financing, such as mortgages or home equity lines of credit, but it can be more competitive with credit card loans and personal lines of credit. The underwriting process for PACE financing is driven by property information, mainly the property assessment history, rather than the typical credit metrics used for underwriting retail credit, like credit score (see Fig. 7.3 for detailed credit criteria). As a result, legitimate concerns exist that PACE financing may advance loans to homeowners lacking the ability to repay.

Given the aforementioned complexities and concerns around PACE financing practices, new legislation to regulate the PACE financing market has been introduced in California, the most vibrant residential PACE financing market.

The Energy Information Administration (EIA) has made available a breakdown of the different ownership structures for solar systems across the US as of September 2016. The EIA reports that, as of September 2016, distributed solar installations, which

Factor	Evaluation Criteria
FICO Score and Loan to Value Analysis	• FICO Version 2 is the most commonly used FICO model that predicts a borrower's potential for becoming ninety (90) days delinquent using 10 different scorecards • Applicants with scores of 700 and above are automatically approved • Applicants with scores of 680-699 are further evaluated and approved if: – Loan to home value ratio is 75% or less (determined by adding balance on 1st mortgage and highest balance on any home equity loans and dividing by estimated home value using Zillow) – Enrolled for automated ACH payments • Generally, all three bureaus are sampled. Underwriting criterion varies by the developer (e.g. best of 3, worst of 3, first to report, etc.)
Bankruptcy Review	• Applicants who have filed for bankruptcy within the last 5 years carry higher credit risk and are declined
Mortgage Review	• Applicants with late mortgage payments reported on their credit report carry higher credit risk and are declined
Title Review	• Typically, through mortgage review
Fraud and Identity Theft Prevention	• Protection against Fraud, Identity theft, SSN abuses, etc.

Fig. 7.3 Typical underwriting criteria for a residential solar customer.

[4] See footnote 3.

they define as all solar installations smaller than 1 MW, totaled 12.3 GW, of which approximately 30% (or 3.7 GW) were TPO.

NREL reports that between 2012 and 2014, TPO residential solar systems represented 62%–72% of the US market, but that TPO market share had decreased by the beginning of 2015, while PV loans and cash purchases grew.[5] By 2017, the TPO market share had fallen to 41% of annual installations, while loans comprised 33% and direct purchases made up the remainder.

Despite the strong economics of residential solar in many locations, penetration of the residential solar system model has been largely limited to higher income households and communities. This is primarily due to the requirement that financing parties (especially tax equity investors) maintain high credit quality of their underlying retail customers, and, to a smaller degree, due to economies of scale owing to larger rooftops. Financing residential systems for lower credit retail customers has always been challenging.

Regardless of the high credit score requirements imposed by financing parties, the retail credit risks associated with residential solar products are arguably lower than those associated with other retail credit products such as car loans, credit card loans, etc. As depicted in Fig. 7.4, households tend to prioritize utility payments more so than other credit products. Developers usually have the option to turn off a system for a customer in default in addition to resorting to the collection agencies to recover the dues. Both of these factors further limit the incentive for a non-payment by the retail customers.

While there is limited public data on consumer defaults and recovery rates for the residential solar sector, anecdotal evidence supports the aforementioned thesis regarding the low defaults for residential solar customers. In 2018, NREL concluded that loans backed by residential solar systems are performing well, as compared across 6770 loans covering more than $186 million in aggregate lending across multiple

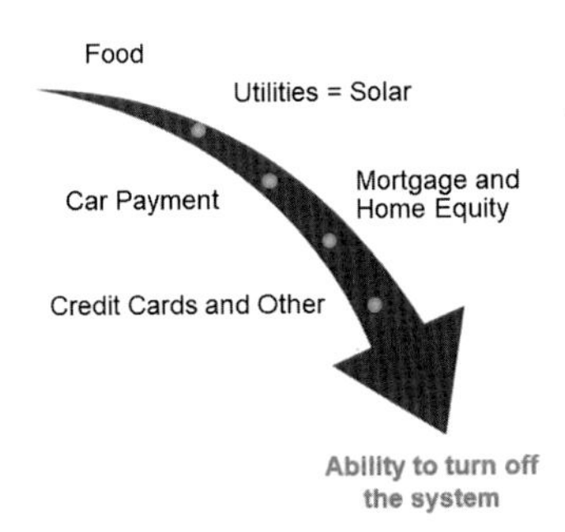

Fig. 7.4 Payment priority for a typical US household.

[5] Schwabe, P., & Feldman, D. (2018). *Solar lending practices by community and regional financial institutions*. National Renewable Energy Laboratory. Retrieved from https://www.nrel.gov/docs/fy18osti/71733.pdf.

lenders. However, these results must be put into the context of the high credit quality of borrowers, early state of the loans, and the recent robust economy.[6]

The previously mentioned inequality in access to residential solar has led to peculiar public policy questions.[7] Specifically, regulators in the US have permitted utilities to recover transmission and distribution costs with a separate charge calculated based on actual energy usage, even though transmission and distribution costs are fixed and do not vary significantly with actual energy usage. Consequently, the transmission and distribution cost charged to households with residential solar is smaller than a non-solar customer, since net electricity consumption for the solar customer decreases once the residential solar system is in place. As an increasing number of households, which tend to be more affluent, migrate to residential solar systems, the burden of transmission and distribution costs gradually shifts to the non-solar households. This raises a question around the fairness of households without solar, especially lower income households, subsidizing households with residential solar. This debate has been most prevalent in Arizona, which became the first state to assess a fee for residential and commercial customers with solar installations in an effort to neutralize the effect of solar installations on non-solar utility customers. As residential solar installations become more widespread, this issue may grow in importance.

Community solar

Community solar refers to an arrangement in which a pool of retail electricity consumers subscribe to a centrally located solar energy system, pay for a portion of the energy generated by the system, and receive a credit on their electric bill for the portion for which they paid. The arrangement is similar to residential solar in that the electricity consumption by the consumer subscribing to the system (the "subscriber") is offset by the portion of energy generated by the centralized system purchased by the subscriber. All subscribers must be located in the vicinity of the solar system and use the same distribution network that would otherwise serve the subscribers. Managing the credit and other risks associated with subscribers is therefore critical.

Community solar has been implemented differently across the US due to different state regulatory policies. In its simplest form, a developer goes through the usual development and construction process for a solar project. During that process, the developer solicits subscribers for the energy from the project either directly or through channel partners, and works with the local utility to set up the bill-crediting mechanism for each subscriber. Once subscribers sign a contract, their local utility includes a credit on their normal electric bill for their share of the energy generated by the solar project. Subscribers also receive a bill from the owner of the solar project, billing them for their portion of the energy from the project.

[6] See footnote 5.

[7] Wolak, F. A. (2017). *Retail pricing to support the 21st century distribution grid.* Stanford Institute for Economic Policy Research. Retrieved from https://siepr.stanford.edu/sites/default/files/publications/PolicyBrief-Nov2017.pdf.

A simpler model in which the subscriber receives only one bill from the utility and the utility then passes the proportional revenue through to the developer could be possible. Such an arrangement could save administrative costs for developers, facilitate easier financing, and avoid multiple bills for subscribers. However, not all utilities or state regulators have embraced this business model.

Subscribers are often permitted to relocate and remain subscribed to the community solar system so long as they do not move out of the utility area, which would deem them ineligible. In the event a subscriber does leave the utility area, the solar project owner must replace that subscriber with one of similar credit quality. Project owners and their channel partners typically maintain a subscriber wait list to mitigate the risk of revenue shortfall that would result from subscriber relocation with inadequate replacement. Yet, despite this migration, subscriber relocation risk has been a major hindrance to the financing of community solar projects.

There are various pricing models for community solar subscriber PPAs. Subscribers can pay a fixed energy payment, which may escalate over time; pay the retail rate with a fixed adder; pay a discount to the retail rate; or pay a retail rate with a floor. The contract terms can also vary and have been as short as six months and as long as 20 years. Subscribers prefer short-term contracts that are linked to retail rates but such contracts are difficult to finance, as financiers prefer longer term contracts with fixed pricing.

Oftentimes, to manage the subscriber relocation risks and to enhance the financeability of the community solar project, a sponsor may seek a large commercial or industrial customer for a significant portion of the project subscription as an anchor customer.

Community solar has gained popularity over a short period. SEIA estimates that 1387 MW of community solar has been installed in the US through 2018. SEIA also states that there are at least 19 states, plus the District of Columbia: that recognize the benefits of community solar and have implemented policies to encourage growth of the program.

The community solar business model addresses a huge latent demand for access to solar for customers who otherwise cannot access it. This lack of access particularly applies to retail customers who rent their homes as well as real estate owners who simply lack adequate rooftops for solar installations. In this regard, community solar enables more equity among customers who desire access to the economic and environmental benefits of solar power. Furthermore, since community solar projects are centralized and typically large, substantial economies of scale can be achieved while retaining the benefits of distributed solar. Unlike the residential solar model described earlier, which effectively eliminates the role of the utility, utilities have a role to play in the Community Solar model. As a result, the transmission and distribution costs are shared more evenly by the utility customers.

Despite strong drivers, growth of community solar remains constrained by limited policy impetus and the complexity with financing the community solar projects.

Community choice aggregation

Community Choice Aggregator (or "CCA") is a non-profit public agency set up by a municipality, or a group of municipalities, to serve the energy needs of a combined load of the residents and commercial businesses within its political boundaries. Once the CCA is in place, the local utility no longer has the obligation to procure electricity for the customers served by the CCA, but it still provides transmission, distribution, and billing services. The CCA entities are created according to state legislatures. The state of Massachusetts is the pioneer of the CCA model and passed legislation in 1997 to enable formation of the Cape Light Compact to serve towns in Cape Cod and Martha's Vineyard. Since the passage of the law in Massachusetts, the CCA model has been used in California, Illinois, New Jersey, New York, Ohio, Rhode Island, and Virginia. Five more states have since introduced legislation to approve CCAs, making it a dominant mechanism to promote renewable energy across the country.

The CCA model has been instrumental in driving down energy costs for its customers. The Northeast Ohio Public Energy Council (NOPEC) serves over 500,000 customers and estimated that in 2016, its customers had saved more than $215 million since its inception in 2000. In certain markets, consumers can get electricity at a lower rate than even utility procured renewable energy, due to collective bargaining power. Larger CCAs that span multiple municipalities and counties may also encourage construction of local projects, which can lead to local job creation and economic development.

A key feature of the CCA program is the opt-out provision. Where the CCA model exists, the CCA is the default energy provider for electricity customers in the applicable service area; once a CCA begins to provide service, residents and businesses are automatically switched over from the utility service to the CCA unless the customer opts out of the CCA and maintain service from the utility. The lowest participation rate for opt-out programs that offer renewable energy is around 75%, whereas the highest participation rate for opt-in programs (those that require affirmative consent for participation) is around 25%. Moreover, the national average opt-out rate is only 3%–5%. The statistics help explain the rapid proliferation of the CCA model.

Despite the strong momentum behind CCA, there are hindrances to the growth of the CCA business model, primarily the CCA credit profile. CCA entities are not guaranteed by the municipal agencies that form them. Furthermore, for the most part, CCA is a pass-through entity, with revenues that include the receipts collected from customers and expenses. As a result, financial institutions struggle to finance projects with CCA offtakes. In order to circumvent the financing difficulties, financiers have incorporated certain cash sweep provisions in the recent deals that are triggered if the CCA entity fails to satisfy certain credit metrics. As of June 2019, Moody's

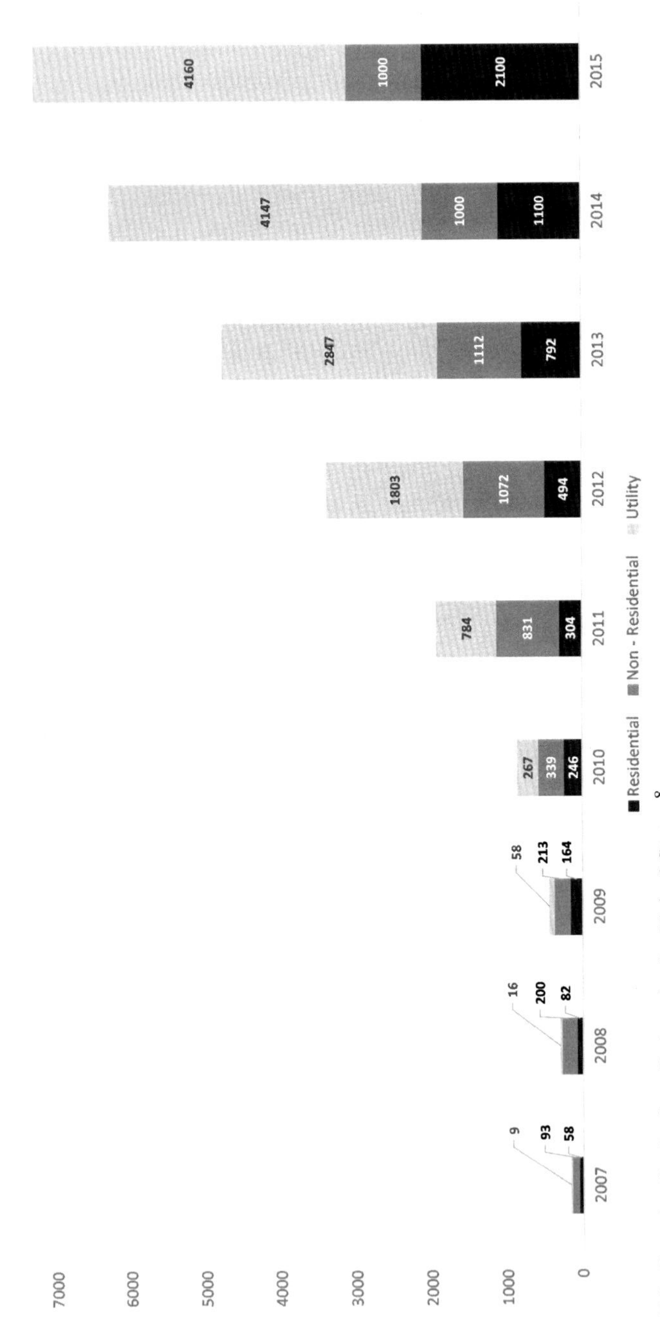

Fig. 7.5 Growth of solar installations in the United States.[8]

have rated two CCA entities – the Marine Clean Energy (rated at Baa2) and the Peninsula Clean Energy Authority (rated at Baa2), both located in California.

Similar to the community solar business model, the CCA model can help customers access renewable energy. With proactive policy support at the local level, the CCA model could also increase the penetration of renewables in the US.

Rooftop solar installations for commercial and industrial applications

Rooftops for commercial and industrial (C&I) companies are ideal for solar installations. As corporations across the United States increase their sustainability commitments, installing rooftop solar for these corporations could have significant economic merits. Unfortunately, as illustrated in Fig. 7.5, the growth of C&I rooftop solar sector has been stagnant in comparison to the utility and residential sectors; this is unlikely to change in the near future. There are several reasons for the relatively low growth of the C&I sector:

- Power purchase agreements and financing contracts are not standardized. The residential solar sector avoided this obstacle because homogeneity exists across installations. Unfortunately, each C&I installation must go through individualized design, construction, and financing, adding significant time and resource commitment.
- The credit profile for C&I hosts can vary significantly. Financing parties typically require credit ratings for solar project offtakers. This is not a problem for utility-scale projects because most of the offtakers, including utilities, are formally rated and, even if unrated, credit concerns can be mitigated. Residential solar has avoided credit issues because retail consumer credit is well understood. However, many C&I offtakers lack both credit rating and general understanding about what it means to be an unrated C&I entity. Unfortunately, the highest demand for C&I solar would come from non-rated C&I offtakers, such as schools, churches, malls, warehouses, etc., but neither rating agencies nor developers have created a credit metric to open these C&I offtakers to financing.
- The economics of the C&I rooftop solar are typically constrained because C&I offtakers are price conscious, despite their commitment to sustainability. The lack of standardized processes for constructing and financing C&I rooftop solar combined with the low PPA rates and expensive financing renders overall financing a significant challenge.
- Finally, the developer space for the C&I space remains fragmented. No single developer or sponsor of note has managed to garner critical mass across the US The situation is not likely to change because the barriers to entry for C&I rooftop solar are likely higher than those for residential solar.

[8] Wood Mackenzie Power and Renewables; Solar Energy Industries Association (SEIA). March, (2019). *U.S. Solar Market Insight.*

Renewable energy in power markets

Renewable energy projects such as wind farms, solar PV facilities, geothermal, and hydro power are all dedicated to the production of electricity. Historically, the electric sector in many countries was the domain of large, vertically-integrated utilities that were heavily regulated or state-owned. In the vertically-integrated monopoly model, one company or entity would be responsible for the generation, transmission, and distribution of electricity in a region, and the price paid by the consumer bundled together these elements.

The restructuring of the power industry in many countries has created scope for competition in the generation sector, allowing for independent generators (not owned by a vertically-integrated utility or government agency). These may be conventional fossil-fuel generators (e.g. coal or natural gas-fired) or renewable facilities such as wind farms and geothermal plants. Many project-financed renewable generation projects operate in the context of these restructured, competitive power markets. A basic understanding of how these work is thus critical for understanding how renewable projects work in practice and their market risks.

Electricity markets are some of the most complex markets ever implemented. The economics and implementation of these markets could easily fill a full book. The objective of this section is therefore only to provide a high level understanding of how renewable projects operate within these markets to aid in understanding project economics, risks and contractual structures.

Fig. 8.1 shows the operating functions within a restructured electric industry. Multiple generating companies compete to sell electricity, under long-term contracts or in spot markets. A regulated transmission company owns the high-voltage transmission grid which moves power across a region. Finally, regulated distribution companies operate the local distribution systems (e.g. the power lines running down the street) and deliver the power sold to customers.

This is of course a highly simplified description of a complex regulatory structure. In many countries, for example, the transmission company only physically owns and maintain the high voltage transmission grid, short-term operations and planning may be entrusted to a "system operator", independent of other market participants. These operators are often referred to as independent system operators (ISOs) in the United States and Transmission System Operators (TSOs) in Europe, although the roles in each case are not exactly the same This system operator will ensure that the system operates reliably and typically also runs a spot market for electricity, used to clear supply and demand on a continuous basis. This is illustrated in Fig. 8.2.

The commercial arrangements run in parallel, as also shown in Fig. 8.2. Customers buy electricity from electricity retailers (a competitive function in some systems and regulated in others). These retailers buy their power under contract from the generators

Renewable Energy Finance. https://doi.org/10.1016/B978-0-12-816441-9.00008-8

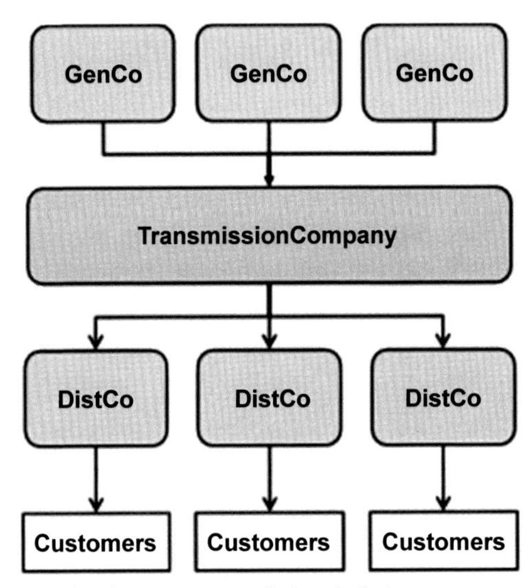

Fig. 8.1 Functional separation in a restructured electric industry.

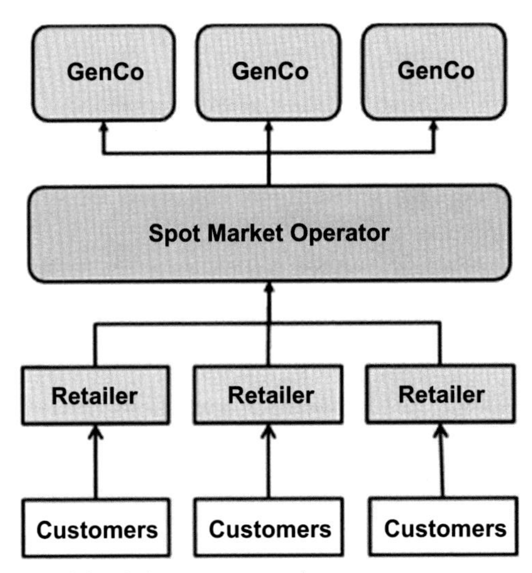

Fig. 8.2 Illustrative flow of funds in a restructured power sector.

or in the spot market operated by the system operator. Where there is a requirement for renewable generation (such as in the renewable portfolio standards described in Chapter 2), at least a defined fraction of the electricity procured by retailers must come from qualifying renewable generators.

Basics of power market design

Demand for electricity varies continuously, by time of day (lower overnight, and higher during waking hours), weather (such as air conditioning loads), and other factors. At the smallest level, the total demand for power varies as individual customers switch on their TV or turn off their lights.

As electricity cannot be easily stored, and demand varies substantially over a day, power markets must be designed to clear frequently. In the absence of other constraints, it would make sense that the lowest marginal cost resources were used first, and moving on to more and more expensive generation to meet higher levels of demand. Electricity markets implement this concept of **least cost dispatch** to ensure that power production is efficient.

In power markets, the "merit order" of generation starts with the lowest marginal cost resources (such as wind farms when the wind is blowing and solar facilities when the sun is shining), and moves up through conventional generation which has identifiable fuel costs, as illustrated in Fig. 8.3. So, for example, the marginal cost of a wind farm is pretty much zero when the wind is blowing – no additional costs are incurred for generating power. While a nuclear plant may be costly to build, the marginal cost of generating each megawatt-hour is very low, so it too is low in the merit order. At the opposite extreme, many systems have natural gas-fired combustion turbines which have high relative fuel consumption per MWh generated and hence high marginal costs. To minimize total costs, power markets generally dispatch the lowest cost generation first, using more and more costly generation as necessary to meet demand in each period.

Power markets must also recognize other engineering constraints on how the power system functions. These include transmission constraints, which limit the flow of power on individual lines (discussed in the next section). Another set of constraints includes the operational characteristics of individual generators. For example, many

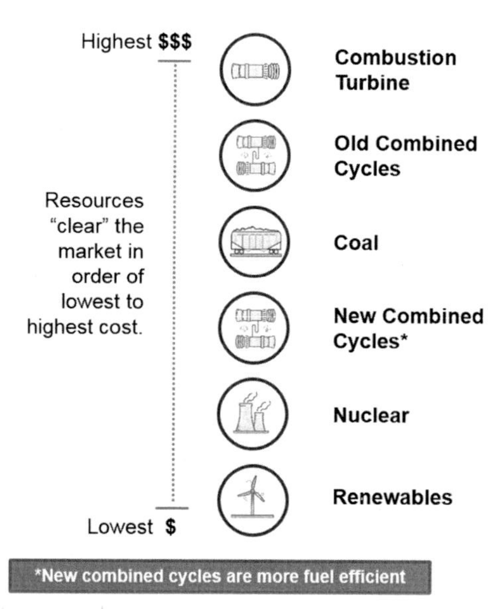

Fig. 8.3 Merit order for generation.

generators cannot be turned on or off instantaneously, but take time (and fuel) to start. The system operator must ensure that all of these constraints are respected in operating the market, so that electrical reliability is protected.

Power markets operate in general like a reverse auction, taking bids to generate (for short periods such as an hour) and using these to clear the market. In simplified terms, the clearing price in a period (e.g. an hour or less) is set by the price of the highest bid needed to meet demand in that period. Since demand is changing frequently, power market prices tend to be volatile even within a day.

Over longer periods, power prices vary not only with demand but also with fuel prices (especially with natural gas prices in many markets), as plants enter or retire from the market, and a host of other factors. Without a hedge, the revenues from a renewable project would vary substantially over time as natural gas and other prices change, even if a project like a wind farm does not directly use fuel. Therefore, some form of hedging of these price risks is typically necessary to finance a new project.

Most sources of renewable energy, of course, also depend on external factors, such as insolation (the amount of sunlight reaching the ground), wind speeds, or the amount of water available in a river. As such, many renewable resources are often referred to as "non-dispatchable" as they are not available on demand. In most systems, renewables will tend to run when available, absent transmission constraints or insufficient demand.

Fig. 8.4 illustrates in a highly stylized fashion some of the relevant time horizons in power markets. Transmission and reliability planning (which are not market functions

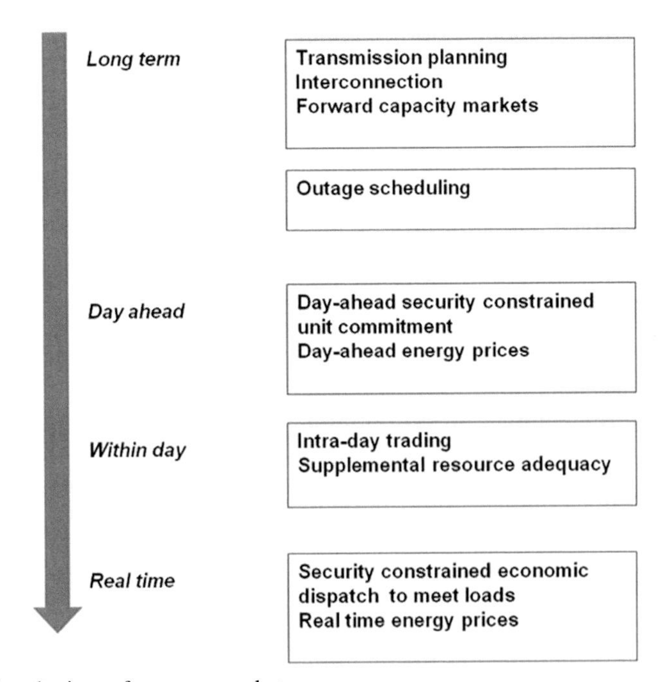

Fig. 8.4 Time horizons for power markets.

exactly) must be done well in advance to ensure that the bulk power system will be capable of meeting future demand. This may be done years in advance. Closer to the operating time, system operators help coordinate outage scheduling of plants to ensure that sufficient capacity is available while allowing for needed plant maintenance.

Given the dynamic constraints on units, most power markets have a day-ahead scheduling process, which allows units to be committed to run the next day to meet load requirements and provides time for units to start, ramp up, etc. This is often coupled with a day-ahead energy market, in which generators bid to be committed, and which produces hourly clearing prices for the next day. A system is usually required to ensure that transmission security requirements are met – e.g. generators are not being scheduled where they cannot run due to transmission constraints. In a full **locational marginal pricing** (LMP) market, this process is known as security-constrained unit commitment (SCUC) and the resulting day-ahead LMPs are paid to the generators scheduled to operate in those hours.

After the day-ahead market, the system operator may do a resource adequacy assessment to ensure that enough units are committed to meet reliability requirements. This can be done on a market basis (with additional generators bidding in) or not. If the system operator finds that insufficient generation has been committed at this stage, it can schedule additional units to be ready.

Finally, security-constrained dispatch (SCD) is conducted by a system operator who has final control and responsibility for the reliability of the power system. In real-time, this system operator must dynamically dispatch units by merit order in order to precisely balance supply and demand, while also recognizing the transmission constraints in effect at the time.

Transmission congestion and LMP

The previous section focused on the temporal aspect of power market operations – the need for the system operator to match supply and demand quite precisely, and to reflect the constraints on generators which prevent them from switching on and off (and limit their ability to ramp from one level to another) immediately. In this section, the focus is on the significant locational issues with pricing created by transmission constraints.

Transmission congestion concepts

Congestion as an economic concept exists when current demand for a scarce resource exceeds capacity. One person making use of that capacity thus creates an external cost for other users. In a power system, congestion refers to the use of the transmission system's capacity.

Fig. 8.5 shows a power market with two generators and two loads, connected by a single line with unlimited capacity. The total load is 160 MW in 1 hour, so total generation (from Generator A and Generator B) must total 160 MWh. By merit order,

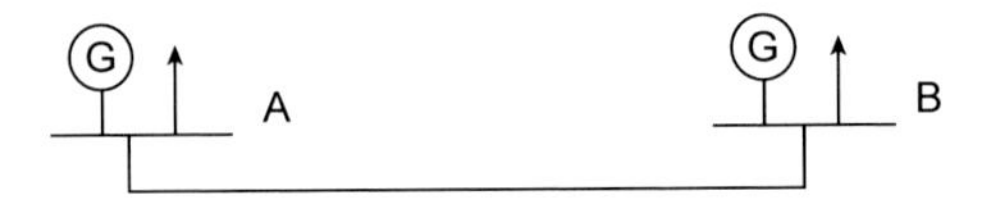

Bus A (Node A)	Bus B (Node B)
Load: 100MW	Load: 60MW
Generation Cost: $10/MWh	Generation Cost: $5/MWh
Generation Capacity: 100 MW	Generation Capacity: 100 MW

Fig. 8.5 A two-node example with no congestion.

the lowest cost solution is to use all of Generator B's capacity first (100 MW), followed by 60 MW from Generator A. As the line is assumed to have unlimited capacity at this point, the location of generation is unimportant. At each node in the simple system, the marginal cost of supply (or price) is equal to $10/MWh, since any additional unit of generation must come from Generator A (as generator B is already operating at full capacity).

Now consider the same situation, but now the transmission line between Nodes A and B has a maximum capacity of 20 MW, as shown in Fig. 8.6. Generator B remains the cheapest source of supply, but the load at Node B is only 60 MW, and the line can only take 20 MW to serve the additional load at Node A. So, Generator B is constrained to 80 MW of output, and the other 80 MWh of generation must come from Generator A at a higher cost.

With transmission congestion, prices between the nodes diverge. Serving an incremental MWh of generation at Node A will cost $10/MWh, since any additional generation must come from higher cost Generator A. Serving an incremental MWh at Node B costs $5/MWh, since that additional supply can come from the cheaper Generator B. The marginal prices at the two locations are thus $10/MWh (Node A) and $5/MWh (Node B), and the congestion cost differential between the two nodes is $5/MWh (calculated as $10/MWh - $5/MWh).

Congestion is very simple with only two nodes, but even with three nodes the marginal pricing effects become much more complicated. In the three-node example in Fig. 8.7, there are again two generators, now with a single load located at a third node.

In the electricity grid, power flows not only on the shortest path, but spreads across the network. On the right-hand side of Fig. 8.7, for example, it is assumed that all lines

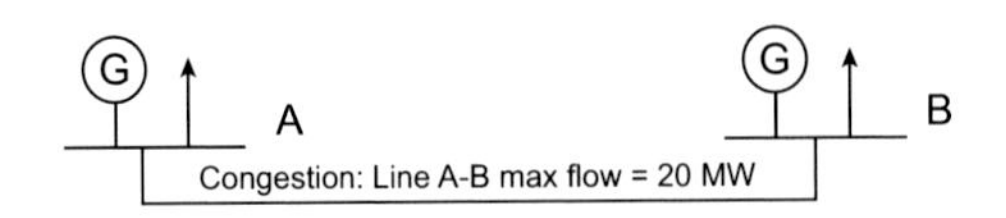

Bus A (Node A)	Bus B (Node B)
Load: 100MW	Load: 60MW
Generation Cost: $10/MWh	Generation Cost: $5/MWh
Generation Capacity: 100MW	Generation Capacity: 100MW

Fig. 8.6 Two-node example with congestion.

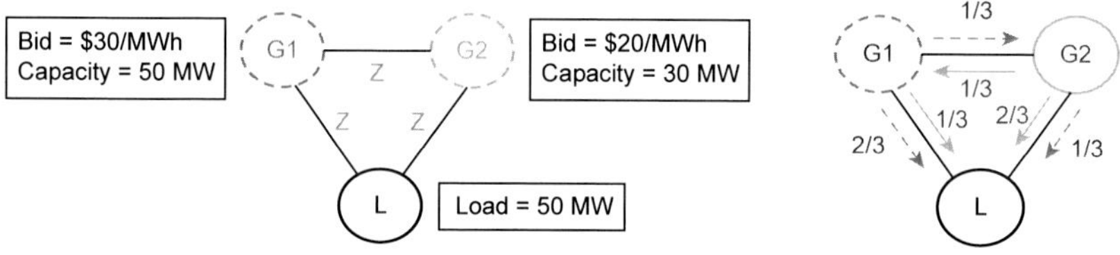

Fig. 8.7 Three-node example.

have the same **impedance**. Hence, flows from Generator 1 to the load do not flow only on the direct line, but also go by way of the Generator 2 node. Under the simple assumptions in this example, 2/3 of the power goes directly from G1 to the load Lm while 1/3 goes via the G2 node. Conversely, for G2, 2/3 of its generation goes directly to L, while 1/3 is transmitted through the G1 node.

With no transmission line constraints or losses, the marginal price in every location (referred to as a "Locational Marginal Price" or LMP) will be same at every node. As shown in Fig. 8.8, all of the cheapest generation (G2) will be used first, with incremental generation coming from G1 at a marginal cost of $30/MWh. The LMP at node L (and everywhere else) is therefore $30/MWh. G2 will produce 30 MW, and G1 will produce 20 MW, to meet the 50 MW of total demand (all at L).

Fig. 8.9 shows the power flows in this simple unconstrained case. As discussed before, 2/3 of generation from G1 goes straight to L ($2/3 \times 20$ MW $= 13.3$ MW) while 1/3 flows through G2 ($1/3 \times 20$ MW $= 6.7$ MW). From G2, 2/3 goes straight to L (30 MW $\times 2/3 = 20$ MW), while 10 MW goes via G1. The table shows the total flows on each line.

What if there is a constraint on the power that can flow on even a single line? This situation is illustrated in Fig. 8.10.

The limit on line G2 – L is 20 MW, less than the 26.7 MW flow under the unconstrained solution seen in Fig. 8.8. Since the load of 50 MW still has to be met, clearly the pattern of generation needs to change (known as a "re-dispatch") in order to ease the congestion on line G2- L.

Consider the shift in generation output shown in Fig. 8.11, where G2 produces 1 MW less (down to 29 MW) and G1 produces 1 MW more (up to 21 MW). The total generation constraint is met (50 MW of generation matches 50 MW of load) but this shift is not enough to resolve the transmission constraint. Under this re-dispatch, flows on G2 – L would still be 26.3 MW, greater than the 20 MW capacity of the line. Thus, a larger re-dispatch is needed to shift more output to G1 (with a higher marginal cost).

Fig. 8.12 shows a feasible dispatch, which meets the load and respects the transmission constraint of 20 MW. Under this dispatch, G2 produces only 10 MW and the higher cost G1 now produces 40 MW. Total generation cost has obviously increased substantially over the unconstrained case presented in Fig. 8.8.

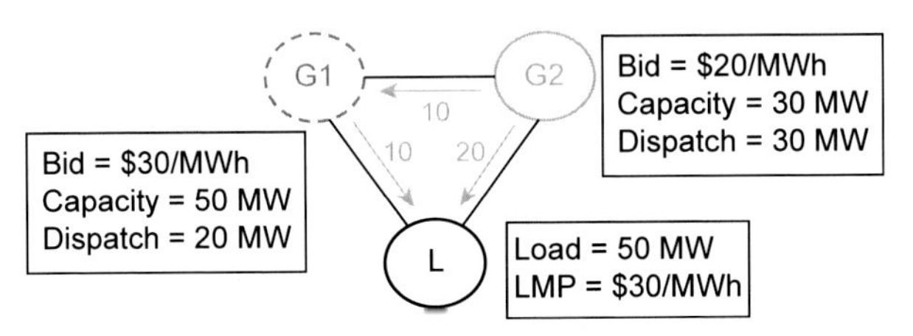

Fig. 8.8 Unconstrained dispatch in 3-node example.

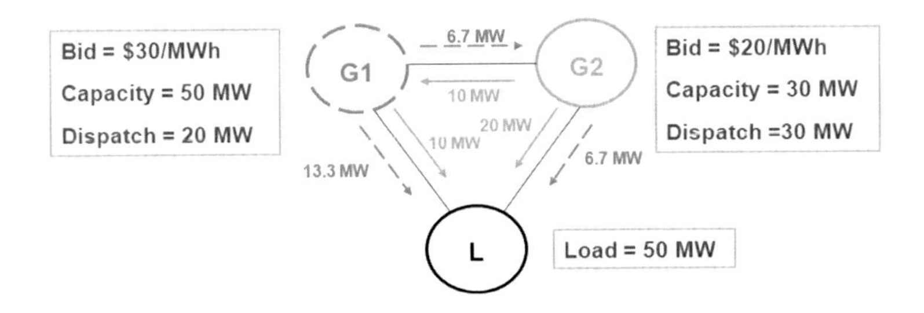

From G1			From G2			Net Line Flow		
Total	To L	To G2 to L	Total	To L	To G1 to L	G1 to L	G2 to L	Total
20.0	13.3	6.7	30.0	20.0	10.0	23.3	26.7	50.0

Fig. 8.9 Power flows in the unconstrained dispatch example.

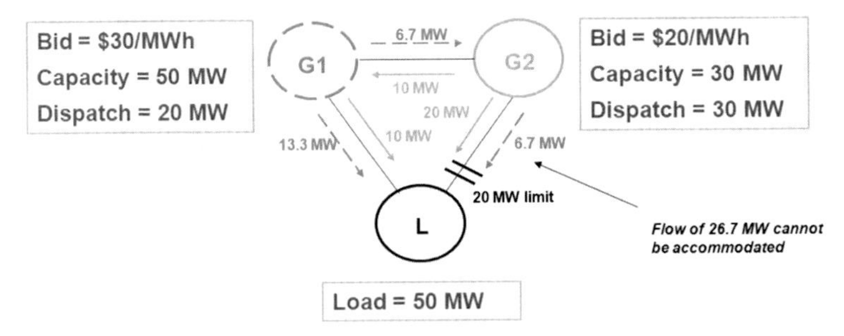

Fig. 8.10 Transmission constraint in a 3-node example.

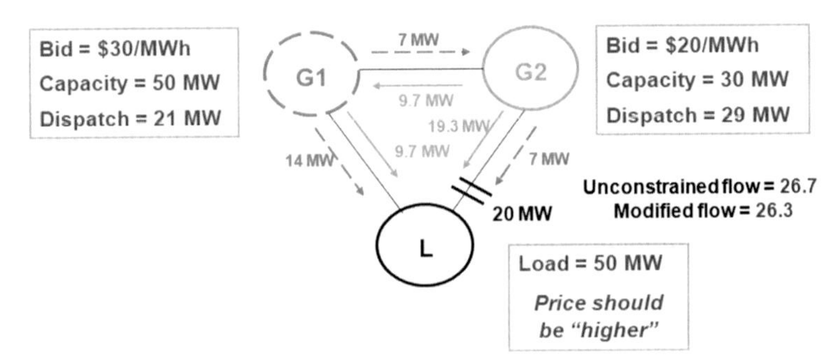

Fig. 8.11 Shifting generation to ease congestion.

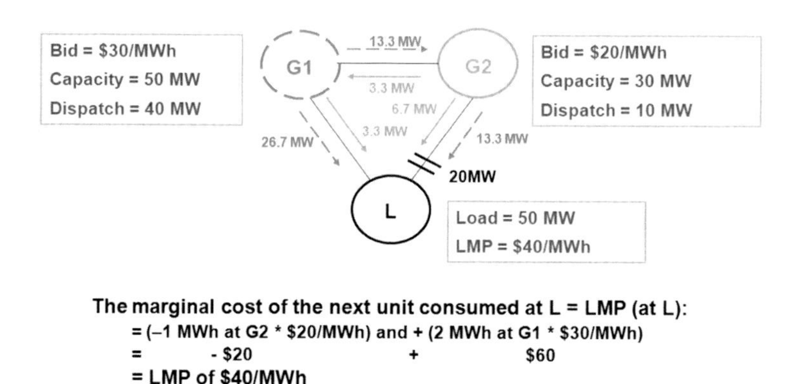

Fig. 8.12 LMPs reflects marginal cost of resolving congestion.

The LMP at every node is equal to the cost of another unit of output (e.g. 1 MW) at that node, while reflecting all energy balance and transmission constraints.[1] The LMP at Node L therefore must consider that 1 MW more must be generated in total. However, given the transmission constraint on line G2 − L, this will involve producing 2 MW more at G1 and 1 MW less at G2. Therefore, the marginal cost of supplying 1 MW at L is (2 × $30/MWh) − (1 × $20/MWh) = $40/MWh.

Note that the LMP at L is higher than the marginal cost of the most expensive generator on the system. This happens because the LMP reflects not only the marginal cost of supplying an additional unit but also the marginal cost of re-dispatching the system. In this way, the external costs of congestion are captured in market prices, which leads to a more efficient set of economic outcomes.

In practice, full LMP market designs capture not only the marginal costs of congestion on the system, but also the marginal losses created by an additional unit. For example, in distant regions of the grid connected by long transmission lines, the marginal loss component (MLC) may be significant, lowering LMPs.[2] In a high demand region, the MLC may be high, increasing LMPs.

The LMP thus reflects the marginal cost of energy (MEC) across the system (the marginal energy cost), the marginal cost of congestion (MCC) which varies by location, and the marginal loss component (MLC). Marginal losses, like marginal congestion costs, also vary by location and can be positive or negative.

The LMP at any location is the sum of the marginal energy, congestion, and loss components, such that:

$$\text{LMP}_i = \text{MEC} + \text{MCC}_i + \text{MLC}_i$$

[1] All of the examples presented here have no losses for purposes of exposition.

[2] The marginal loss component may be positive or negative. In this case, under the usual sign convention, the marginal loss component of the LMP would be negative, lowering the total LMP at this distant generation node.

Here the subscript i represents the location on the grid. Different LMP systems sometimes have different sign conventions for the congestion and loss components, so it is important to check how the LMP is defined in that market before analyzing any specific component.

Examples of LMP markets

In practice, a large interconnected power grid consists of hundreds or thousands of nodes and many transmission lines. For example, Fig. 8.13 illustrates some of the major transmission lines in the PJM market in the Eastern United States, shown by voltage.[3] There are many other lines that are not shown. In the PJM power market, individual nodal LMPs are calculated at over 3000 locations continuously.

In LMP markets, the prices are calculated using sophisticated constrained optimization software that uses bids from generators and loads to clear the market, while representing the maximum flows on lines in a manner similar to the simple three-node example presented in the previous section.

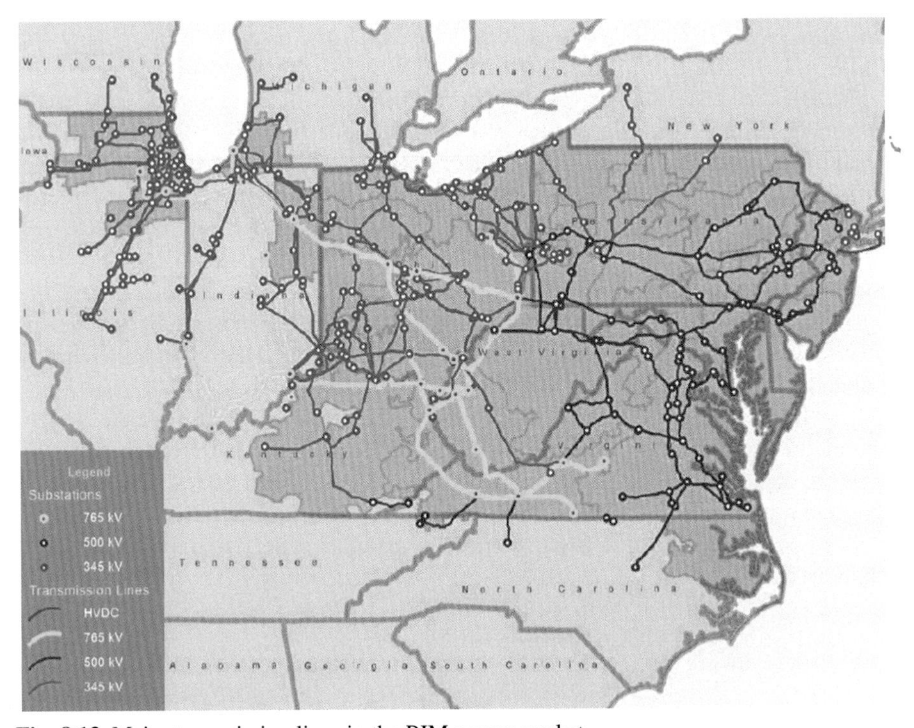

Fig. 8.13 Major transmission lines in the PJM power market.
Source: PJM Interconnection, used by permission.

[3] PJM. (2018). *2017 PJM regional transmission expansion plan*. Available from www.pjm.com.

In full LMP market designs, hourly day-ahead LMPs are calculated at each node, based on day-ahead bids and the day-ahead security-constrained unit commitment (SCUC) process. This process in repeated in real-time during plant dispatch, which produces 5-min real-time LMPs at each node. Prices can change substantially over a day, depending on supply and demand conditions. Fig. 8.14 shows the three LMP components in a region of high demand on a peak day, with high LMPs in the peak demand period in the middle of a hot summer day.

LMPs by definition also vary spatially. There can be regions of higher prices where demand is high and transmission constraints are binding. In regions of surplus generation, also affected by transmission constraints, LMPs can become very low and even negative.

The LMP market design has become widely accepted in the United States, as shown in Fig. 8.15, operating in PJM, New York (NYISO), New England (ISO New England), Texas (ERCOT), the central US (MISO and SPP), and California. LMP-based market designs are also used in New Zealand and other countries.

Other regions of the United States (such as the Southeast and much of the West) continue to operate on more of the traditional vertically-integrated model using physical transmission rights, with open access to the transmission grid mandated by FERC. In these regions, there is some bilateral trading at the wholesale level, but without clearing spot markets.

Zonal markets in Europe and elsewhere

The LMP market design attempts to capture the short-run marginal costs of electric transmission, reflecting re-dispatch costs, reliability constraints, and marginal losses. However, many power markets rely on a simplified representation of the transmission grid in terms of scheduling and market pricing.

For example, some markets act as a single pricing zone, ignoring transmission constraints in setting prices. In these cases, supply and demand bids can be cleared

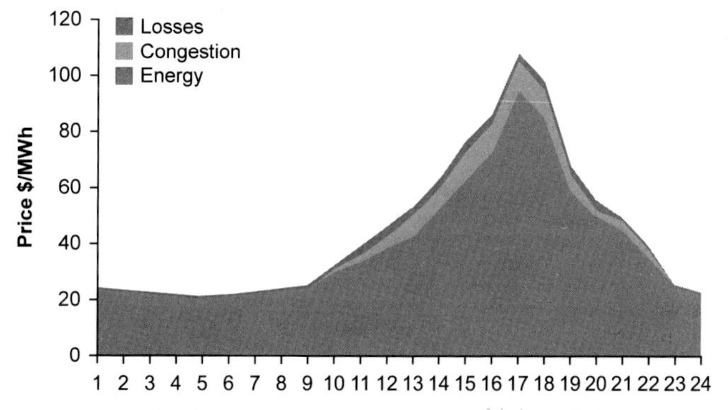

Fig. 8.14 Example of LMP prices and components over a high peak day.

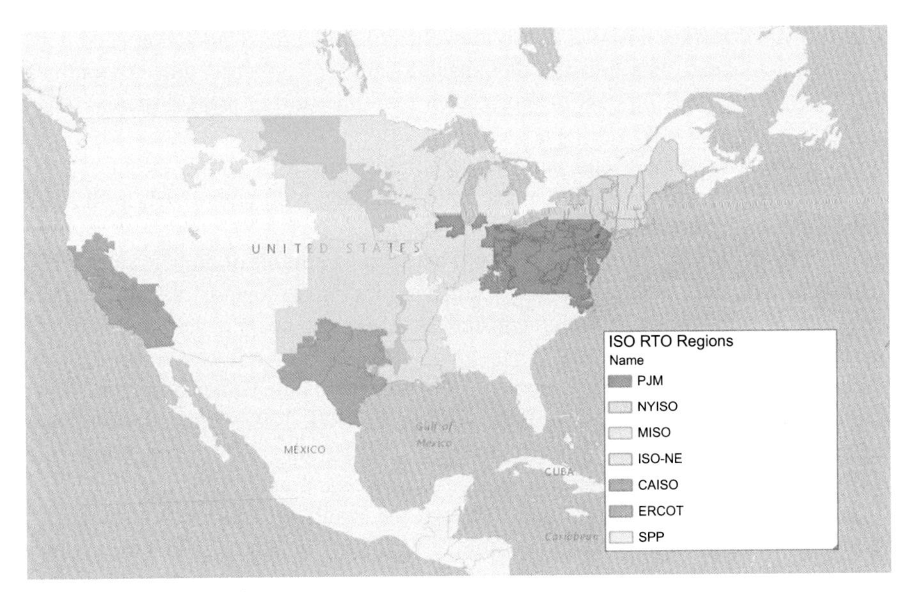

Fig. 8.15 US LMP markets.

directly, although since there are almost always some transmission constraints in actual systems, a process must be included for the system operator to change the unit commitment and dispatch to reflect these as necessary. The use of a single pricing zone where transmission constraints exist creates a range of incentive problems, but it is still widely used in some systems.

In larger systems with recognized transmission constraints, zonal systems combine nodes into zonal groups which are treated in economic terms as being at the same location. As shown in Fig. 8.16, a simplified interconnected system is split into four zones. This creates some administrative simplicities, but at a cost. For example, in an interconnected grid, it is highly unlikely that generators located at nodes A and B would have the same flow impact on Line A-C, where the constraint exists. But, as demonstrated in the previous three-node example, the re-dispatch logic for efficiently handling congestion depends on the system operator knowing which units to re-dispatch based on the consequent flows in the constrained lines.

A zonal market model which couples together many national- or system operator-based markets can be found in the European Union. Most EU countries have markets that clear their individual market and treat all generation and load as being at the same point. A market coupling mechanism calculates the flows between individual zonal markets. While EU market coupling is superior in economic terms to the explicit transmission auctions previously used in the EU, the zonal simplification still creates transmission inefficiencies and incentive problems. As the penetration of intermittent renewables resources increases, these problems may grow.

While this market coupling system does enable international trade, it has limitations. First and foremost, individual TSOs must have mechanisms for resolving the remaining congestion that cannot be reflected in a zonal system. Second, a zonal

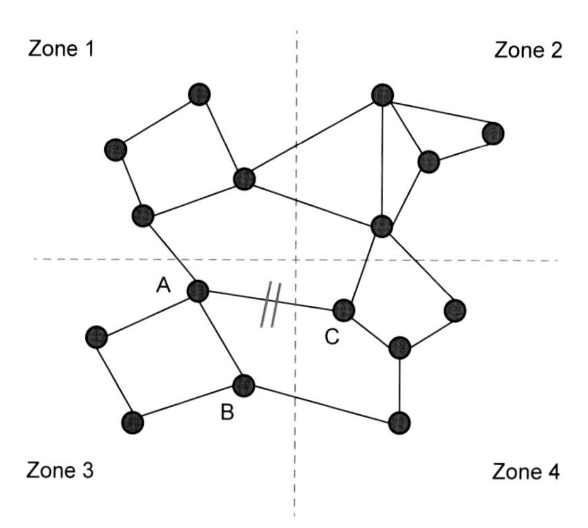

Fig. 8.16 Zonal definitions in a network.

system cannot efficiently define a set of transmission rights that uses all of the transmission capacity completely. If zonal systems have advantages in terms of historical acceptability, the economic costs of inefficient transmission allocation remain a distinct downside.

Capacity and ancillary services markets

To preserve electrical reliability, power systems need sufficient installed generation capacity over various time horizons. This includes sufficient installed capacity on a long-term basis to meet load (since generation capacity cannot be added instantaneously) and for various flexible reserves need to balance supply and demand on short-time horizons. These markets are highly complex and hence this section provides only a brief overview of the issues from a financing perspective. The reader is directed to the references for a more complete description of these markets.

Resource adequacy mechanisms

Capacity market and payment mechanisms seek to ensure that there are sufficient supplies of capacity are available to meet future load reliably. This can be through a tradable market construct or auction (often referred to as a capacity market), where a capacity product is bought or sold (usually on a forward basis), or through a payment to generators who make capacity available, especially during peak periods. One of the earliest competitive power markets in England and Wales had such a capacity payment that paid generators a premium for having capacity available when a calculated Loss of Load Probability (LOLP) factor was high. This was designed to ensure that generators had incentives to make their capacity availability when it was most needed (e.g., peak demand periods), but it also created market power problems.[4]

Currently, many electricity markets have a tradable capacity product which is purchased by load-serving entities (e.g. utilities or retailers which are responsible for supplying power to final customers). Well-known examples include the PJM Reliability Pricing Model (RPM), and ISO New England's Forward Capacity Market (FCM).[5] In Europe, the United Kingdom designed a capacity market for the British system (in England, Wales, and Scotland) but its implementation has been delayed over issues associated with European state aid rules.[6] Other European countries have also developed capacity market mechanisms.

Capacity is typically traded forward so that suppliers can offer new capacity from projects that are not yet built to meet future forecast demands. For example, in PJM, an auction is held to acquire capacity three years in advance of when it is needed. Auction prices are differentiated by location in PJM (under a zonal system) to help create incentives for generation to be made available where it is needed, reflecting constraints on the transmission grid.

Many renewable resources such as wind and solar are intermittent, and depend on external factors (such as wind speeds) to determine their output. Thus, they often cannot be depended on to the same degree as many conventional resources (such as gas turbines) to be able to generate when most needed. To account for this, most capacity market designs derate the capacity from intermittent resources such as wind and solar based on their historical availability at periods of peak demand. As a simple example, such a derating may account for the fact that a solar PV facility will be expected on average to be generating 50% of its maximum capacity at peak periods, while a wind farm (whose output may be higher in off-peak periods) gets an even higher derating.[7] To the extent that a project is covered by a PPA, any capacity value from the unit is typically sold bundled with energy, renewable credits, and other attributes, and the project will be insulated from variation in capacity market prices.

Some other markets do not have capacity markets, and rely on scarcity pricing in peak periods to provide the incentives to invest in sufficient generation. The ERCOT market in Texas is a well-known example of an energy-only market.[8] Other energy-only markets can be found in Australia (the National Electricity Market), Europe, and New Zealand.

[4] Newbery, D. (1995). Power markets and market power. *The Energy Journal, 15*(3).

[5] The PJM RPM capacity market is described in PJM, "RPM 101: Overview of Reliability Pricing Model" available at www.pjm.com. A set of detailed "FCM 101" presentations describing the ISO-NE FCM market is available at https://www.iso-ne.com/participate/training/materials.

[6] McNamara, F. (2014). Capacity market. Department of Energy and Climate Change. Retrieved from https://assets.publishing.service.gov.uk/government/uploads/system/uploads/attachment_data/file/335760/capacity_market_policy_presentation.pdf.

[7] These are only simple examples, and various markets have complex rules for establishing the capacity rating of renewable and other resources.

[8] Bushnell, J., Flagg, M., & Mansur, E. (2017). Electricity capacity markets at a crossroads. UC Davis Energy Economics Program. Retrieved from http://deep.ucdavis.edu/uploads/5/6/8/7/56877229/deep_wp017.pdf.

Ancillary services markets and payments

To operate the power system reliably, grid operators need to be able to call on flexible generation to meet outages, change output to reflect short-term variations in supply and demand, and supply other technical requirements. In the context of electricity markets, these are often referred to as ancillary services. In some markets the grid operator procures these ancillary services under contract which are then recovered from loads in transmission costs; this approach is common in the European Union for example. In the US ISO markets, sophisticated markets have been designed (which parallel the day-ahead and real-time energy markets discussed earlier) to procure and price these ancillary services products, which include regulation (units that can ramp up and down to adjust their output levels in response to an automatic generation control (AGC) signal) and operating reserves. Operating reserves are generally classified as spinning or synchronized reserves that actively generate and can change their output as needed; and non-spinning reserves, which are on standby, can start generating within a defined short period of time (e.g. 10 min).

Most large-scale renewable projects are not well suited to providing these ancillary services. Hence, this is generally not a major revenue source for renewables projects. As will be discussed in Chapter 12, however, some forms of energy storage technologies are well suited to providing these ancillary services and contracts to provide such services and may be a major revenue stream for these storage projects.

The future of energy markets

As is apparent from the discussions above, electricity markets have largely been designed around the characteristics of fossil fuel-fired thermal plants. As renewable energy and storage penetration increase, the constraints that have shaped electricity market design (re-dispatch costs for congestion, for example) will change, and new concerns such as planning for reliability in systems with large amounts of intermittent renewable generation will come to the fore. It also appears likely that in many systems the amount of distributed generation (such as rooftop solar) will also increase substantially. Many market participants will be generating as well as using electricity, and market pricing, settlement and regulation will need to adapt.

Managing transmission costs and risks for renewable projects

Large-scale renewable energy projects such as wind farms and large utility-scale PV facilities must be connected to the transmission grid to deliver their product to customers. Connecting to the grid is a major cost for large renewable projects, often costing many millions of dollars. Once connected, renewable and other generators are also charged to move power across the grid. Investors in renewable projects therefore should be aware of these costs and how they can change over time, which encompasses a complex set of engineering, regulatory, and economic considerations. This chapter seeks to provide a brief introduction to the often highly complex issues which arise with transmission, and which can vary substantially between jurisdictions and over time.

The electric transmission sector is highly regulated in almost all countries. The costs and risks to a new or existing renewable energy project are often affected by complex rules set by governments and regulators, and implemented by transmission system operators. Given the complexity and changing nature of these rules, this chapter will seek only to provide an introduction to some of the core concepts regarding transmission regulation and its impact on renewable projects, and how these are implemented in a few illustrative systems around the world.

Connecting new generation projects to the grid

The first issue arising is how a new project will connect to the transmission grid, and how much this will cost the project sponsor.[1] A project developer, conscious of the costs of grid interconnection for a project, will make the feasibility and cost of the grid connection a major factor in selecting a site for a new project. Not only must the land and resource (e.g. wind or sun) exist at the site, but it must be feasible to connect the site at reasonable cost (and with reasonable likelihood of regulatory approvals) to the existing grid. For this reason, renewable project developers often employ engineers and consultants with detailed knowledge of the grid configuration in order to analyze where new projects are most feasible from a transmission perspective.

Proposing a new interconnection

Assume, for example, that a wind developer has identified an attractive site for a new wind farm. Customers in the region seeking to sign PPAs for new wind farms, the wind

[1] Some smaller projects may connect at distribution level in some countries but many of the same economic considerations will apply

Renewable Energy Finance. https://doi.org/10.1016/B978-0-12-816441-9.00009-X

resource is attractive, and the environmental siting issues appear tractable. There is a major transmission line relatively close by, only 10 km (6.2 miles) away. The developer, however, cannot just connect to the grid as it sees fit. The local transmission company or host utility must first determine if such a connection is technically feasible and what costs a new generator connected in this area will impose on the system. These costs are often much larger than the costs of the line connecting the project to the nearest line on the grid, and may include upgrading elements of the transmission network far from the project. Since, in many countries, a new generator is responsible for paying these interconnection costs, this can be a major driver of the economics of projects.

Determining interconnection costs

The first step is typically for the developer to make an application for interconnection, which triggers an initial engineering feasibility study by the grid operator to ensure that an interconnection is even possible in this area of the grid.[2] The feasibility study may include initial power flow and reliability studies. The initial application also enters the project into the "transmission queue," the set of projects that are proposed in a transmission operator's region. Since new generation projects pay for the incremental costs of connecting to the grid, the position in the transmission queue can be important as interconnection capacity is limited - later projects may have to pay additional costs if interconnection capacity in that region is already used by earlier generation projects. The feasibility study will often provide the developer with an initial review of the potential for interconnection and its costs.

If the developer wishes to take the project to the next stage, a system impact study will be required. In this phase, the transmission operator will make a detailed assessment of system constraints affected by the generator queue request, and identify the local attachment facilities and their needed upgrades, and the broader system upgrades necessary to connect the new wind farm. At this stage, the developer will receive a more precise cost and lead time estimate as well, which is important for capital budgeting.

Finally, if the project continues to go forward, the transmission operator will need detailed engineering and design work, and develop the schedule to complete the new plant interconnection. All of these studies require the project developer to sign agreements to conduct the analysis and design work. If the project goes ahead, the project developer or sponsor will need to sign an agreement authorizing the construction of the facilities and detailing payment terms.

Grid interconnection costs can be substantial for new renewable projects, sometimes in the hundreds of dollars per kilowatt connected (or even more). Given the magnitude of these costs (and their potential variability, which is generally outside

[2] This discussion is based loosely on the PJM new generator interconnection process, which is described in detail in PJM Interconnection Inc. (July 26, 2018). *PJM manual 14A: New services request process*. Other transmission system operators will have different processes.

of the direct control of the project sponsor), budgeting for these costs with appropriate contingencies is critical to ensure successful project economics.

Continuing transmission costs

Once a renewable energy project is built and interconnected, it can, like other generators, face ongoing transmission costs. In several European Union countries, for example, a fraction of total transmission costs is recovered from generators through tariffs. These can range from the small (a few percent of total costs in France) to the substantial (30% or more in some Scandinavian countries). In some countries, transmission tariffs differ by season and time of day to reflect peak usage on the electric grid. In a few countries, such as Great Britain and Norway, the generation fixed charges vary by location as well.[3]

The UK zonal system

As a specific example, Great Britain has a relatively complex locational transmission tariff for generators, which is designed to provide incentives for generators to locate in regions where more supply is needed, and less in areas where there is surplus generation. This structure is consistent with the overall design of the British power market, which does not include a direct locational component in electricity prices (such as the LMP prices in the United States). With no locational element of Transmission Network Use of System (TNUoS) charges, there would be no incentive to locate where generation is most needed. As such, there is no need for transmission rights in Great Britain to sell power into the domestic market, as all generation in short-run market terms is treated equally, regardless of location.[4]

Total transmission costs are split between generators and loads in Great Britain, with a significant allocation to large generators. National Grid ESO, which acts as the system operator for the British system, uses a complex methodology to allocate costs between generators and loads, then split these costs based on location. At present, there are 27 generation charging zones.

There is substantial variation in the TNUoS charges by location. For example, as shown in Fig. 9.1, large generators in northern Scotland and some other regions with transmission constraints pay larger TNUoS charges. Some regions with large little generation (such as Central London and Southwest England) receive negative charges for connection, recognizing, at least partially, the value of generation to the grid in these regions. Generators pay TNUoS based on their Transmission Entry Capacity (TEC), which is typically fixed from one year to another, absent unit upgrades, retirements, etc.

[3] ENTSO-E. (May 2018). *ENTSO-E overview of transmission tariffs in Europe: Synthesis 2018.* Available from www.entsoee.eu.

[4] In most European and other countries (some large countries being exceptions) all generation is treated as being in a single "zone" in market terms, so transmission rights must be acquired by generators only if selling to a neighboring market.

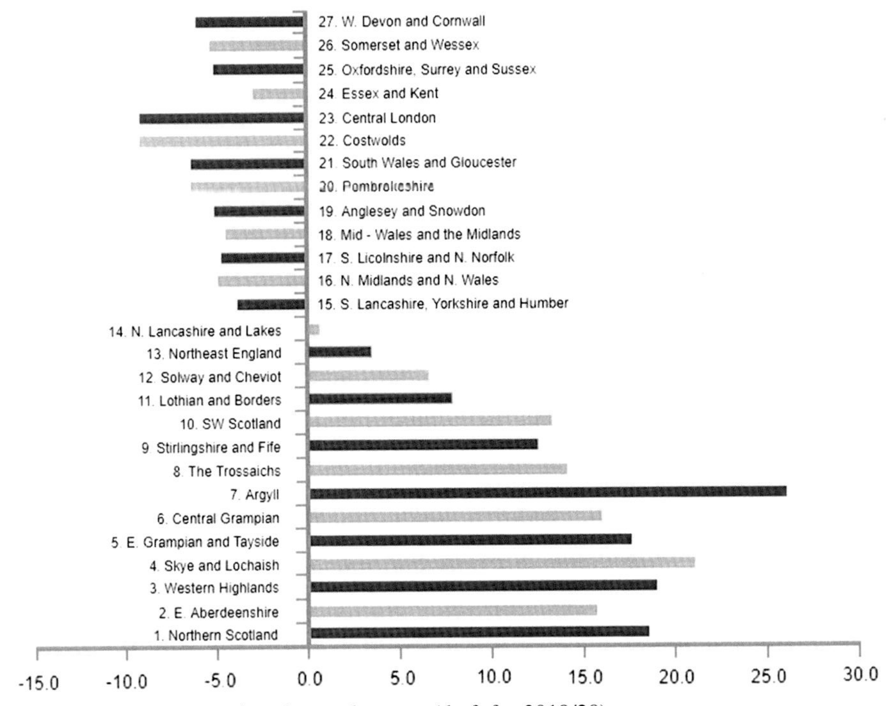

Fig. 9.1 TNUoS generation charges by zone (draft for 2019/20).
Source: National Grid ESO.

TNUoS charges to generators, like those in most entry capacity-based systems, are set by regulators and system operators. For project investors, however, it can be important to understand how these costs may change over time. In the British system, for example, the transmission operators have their overall revenue requirement set periodically by the regulator; as this changes, tariffs charged to generators can change. Also, the process and inputs used to allocate costs and create zonal TNUoS charges can change over time, which could create relative additional costs or savings to a particular generation project over its lifetime, and hence affect project cash flows.

US Physical transmission rights-based systems

Regions of the US outside of the ISOs provide an illustration of how physical transmission rights systems work. In these regions, there is generally a transmission provider which is part of the local franchise utility, but which is required to provide transmission services to all system users on an open access basis. This is done through acquisition of physical transmission rights, rather than the LMP systems discussed in the previous chapter.

Fig. 9.2 Sale of power to a distant user in a physical rights/contract path system.

In simple terms, physical transmission rights systems basically operate on an open access reservation system. To use the transmission grid to sell power to a customer, for example, the generator needs to reserve capacity on a specific path between its interconnection point and the point where the customer will receive the power. This is often referred to as a contract path.[5] To ensure that capacity will be available and that the customer can receive its power, the generator (or someone doing the scheduling on its behalf, in our example) will need to reserve "firm" transmission service along the path.[6]

Fig. 9.2 provides a highly stylized example of a sales transaction between a wind farm and a purchasing utility (purchasing under a PPA). The actual sale of the power could occur at multiple points along the contract path. In many cases, the renewable project company will sell its power at its point of interconnection into the grid (sometimes referred to as "at the busbar"), and the purchaser will be responsible for making the transmission arrangements to get the power to where it is needed. Another option is that the PPA may specify that the power will be delivered at the customer's location, in which case the project will be responsible for ensuring that transmission service is available and for paying the monthly cost. Finally, sometimes buyers and sellers will agree that the power may be delivered to an intermediate market hub – a location where power contracts are commonly traded and where hedges may be available. In this case, the renewable generator will be responsible only for arranging transmission from its location to the hub.

However, there are still risks to be considered to the project sponsor and investor. First, if the renewable project is not selling power at its location, it will typically require long-term firm transmission service to the delivery point. Under FERC rules, if insufficient transmission capacity on a contract path is available, then the transmission provider can charge for engineering upgrades necessary. This imposes additional capital costs and lead times on the project.

Second, the monthly transmission charges will generally be linked to the utility transmission provider's FERC cost-of-service-based tariff. Under cost of service regulation, the transmission provider is allowed to recover its operating and capital costs (including return on and of capital invested in providing service) as part of its regulated

[5] The contract path is a regulatory concept that has no basis in actual power flows. The power in the transmission network does not flow down a single path in an interconnected grid system.

[6] U.S. transmission providers also offer "non-firm" service but this is on an "as available" basis.

annual revenue requirement.[7] As this annual revenue requirement changes, the transmission tariff will change, which will change the transmission costs of the project company and hence affect project cash flows. While the project company may have little alternative but to pay the tariff rate once its PPA is set, it can be important to understand how these transmission tariffs may change over time.

Dynamic pricing systems

The LMP transmission pricing systems described in Chapter 8 provide the best example of transmission risk management for renewables generation. With certain simplifications, many of these same lessons can be applied to inter-TSO transactions in coupled zonal systems, such as those used in much of the European Union.

Under dynamic pricing systems, some components of transmission costs are not fixed, but depend on the usage of the transmission system and hence vary continuously.

Transmission risk in LMP systems

In an LMP system, energy prices vary by location on the grid, and reflect dynamic constraints on moving power across the grid (congestion) and losses. As the name "locational marginal pricing" suggests, the objective of LMP prices is to create local prices at each node (or zones for loads) that reflect the marginal impact of an additional unit of generation or demand at each location on the grid.

In this system, a generator receives its local nodal price for its output, and a load pays this price (or a zonal price derived from local nodal prices). When there is congestion on the system and losses, LMPs at the generation and load points will differ, creating a basis differential in prices. Unless hedged, one party or another (the generator or its customer) bears the basis risk, which changes frequently and can be highly volatile.

Fig. 9.3 shows the congestion price components over one day at two nodes: one with high wind availability, and another closer to load centers. As can be seen, the differences in congestion prices (which usually make up the largest part of the differences in LMPs between locations) vary substantially throughout the day, from approximately \$1.5/MWh around 9 a.m. to over \$7/MWh overnight.

Managing LMP transmission basis

If the PPA or price hedge is set based on the LMP at the generator location (busbar), the customer bears the transmission basis risk. As it must purchase power at its location to serve its load, this LMP will be different from the generator busbar LMP. If the PPA or price hedge is set at the customer's local price, then the generator bears the basis

[7] For a more detailed explanation of cost of service regulation and U.S. utility finance, see Morin, R. (2006). *New regulatory finance*. Public Utilities Reports.

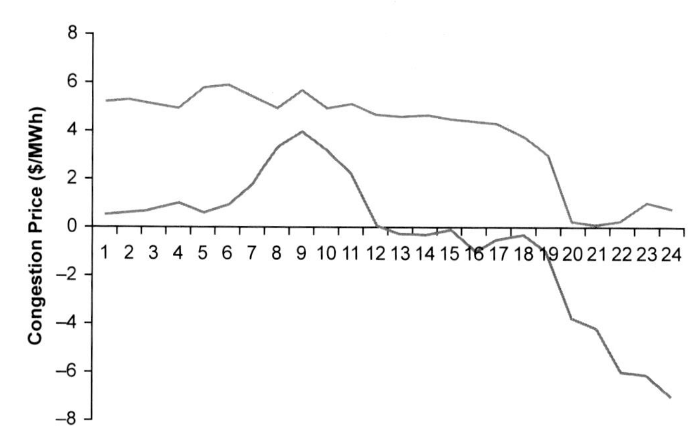

Fig. 9.3 Example of congestion price differences at two nodes across a day.

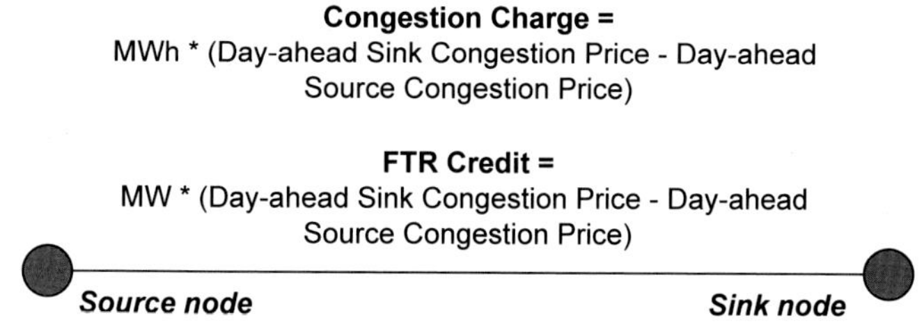

Fig. 9.4 Congestion charge and FTR credit.

risk. Finally, as discussed before, if the PPA or hedge is set at a market hub, then the generator and customer each bears the risk between its location and the hub.

Under LMP market designs, the primary means of hedging these locational basis risks are financial transmission rights (FTRs).[8] FTRs are financial instruments, which merely hedge congestion price differences between two locations. They do not affect or provide the holders the right to the physical flow of power.[9]

An FTR is defined by a source node and a sink node, which then determines the locational congestion price components as these vary by node. For example, an FTR might be defined for 100 MW of capacity for each hour in a year between Node 1 and Node 2. As shown in Fig. 9.4, the payoff to the FTR holder in each hour is the difference in congestion prices at the two nodes multiplied by the capacity of the FTR (100 MW). So if the congestion price component of the LMP was $3/MWh at Node 1 and $7/MWh at Node 2, the payoff of the FTR would be ($7/MWh − $3/MWh) × 100 MW or $400 in that hour. FTRs are longer-term hedges, effective for

[8] These are called Transmission Congestion Contracts (TCCs) in the NYISO system and Transmission Congestion Rights in ERCOT in Texas. The basic concepts are the same.

[9] For a full discussion of FTR issues see Rosellón, J., & Kristiansen, T. (Eds.). (2013). *Financial transmission rights: analysis, experiences and prospects*. Springer-Verlag.

months or years, but the concept is the same. The total payoff of the FTR is just the sum of the individual hourly payoffs.

Note that the FTR payoff or credit mirrors the congestion charge (cost of moving power due to congestion from one node to another), as shown in Fig. 9.4 as well. For a fixed quantity of power, an FTR defined against the relevant nodes thus provides a "clean" hedge against locational congestion price differences.

FTRs are sold by US independent system operators in centralized auctions. Generators, utilities, or traders can buy FTRs to hedge transmission risks or to speculate in relative congestion prices.[10] Over time, one might expect that the prices paid in the FTR auctions would tend to converge toward the payoffs from holding the FTRs.[11]

FTR payoff example

Fig. 9.5 provides another example of hedging basis risks using FTRs. Assume that the wind farm continues to generate 100 MW at its (source) node, and the customer buys it at a PPA price of $28/MWh at the source node. The customer then sells the power to the ISO at the generation node and receives the hourly price at that location, which is $30/MWh in this example.

The customer actually needs the power at a distant node, where, due to congestion and losses, the LMPs are higher ($38/MWh). Since it is buying power at one location

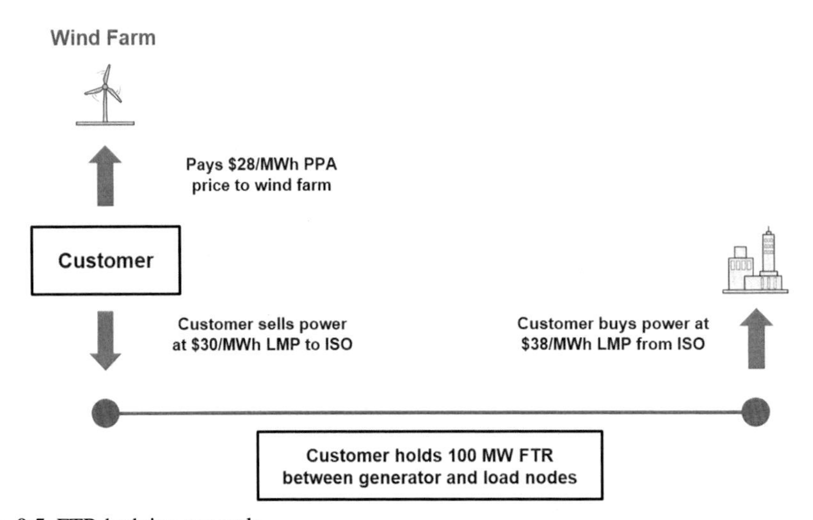

Fig. 9.5 FTR hedging example.

[10] Adamson, S. & Parker, G. (2013). Participation and efficiency in the New York financial transmission rights markets. In J. Rosellón & T. Kristiansen (Eds.), *Financial transmission rights: Analysis, experiences and prospects*. Springer-Verlag.

[11] Adamson, S., Noe, T., & Parker, G. (2010). Efficiency of financial transmission rights in centralized coordinated auctions. *Energy Economics, 32*, 771–778.

LMP Components

Gen node:	
Marginal energy cost	$35/MWh
Marginal Congestion Cost	-$6/MWh
Marginal Loss Component	$1/MWh
Total LMP	*$30/MWh*

Load node:	
Marginal energy cost	$35/MWh
Marginal Congestion Cost	$1/MWh
Marginal Loss Component	$2/MWh
Total LMP	*$38/MWh*

Paid to Wind Farm

PPA Price * Output

$28/MWh * 100 MWh = $2800

Paid to Customer by ISO

Gen LMP * Output

$30/MWh * 100 MW = $3000

Paid by Customer to ISO

Load LMP * Output

$38/MWh * 100 MW = $3800

FTR Payoff to Customer

FTR payoff = [$1/MWh - ($-6/MWh)]* 100 MW = $700

FTR cost = $5/MWh * 100 MWh = $500

Net FTR payoff = $200

Fig. 9.6 Example of FTR payoff.

and using it at another, the customer faces a substantial basis risk. To help hedge this risk, the customer also purchases a 100 MW FTR between the two nodes at a cost of $5/MWh.

Fig. 9.6 shows the total payoffs to each party under the PPA and the FTR. Of special note is the difference between the LMP revenues to be paid to the customer for the power ($3,000) at the gen node and the price the customer pays at the lode node ($3,800). This amount is not completely hedged by the FTR, since the FTR, by definition, is defined against the difference in the congestion components (which is $7/MWh) and not the LMPs (which also include the marginal loss component difference).

FTRs provide a means for hedging basis risks but these have limitations for renewable energy projects in particular:

- As shown in the above example, FTRs only hedge the congestion component in LMPs and not the marginal losses, so they are fundamentally an incomplete hedge.
- FTRs only hedge risks when output matches the FTR quantity. This was true in the example but is rarely the case in practice. Renewable projects have uncertain levels of output and rarely run 24 hours a day. It is therefore difficult to hedge their output with FTRs that are defined on a 24 hour basis.
- FTRs are difficult to price, which raises substantial transaction costs in the auctions.

Usually, renewable energy projects are built against long-term PPAs and hedges. But FTRs are generally not issued for such long terms by ISOs, so it is likely that a hedge might be replaced for the duration of the project, increasing long-term transmission price risks.

Transmission risks in zonal systems

Similar dynamic risks exist in zonal systems; only the prices vary by zone and not by node. So, for example, if a renewable generator in one zone sells to a customer in another zone, it faces the risk that zonal prices in the customer zone will be higher, creating a price risk. Dynamic zonal systems also usually have inter-zonal hedge contracts (similar in concept to the FTRs described above) to help manage these risks. The limitations on FTRs described above in managing these risks also generally apply to these inter-zonal transmission contracts as well.

Alternative off-take strategies and managing merchant risks

10

The most common traditional project finance structure for renewable energy in many countries has been based on long-term power purchase agreements (PPAs) with utilities. PPAs with utilities have been around for many years, and these PPAs can provide price stability with a creditworthy counterparty to support investment. However, the renewable energy markets have continued to evolve and, in many regions, the penetration of renewable generation has come close to or exceeded government mandates for renewable purchases, such as the renewable portfolio standards described in Chapter 2. This makes new long-term PPAs hard to come by. For this reason, and to meet new demands for renewable energy, project developers have turned to alternative off-take strategies to make their project bankable.

This chapter explores some of the more common alternative off-take strategies for renewable projects. It also discusses how remaining merchant risks (for power not hedged under a PPA or otherwise) are viewed in a project finance structure. Most of the discussion in this chapter will be focused on wind generation. However, the concepts are general and equally applicable to solar projects.

Corporate PPAs

Over the past few years, major companies in the United States and elsewhere have stepped up their direct purchases of renewable energy. These corporate purchases have helped to fill the void created by the limited demand for PPAs from traditional utilities. Under these structures, companies (often large multi-national corporations) have been increasingly procuring electricity directly from renewable energy sources through bilateral contracts to satisfy their own electricity consumption needs.

Companies' strategic shift to procuring renewable energy has generally been driven by two main considerations. First, it helps achieves their corporate sustainability goals, with corresponding impacts on employee morale and brand image among their partners and customers. Second, the cost of electricity from renewable energy sources has often dropped to the point that companies can benefit from the savings generated by procuring electricity directly from renewable resources instead of the traditional utilities.

The burgeoning power demand from data centers is making electricity costs an important component of the overall cost structure for technology companies. Large technology companies have responded in many cases by rapidly increasing their renewable energy purchases. In less than a decade, corporate PPAs for renewable energy have grown from zero to more than 13 GW in the United States alone. For example, Microsoft claims to be one of the largest players in this market, starting

Renewable Energy Finance. https://doi.org/10.1016/B978-0-12-816441-9.00010-X

from a 110 MW wind project in Texas in 2013 and to now possessing a portfolio of more than 1.2 GW (1200 MW) in six states and three continents.[1]

Controlling the cost of electricity has become such an important consideration that some technology companies are moving data centers closer to power generation sources. While the proliferation of corporate PPAs has provided a favorable avenue for renewable project developers to secure long-term off-takes for new projects, corporate PPAs are not without additional hassles or risks for developers.

First, many of the corporate purchasers are highly cost conscious, and are entering into PPAs for economic reasons. As a result, corporate buyers are often reluctant to lock in power prices for terms longer than 10–15 years. The corporations are often concerned that if they lock in prices over longer-terms, they may miss any potential cost savings generated by future technological innovations.

Second, corporations have a materially different credit profile than traditional utilities. Since the utilities are regulated and their obligations for power purchases are recovered through the regulatory mechanism, a regulated utility usually serves as a better counterparty from credit standpoint than a typical corporation with the same credit rating. To make matters worse, corporations are often averse to providing sizable corporate guarantees as a credit support to back their obligations under the PPAs. The amount of credit support provided by corporate purchasers may be as little as one to two years' worth of revenues under the PPA; this is not consistent with the actual risks that a renewable energy project is exposed to, as calculated using counterparty credit exposure tools.

Third, corporate PPAs often are structured for delivery (and settlement) at a hub location instead of the point of interconnection. This makes sense for the corporate buyer as their electricity demand is usually distant from the renewable energy project itself. In addition, corporate purchasers often have little or no expertise in structuring the transmission contracts to wheel the power from the project to their consumption point. The use of a hub location, rather than the project location, for energy purchases under a corporate PPA exposes the project to transmission risk.

Finally, over time, corporate purchasers of renewable energy have learned to offload many unwanted risks back to renewable energy companies. For example, under current settlement mechanics, a corporate PPA may more resemble a collar than a swap, which implies that a renewable energy project with a corporate PPA loses the benefit of the fixed price protection if the settlement prices at the reference hub fall below a floor price; however, the project gets to keep the upside if the settlement prices increase beyond a certain level. If the sponsor's objective is to get a fixed price contract through a corporate off-take, a collar mechanism significantly negates the price protection afforded by such PPAs. Fig. 10.1 illustrates the realized prices after giving effect to a forward/swap contract as compared to a collar.

Despite these shortcomings from the perspective of project developers, corporate PPAs have been a major driver for the growth of renewable energy projects in the United States in recent years.

[1] Microsoft Press Release dated October 16, 2018.

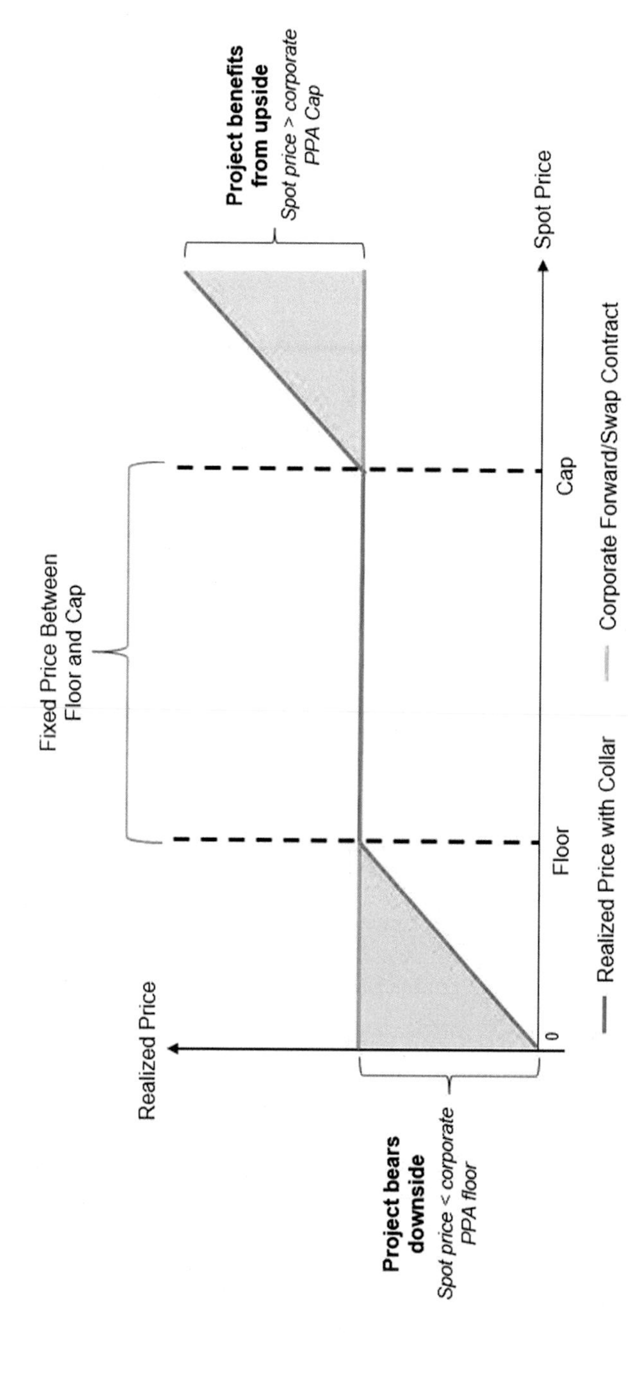

F g. 10.1 Comparison of realized price between a forward or swap contract and a collar structure.

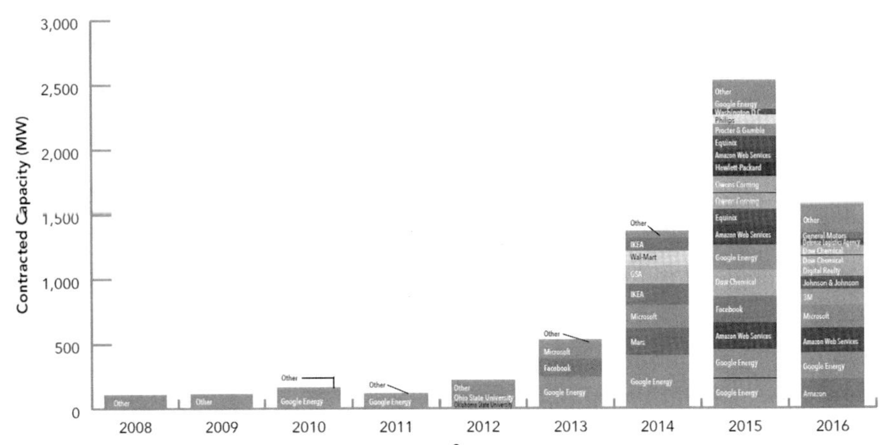

Fig. 10.2 Growth in corporate PPAs in the US.[2]

Fig. 10.2 shows the quantity and total capacity of PPAs signed by major corporate purchases in the United States from 2008 to 2016.

The European Union is also starting to see an increase in corporate PPA activity for renewable energy procurement. Developers are turning to corporate PPAs as government subsidies or contracts have become less available or generous in many EU countries.[3]

Commodity hedges

Structuring hedges for renewable projects is made difficult by the intermittent output of the projects themselves and the financial structure of project companies. Banks have been actively brokering and dealing various derivatives products for electricity (usually in the form of forward swaps) since the restructuring of the power sector began in the 1990s. But these typically have a flat profile, covering all hours with equal output (an "around the clock" contract) or covering a set of peak hours (again with equal output in each of these hours).

Dependent on the weather, a renewable project, such as a wind turbine or a PV facility, has a variable and unpredictable output profile. This makes the standard traded financial hedges in power markets difficult to use with renewable projects. There is always a mismatch between the fixed quantity in the hedge and the output of the plant, which introduces a volumetric risk.

[2] American Wind Energy Association Annual Report. (2016).

[3] Deign, J. (2018, September 4). *Europe on the cusp of a 'corporate PPA revolution'*. Retrieved from https://www.greentechmedia.com/articles/read/windeurope-europe-on-verge-of-corporate-ppa-revolution#gs.z87e3d.

Most commodity hedges are offered at market hubs, which allow the bank or other counterparty better opportunities to manage its risks. However, as with the corporate PPAs discussed previously, this introduces a basis risk between the project and the hub location, which in itself can be hard to manage. As discussed in Chapter 9, transmission basis can be managed through various, albeit imperfect, means.

Renewable energy projects are also susceptible to transmission curtailment.[4] Transmission curtailment arises when severe transmission grid congestion or other reliability issues force the system operator to reduce certain generation to maintain the reliability of the system. If not structured properly, a transmission curtailment may expose the renewable energy project to liability for liquidated damages resulting from the non-delivery of the power under the hedge.

Commodity hedges typically require substantial credit support or guarantees. From the perspective of a bank desk offering a fixed price hedge, for example, the project company counterparty could default and leave the bank with a large exposure. Non-recourse project companies, however, have very limited assets, and these are pledged to lenders and other parties. From a bank's perspective, writing a commodity hedge involves an implicit credit exposure to the counterparty. Traditional credit support mechanisms (such as guarantees) are often unavailable as sponsors typically have limited access to capital or balance sheets on their own. Lien-based structures for credit support may have their own issues – for example, a lien-based credit support necessitates forbearance and cure rights for the benefit of the tax equity investors, which may make structuring the financing difficult.

There is also the timing of commodity hedges. The sponsor needs the hedge to start when the project begins operating and producing power, but there is always some uncertainty around the in-service date for a new project. Under these circumstances, however, the bank or other hedge provider is exposed to construction risk with respect to the timing of the hedge, which affects its ability to offset or otherwise manage the risks.

Finally, as will be discussed in this chapter, there are often price risks later in a project's life – for example, at the end of an initial 10- or 15-year corporate PPA. These risks are especially difficult and costly for sponsors to hedge. Banks and other hedge providers face liquidity constraints in offering long-dated hedges and few want to offer hedges with such a long-term exposure.

Types of hedges

Despite these challenges, a reasonably vibrant market for hedges has evolved in which investment banks and commodities trading houses provide power price hedges for renewable energy projects. There are two common types of hedges offered in the market: physical hedges and financial swaps. These have different legal and operational

[4] Both corporate PPAs and utility PPAs are also susceptible to the transmission curtailment risks. However, the utilities are willing to bear a greater share of the risk due to their expertise in the transmission markets.

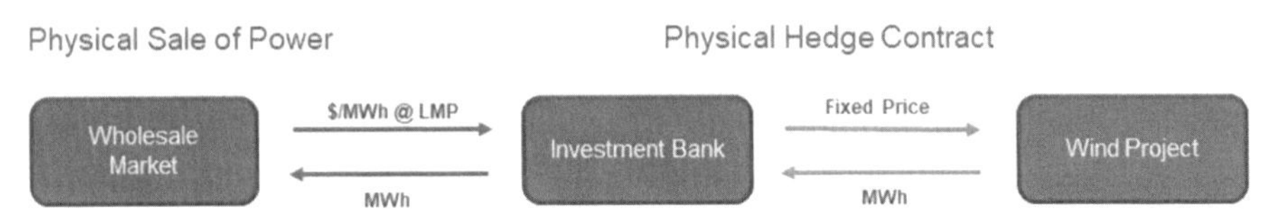

Fig. 10.3 Physical hedge contract.

structures, but both offer the project company a significant degree of price certainty that can help support a project's financing.

Under a physical hedge structure, as illustrated in Fig. 10.3, the renewable energy project company contracts to deliver the electricity produced at a specified location, and the bank or other hedge provider makes a periodic payment for the power delivered at a fixed price. The project company takes the basis risk between its location and the delivery point (often a hub point).

A financial hedge or swap operates around payments in the electricity spot market (for example, an ISO's electricity market). In this arrangement, as illustrated in Fig. 10.4, the project company sells all of its power into the spot energy market, and receives that price (e.g. an hourly price) from the ISO. Specifically, the project company sells power to the ISO at a defined location (for example, at a specific node in an LMP market), so power is priced at the relevant LMP. Separately, under the financial swap contract, the bank or hedge provider and the project company then make payments (a "fixed for float" swap) linked to the actual LMP at the project's location. When the LMP is above the fixed hedge price, for example, the project company makes a payment to the bank, calculated as the LMP minus the hedge price times the quantity of the hedge. When the LMP is below the hedge price, the bank pays the difference to the project company. In this way, the project company receives a final fixed price for its output.

Key elements of hedge contracts

A hedge contract is typically executed when a **Notice to Proceed (NTP)** is issued, concurrent with the financial close for tax equity and construction financing. The timing of the hedge contract execution is driven by the credit support requirements. During the construction phase, the financial institutions have a higher credit exposure to the counterparty, because, as discussed in earlier chapters, the risk of default is highest during the construction phase. Therefore, the financial institutions usually insist on a credit support in the form of a letter of credit (LC). Once the project reaches commercial operation, the financial institutions are willing to accept a first lien on virtually all

Fig. 10.4 Financial hedge contract.

of the assets of the project company in order to satisfy the credit support requirements. If the construction financing closes contemporaneous with the execution of the hedges, the project sponsor can source the LCs from the construction financing. On the contrary side, if the hedge is executed before the construction financing closes, the sponsor will need to provide LCs on its own.

Hedge contracts will generally have a delayed start, such that the hedge does not take effect until a few months after the project is operational. This ensures that the project does not incur any liquidated damages for the non-delivery if there are any construction delays.[5] However, as a result, some initial project electricity output may remain unhedged.

Hedge contracts (whether physical or financial) typically settle with reference prices at a liquid trading hub. Furthermore, the hedges are usually sculpted to the shape of the electricity generation profile using the on- and off-peak trading blocks available in the market. Using a liquid hub price may allow the hedge provider to offer a hedge at a better price, but will typically leave residual basis risk (from the project location to the hub location). These risks must be modeled and managed by the project sponsor and lenders.

As a renewable project will not produce a fixed amount of energy in a given time period, the hedge must be sized against the expected variability in project output. A common structure sizes the hedge to the 1-year p-99 production level of the project to ensure there is slack to account for generation variability. This is important because the financial institutions want to ensure that the project can reliably deliver the contracted electricity production. Otherwise, the project company may incur liquidated damages due to a failure to generate the contracted generation volume.

Depending on the project specific characteristics, the 1-year p-99 production often ranges from 75% to 85% of the expected generation for a wind project and 80%–90% of the expected generation of a solar project. Consequently, the power hedge will likely expose the project company to merchant price exposure (that is, to variation in power prices in the market) for the remaining 10%–25% of the unhedged electricity generation.

Tracking accounts in hedge contracts

Even with a power price hedge, the project may still be exposed to residual risks due to the intermittent nature of the renewable energy generation and the structural complexities outlined previously. For example, in any given time interval (e.g. minute, hour, day, etc.), the actual generation from a wind farm will likely vary depending on the actual wind speed and project specific attributes. Therefore, there will be instances when the actual generation output could be lower than the hedged output and vice-versa. Also, other events such as scheduled maintenance, forced outages or

[5] Liquidated damages arise if t a project fails to deliver electricity output or delivers less than the contracted output. In such situation, the hedge counterparty may require a payment in the amount of the payment the hedge counterparty incurs in order to cover the shortfall. Depending on the market prices, the liquidated damages can be significant.

transmission curtailments may affect output. Therefore, renewable energy projects will almost invariably have periods in which the project output is below the amount of generation contracted under the commodity hedge. This will expose the project to financial penalties in the form of liquidated damages resulting from the under-deliveries.

It is not in the interest of the project sponsor or the hedge provider (such as a bank) to set up an otherwise well-operating renewable energy project for an event of default due to the shortfall in wind generation or other physical realities of renewable energy projects. Therefore, as a part of the commodity hedge, banks and other hedge providers often incorporate a financing mechanism called a tracking account to help alleviate any temporary shortfalls in generation for the renewable energy project.

A tracking account is essentially an add-on working capital facility as a part of the commodity hedge contract, which can be drawn to cure any liquidated damages caused by temporary generation shortfalls. The amount drawn under the tracking account equals the amount of generation shortfall in a given time interval multiplied by the reference price (at which the commodities hedge settles). Any amount drawn under the tracking account increases the borrowing under the tracking account. The borrowings drawn under the tracking account are repaid in subsequent time intervals by the excess revenues generated by the project. These excess revenues are calculated as the positive difference between the actual generation from the project and the hedge quantity multiplied by the actual price the project receives at its location.

The tracking account mechanism is premised upon the simple logic that while the production from a renewable energy facility may vary widely even over a short period (e.g. 1 hour), over a longer period of time (e.g. a year) the generation amount is more predictable and stable. In other words, over a longer period, the variations in generation output from a renewable energy resource will tend to return to expected values. Consequently, so long as a fraction of the generation output is kept unhedged, the borrowings under the tracking account should eventually reduce to zero. As explained before, commodity hedge providers typically size the hedges at 1-year p-99 generation output, which usually ranges between 75% and 90% of the generation output depending on the project specific parameters and the nature of the renewable energy resource, Therefore, given the amount of slack (10–25% of the expected generation) over the term of the hedge, the borrowings under the tracking account should tend to zero.

Borrowings under the tracking account accrue interest at a pre-agreed rate similar to a working capital facility. Tracking account borrowings are available only through the term of the hedge contract, and any outstanding borrowing amount under the tracking account at the end of the hedge term is often due and payable at the expiration of the hedge. However, commodities hedge providers routinely allow the renewable energy project sponsors to repay any outstanding tracking account balance in equal monthly/ quarterly installments over a 2- to 3-year period upon the expiration of the hedge term.

The simple tracking account mechanism has become an integral part of commodity hedges for renewable energy projects and has helped to increase the ability to finance these projects. In recent years, this mechanism has been successfully expanded to cover transmission basis differential risk as well. Incorporating transmission basis differential risks into a tracking account requires a significant increase in the amount of the working capital facility. For example, a $5 million tracking account facility might be sufficient for a typical 200 MW wind project to cover temporary generation shortfalls alone, but it may need to be increased to $20 million if the tracking account is also used to cover transmission basis differential risks as well. The tracking amount quantity is a function of the duration of price deviations from initial expectations, and these can be persistent for transmission basis, as new transmission upgrades (which tend to lower transmission basis differentials), for example, are often slow to materialize.

To illustrate the functioning of the Tracking Account, consider an example of a small wind project with a hedge contract for 100% of the capacity of the project at 200 MWh per month at a price of $100/MWh. Assume that the Tracking Account does not have any cap and accrues interest at the rate of 5% per annum. For the sake of simplicity, also assume that the power prices are normally distributed and the prices for the hub at which the hedge contract settles have a mean of $100/MWh and standard deviation of $10/MWh. Similarly, the node prices have a mean of $99/MWh and a standard deviation of $10/MWh. The hub and node prices are assumed to be correlated with a correlation coefficient of 80%. Finally, assume that the output of the wind farm also follows normal distribution with a mean generation of 100 MWh and standard deviation of 5 MWh.

The table in Fig. 10.5 provides monthly calculations for this Tracking Account. Columns 1 to 3 provide a randomized sequence of wind generation, hub prices, and node prices respectively using the statistical parameters stated above.

For the first month, the Tracking Account balance starts with zero. The project output is 208.28 MWh. The hub price and the node price are $105.19/MWh and $91.22/MWh respectively. Accordingly, the Realized Revenue for the first month is:

$$\begin{aligned}\text{Realized Revenue} &= \text{Project Output} \times \text{Node Price}\\ &= 208.28\ \text{MWh} \times \$91.22/\text{MWh} = \$19000.17\end{aligned}$$

Fixed Obligation refers to the fixed payment due to the project from the hedge counterparty for a given month.

$$\begin{aligned}\text{Fixed Obligation} &= \text{Hedged Contract Quantity} \times \text{Fixed Price}\\ &= 200\ \text{MWh} \times \$100/\text{MWh} = \$20,000\end{aligned}$$

The Floating Obligation that the project owes to the hedge counterparty for the first month is calculated as follows.

$$\begin{aligned}\text{Floating Obligation} &= \text{Hedge Contract Quantity} \times \text{Hub Price}\\ &= 200\ \text{MWh} \times \$105.19/\text{MWh} = \$21307.89\end{aligned}$$

With a typical financial hedge without a Tracking Account, the project would owe the positive difference between the Floating Obligation and Fixed Obligation to the counterparty. For the first month, this would be $1,307.89. With the Realized Revenue

Month	Output (MWh)	Hub Price ($/MWh)	Node Price ($/MWh)	Realized Revenues	Fixed Obligation	Floating Obligation	Beg Tr Acct Bal	Addition	Subtraction	Interest	End Tr Acct Bal	Revenues after TA Adjustment	Revenues w/o Tr Acct
											$0.00		
1	208.28	105.19	91.22	$19,000.17	$20,000.00	$21,037.89	$0.00	$2,037.72	$0.00	$0.00	$2,037.72	$22,037.72	$17,962.28
2	201.07	112.32	112.40	$22,600.51	$20,000.00	$22,463.90	$2,037.72	$0.00	$136.61	$8.49	$1,909.60	$19,871.88	$20,136.61
3	196.82	84.89	91.47	$18,002.15	$20,000.00	$16,978.73	$1,909.60	$0.00	$1,023.41	$7.96	$894.14	$18,984.54	$21,023.41
4	192.74	91.59	85.22	$16,424.55	$20,000.00	$18,318.16	$894.14	$1,893.62	$0.00	$3.73	$2,791.48	$21,897.34	$18,106.38
5	202.96	99.47	96.11	$19,506.74	$20,000.00	$19,893.55	$2,791.48	$386.81	$0.00	$11.63	$3,189.93	$20,398.44	$19,613.19
6	194.94	87.78	90.24	$17,591.75	$20,000.00	$17,555.15	$3,189.93	$0.00	$36.61	$13.29	$3,166.61	$19,976.69	$20,036.61
7	203.54	112.78	106.37	$21,650.01	$20,000.00	$22,555.51	$3,166.61	$905.50	$0.00	$13.19	$4,085.30	$20,918.69	$19,094.50
8	191.39	71.12	86.33	$16,522.78	$20,000.00	$14,223.40	$4,085.30	$0.00	$2,299.37	$17.02	$1,802.95	$17,717.65	$22,299.37
9	197.07	96.32	102.34	$20,168.35	$20,000.00	$19,263.13	$1,802.95	$0.00	$905.22	$7.51	$905.24	$19,102.29	$20,905.22
10	205.82	112.02	108.06	$22,240.90	$20,000.00	$22,403.38	$905.24	$162.48	$0.00	$3.77	$1,071.49	$20,166.25	$19,837.52
11	200.35	102.57	90.96	$18,223.78	$20,000.00	$20,513.21	$1,071.49	$2,289.43	$0.00	$4.46	$3,365.39	$22,293.90	$17,710.57
12	197.42	105.74	109.14	$21,547.11	$20,000.00	$21,148.90	$3,365.39	$0.00	$398.21	$14.02	$2,981.20	$19,615.81	$20,398.21
13	197.50	109.99	107.39	$21,209.61	$20,000.00	$21,997.23	$2,981.20	$787.61	$0.00	$12.42	$3,781.24	$20,800.04	$19,212.39
14	204.92	85.03	86.09	$17,642.01	$20,000.00	$17,006.63	$3,781.24	$0.00	$635.38	$15.76	$3,161.61	$19,380.38	$20,635.38
15	208.75	122.00	115.64	$24,140.32	$20,000.00	$24,400.77	$3,161.61	$260.45	$0.00	$13.17	$3,435.24	$20,273.62	$19,739.55
16	192.99	92.62	96.03	$18,532.00	$20,000.00	$18,524.60	$3,435.24	$0.00	$7.40	$14.31	$3,442.15	$20,006.92	$20,007.40
17	191.20	83.80	90.48	$17,299.51	$20,000.00	$16,760.72	$3,442.15	$0.00	$538.79	$14.34	$2,917.71	$19,475.56	$20,538.79
18	199.83	79.53	84.20	$16,825.76	$20,000.00	$15,906.05	$2,917.71	$0.00	$919.71	$12.16	$2,010.16	$19,092.45	$20,919.71
19	195.15	90.01	89.12	$17,391.46	$20,000.00	$18,002.59	$2,010.16	$611.13	$0.00	$8.38	$2,629.66	$20,619.50	$19,388.87
20	201.75	99.21	99.85	$20,144.98	$20,000.00	$19,842.78	$2,629.66	$0.00	$302.21	$10.96	$2,338.41	$19,708.75	$20,302.21
21	198.78	95.89	85.16	$16,927.82	$20,000.00	$19,178.34	$2,338.41	$2,250.52	$0.00	$9.74	$4,598.67	$22,260.26	$17,749.48
22	203.48	108.04	107.46	$21,865.28	$20,000.00	$21,608.69	$4,598.67	$0.00	$256.59	$19.16	$4,361.24	$19,762.57	$20,256.59
23	197.64	87.16	89.11	$17,611.58	$20,000.00	$17,432.17	$4,361.24	$0.00	$179.41	$18.17	$4,200.00	$19,838.76	$20,179.41
24	206.32	118.12	109.77	$22,648.46	$20,000.00	$23,623.60	$4,200.00	$975.14	$0.00	$17.50	$5,192.63	$20,992.64	$19,024.86
Standard Deviation				**$2,318.58**								**$1,122.39**	**$1,124.92**

Fig. 10.5 Example of a tracking account (simulated prices and quantities).

of $19,000.17, the project would then net $17,962.28 for the first month – much lower than the $20,000 that the project would have earned with a hedge had there been no basis differential. However, with a Tracking Account, the project would be able to "borrow" the positive difference between the Floating Obligation and the Realized Revenue – most of which is driven by the close to $14/MWh basis differential. Consequently, for the first month, the project can draw $2,037.72, thereby yielding the total revenue of $22,037.72.

In the second month, the project performs better than the mean value (201.07 MWh vs. 200 MWh). Moreover, the node price is also at a premium relative to the hub price ($112.40/MWh vs. $112.32/MWh). Consequently, the project generated excess revenues and can repay the Tracking Account balance in the amount of $136.61. Please note the Tracking Account balance accrues an interest of $8.49 calculated at 5% rate on the balance of $2,037.72 as of the end of the first month.

The rest of the calculations in Fig. 10.5 follow the same pattern. A few observations can be made about the usefulness of the Tracking Account based on the standard deviations calculated at the bottom of the table. In the absence of the hedge, the project is subject to a greater uncertainty in revenues owing to the variations of production output and market prices. The financial hedge can materially help reduce the volatility of the revenues and the Tracking Account can help smooth out the variability in the cash flows further.

Fig. 10.5 is meant to illustrate the Tracking Account concept in a simplistic manner using normally distributed price and production output. However, the framework can be expanded in a Monte Carlo simulation to estimate the various statistical parameters for the Tracking Account using complex stochastic processes for the input variables.[6]

The proxy revenue swap

A proxy revenue swap (PRS) is effectively a weather derivative product, generally offered by insurance companies interested in underwriting weather risks. Since these risks (e.g. wind speeds) are generally uncorrelated with other risks in their portfolio, this product can offer attractive diversification for an insurer. Since its introduction in 2016, proxy revenue swap products have gained popularity among project sponsors, and a variation of the product has been used by corporate off-takers to offload some of the weather risk in their portfolio of PPAs.[7]

[6] Akin Gump has made a Tracking Account calculator available, which illustrates the one-time swap settlement for a commodities hedge employing Tracking Account. (https://solutions.akingump.com/swapsettlement/).

[7] Microsoft announced on October 2016 that it signed three separate Volume Firming Agreements (VFA) with Allianz, in conjunction with their partners at Nephila, covering three wind projects in the U.S. in Texas, Illinois, and Kansas, totaling almost 500 MW

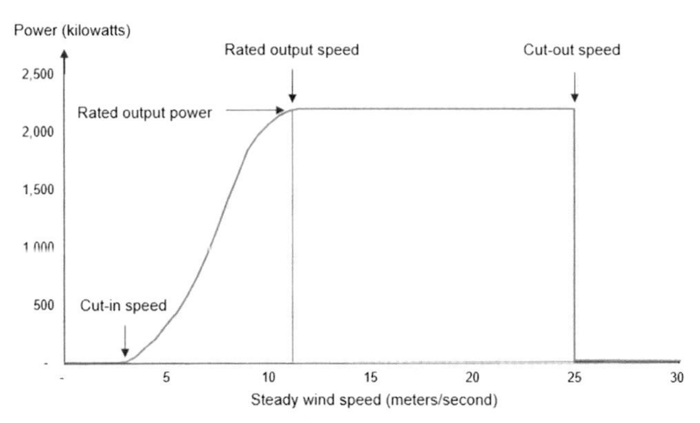

Fig. 10.6 The S-curve for a wind turbine.

Structure of a proxy revenue swap

Fig. 10.6 shows a typical S-curve for a wind turbine with a rated output of 3 MW, which may have a "cut-in" speed of 4 m/s and "cut-out speed" of 15 m/s. The cut-in speed is the threshold wind speed below which a turbine may remain still and not generate any electricity. The cut-out speed is the wind speed at which the turbine may cease to operate to prevent damage due to excessive stress caused by high wind speed.

Fig. 10.7 illustrates the probabilistic version of the wind speed for a typical wind farm. In this figure, one may observe that the mean speed for the wind site under study is 7.25 m/s.

Using the data in these two figures together, one can say that an insurance company may be willing to provide a proxy generation swap with an average speed of 7.25 m/s, which equates to an output of 1.451 MW per turbine or 12,710.76 MWh per turbine. If the market power price for a 10-year swap is $25/MWh,[8] the wind site under study should fetch $317,769 per turbine in annual proxy revenues for the term of 10 years. Consequently, a 300 MW wind farm with 100 turbines with a rating of 3 MW per turbine would fetch an annualized revenue of $31.78 million for the wind farm.

Insurance companies usually charge an up-front and periodic premium for entering into the PRS. The premium payment can be settled against the proxy revenues due from the insurer.

[8] Please note that, all else being equal, the fixed price available under the PRS would always be lower than that available in the market for a commodities hedge because the contracted quantities under a commodities hedge are fixed at the outset. Therefore, if the project does not deliver the contracted quantities, there are liquidated damages. The Tracking Account mechanism can relieve the stress caused by the potential underperformance. Nevertheless, PRS works differently and assumes the wind generation risks. As a result, the insurance company providing a PRS is assuming the covariance risk that the wind generation from the wind project may be lower when the market prices are higher. Consequently, the fixed price under the PRS has to be lower than that under the commodities hedge, all else being equal, for this incremental risk.

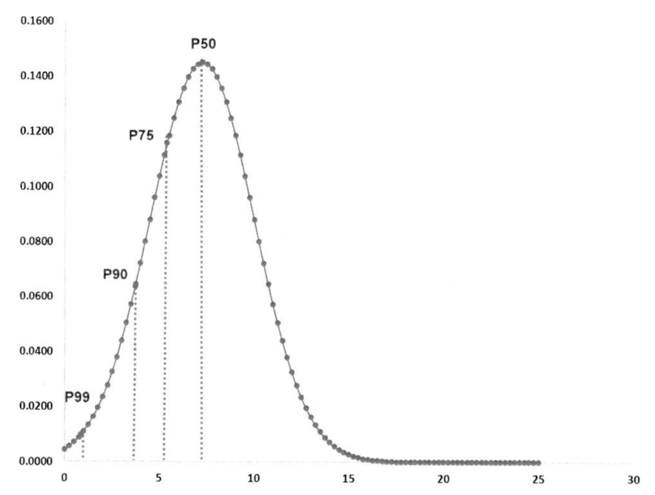

Fig. 10.7 Probability distribution for wind speeds.

Ultimately, project sponsors are interested in hedging revenues to be received by the project, and not volumes. Of course, if there is a fixed price hedge (a commodity hedge or corporate PPA, for example) for the output, a volumetric hedge translates directly into a revenue hedge. In the full proxy revenue swap structure, once the wind generation profile is generated for the site, the insurance company can use forward prices from the power market to construct the annualized "proxy revenue" that a wind farm may generate over the term of the hedge. Since the power markets have any reasonable liquidity only for hub-settled products, the proxy revenues are usually derived using hub-settled power prices. The following diagram illustrates the structure of a PRS for a wind project.

As shown in Fig. 10.8, the wind project company sells power in the market into the local market and receives the local LMP for its output. The proxy revenue (calculated using the manufacturer specified S-curve, realized wind speeds, and the realized power

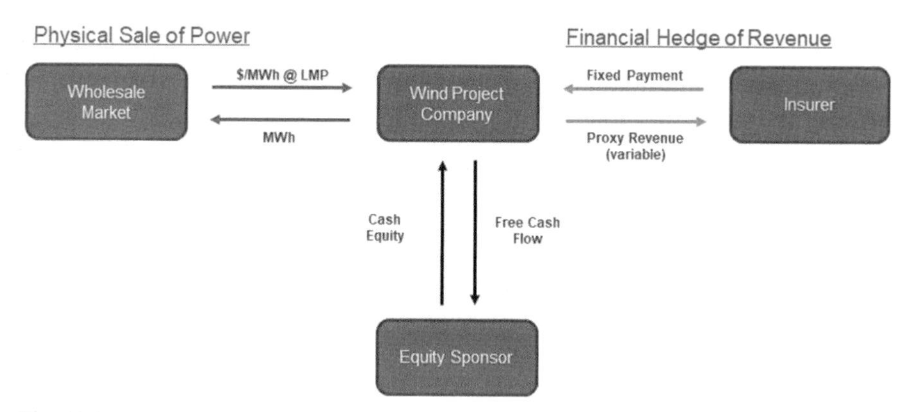

Fig. 10.8 Structure of a proxy revenue swap

prices at the specified reference location over the time interval) is paid to the insurer periodically. In return, the project company receives a fixed payment from the insurer as defined in the proxy revenue swap contract. The proxy revenue swap therefore acts as an (approximate) contract for differences on project revenues, which then stabilizes cash flows to the project sponsor, lenders, etc.

A proxy revenue swap will generally not hedge the revenues for the project perfectly. For example, the actual project S-curve in real time will seldom match the manufacturer's specified S-curve exactly. Also, the project may be unavailable due to scheduled or forced outage or transmission curtailment. Finally, proxy revenues are usually calculated using market hub reference prices at a location specified in the contract, while the project receives revenues based on the power prices at the project's point of interconnection. This exposes the project to transmission basis differential risk. Despite these residual risks, proxy revenue swaps have gained popularity among renewable energy sponsors, as they hedge both power price and generation volume risks for wind projects. This greatly improves the bankability of projects without utility-style PPAs and potentially increases the debt financing a sponsor can raise for the project.

The proxy generation swap and put structures

Given the popularity and limitations of the proxy revenue swap, insurers unsurprisingly offer some alternative structures to their customers. Under the proxy generation swap, as shown in Fig. 10.9, the proxy revenues and payments are calculated using the LMPs at the project location (rather than the hub price as used in the standard proxy revenue swap illustrated in Fig. 10.8). Thus, the proxy generation swap uses the same index price for both legs of the swap and effectively fixes the wind generation from the project.

A proxy revenue put essentially guarantees a floor on the revenues generated by the project. Therefore, if output market prices are higher than the floor price, the wind

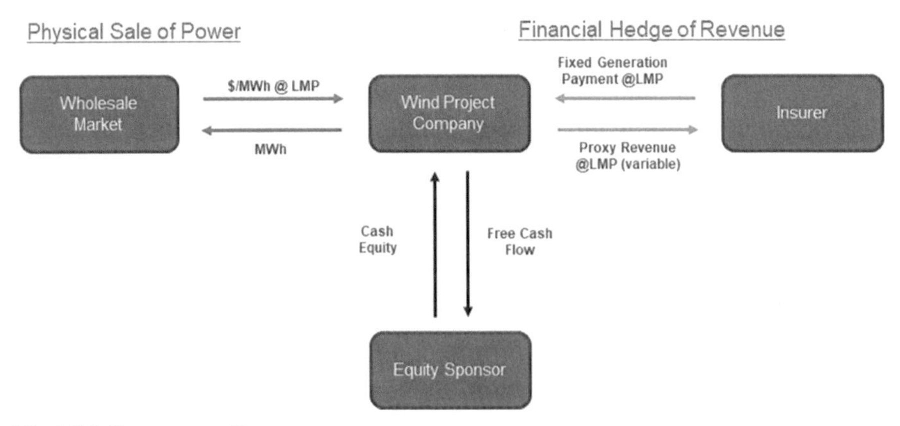

Fig. 10.9 Proxy generation swap.

project company benefits from the upside afforded by the higher power prices. On the contrary, if the power prices fall below the floor price, the project company is protected from the downside. Of course, because the insurer is now taking on a very different risk, the pricing of the up-front premium payment made to the insurer – which is required in all of these proxy revenue structures - will be quite different.

While most of the proxy revenue swaps discussed here were initially designed for wind projects, the insurance companies have been aggressively marketing the same products for solar projects as well.

Comparing project off-take options

The different off-take structures outlined in this chapter have varying characteristics and residual risk profiles for project sponsor companies. Table 10.1 highlights some of the key characteristics of the various off-take structures and compares them to the "traditional" utility PPA structure in which the power is sold at the project busbar to the off-taker, who then takes the price risks. It should be noted that proxy generation swap could be combined with corporate PPAs and/or commodity hedges because the product addresses the wind generation risks, whereas the other two products primarily address the price risk.

This summary provides only a high-level view of some of the many options and variations that exist.

Quantifying and managing merchant price risk exposures

As discussed earlier, renewable energy projects with project finance structures have limited ability to handle large exposure to variations in power prices. These "merchant" price risks directly affect cash flow and subsequently the ability of projects to service debt and meet other obligations. For this reason, projects with large merchant risk exposures are typically financed on balance sheets, relying on standard corporate finance structures. In the European Union, for example, this is relatively common and many new projects are built on balance sheets by large European generation groups. However, even for hedged projects (e.g. those with PPAs or the alternative off-take structures discussed in this chapter) there is often a level of residual merchant price risk. For a project with a utility PPA, for example, the PPA might not cover all of the capacity of the project, leaving some generation (in some periods, for example) exposed to price risks. With the alternative off-take structures, significant residual price risks are very common.

Many corporate PPAs and utility PPAs do not cover the useful life of the renewable energy projects. As a result, the projects are exposed to the merchant price risks for the

Table 10.1 Comparison of various off-take structures.[9]

	Utility PPA	Commodities hedge	Corporate PPA	Proxy revenue swap
Initial transactions	1980s	c.2007–08	c.2010	2016
Counterparty	Utility	Commodities merchant	Corporation	Weather risk investor
Physical vs. financial	Physical	Physical (ERCOT)/ financial	Financial	Financial
Unit price	Fixed	Fixed	Fixed	Effectively varies by volume
Settlement point	Busbar	Trading hub	Busbar or trading hub	Trading hub
Price basis risk	No	Yes	Depends on settlement point	Yes
Market sales	No	Yes	Yes	Yes
Revenue based on wind production?	Yes	Yes	Yes	No
Revenue based on availability?	Yes	Yes	Yes	Yes
Revenue based on efficiency?	Yes	Yes	Yes	Yes
Fixed volumes	None	Hourly (p-99)	None	None
Minimum delivery obligations	Annual minimum production	None	Annual minimum availability	None

uncontracted useful life - once the PPA ends, but while the project is still able to generate electricity.

For a commodity power price hedge, for example, the hedge is typically sized to the estimated 1-year p-99 level of generation. Since generation is expected to be higher, the project is exposed to the merchant price risk for the generation quantity between the actual generation output and the hedged quantity.

[9] Norton Rose Marketing Presentation, Robert Eberhardt.

One subset of merchant risks is transmission basis risk, as discussed in Chapter 9. To the extent that the off-take contracts settle for a price at a location other than the point of interconnection, the project will be exposed to the transmission basis differential. This is a function of the market prices in the two locations (hub vs. project).

Uses of market price forecasts and analysis

For projects that have material merchant price exposures, developing forecasts of future prices is necessary to predict future project revenues. The forecast and analytical process also can provide insight into the drivers of future market prices and the long-term risks facing the project sponsor:

- Merchant power price forecasts ensure that price risks are quantified at the incipient stages of due diligence for projects under development, helping to ensure unattractive projects are filtered out early in the evaluation process
- The merchant price forecasts can be useful in negotiating commodities hedges since the analytic process provides a ground-up fundamentals-based view on future market prices
- The transmission and basis risks can be quantified, and the projections can be adjusted as the project moves through operation phases, transmission upgrades come online, and other changes are made to the transmission system.

Approaches to developing merchant price forecasts

Forecasting future power prices involves considerable uncertainty. As described in Chapter 8, power markets are inherently complex, and market prices are affected by generation costs, transmission system conditions, demand variation, and other factors. In this section we provide a very brief overview of sources of future market prices used in project finance analysis.

Forward market prices

Wholesale power market participants trade electricity under forward contracts, such as swaps. For electricity, unlike some commodities, much of this trading is done on an over-the-counter (OTC) basis, so there is less visibility in market prices than there is in more liquid futures markets (for oil and natural gas, for example). Still, there are price indices available for some periods and some market hub locations from trading platforms such as ICE and through price reporting sources. These can provide some indications of where the forwards markets for future power delivery are trading, but they do have substantial limitations. Volumes are often relatively small and liquidity typically extends for only a few years out. Still, despite these limitations, it is common to use forward market hub prices for power as one indicator of where spot prices may go when these prices are available.

Commercial price forecasting services

It is also possible to purchase forecasts of future power market prices from various suppliers, and these are often used in project financing to assess future merchant revenues. One disadvantage of these data sources is that the inputs driving the price forecasts and the methodologies are often opaque. It is therefore difficult to converge on a single data source and price forecast when various financing parties are involved. Furthermore, since the data sources are widely available for everyone, there is no way to develop a customized forecast that is proprietary to a single party.

Fundamentals-based models

Another way to develop power market forecasts is using fundamentals-based software models, which are also widely used in transmission congestion analysis, generation, and transmission asset valuation, etc. Commercially available models such as Aurora XMP, ProMod, GE MAPS, and others simulate the operation of the spot market, usually on an hourly basis. Some project finance participants maintain their own fundamentals-based forecasting capability, but most develop these forecasts externally with specialized consulting firms.

Since these models simulate the operation of complex spot markets, the data input requirements are high. A few notable examples include:

- Hourly and seasonal electricity demand shapes and future growth and changes
- Fuel price assumptions (especially for natural gas)
- Data on existing generating plants and projected new additions
- A representation of the transmission system, its constraints, and expected transmission upgrades
- Resource profiles for renewable energy sources such as wind and solar

Once the inputs are agreed upon for a particular scenario, these programs can provide hourly forecasts of prices, transmission congestion, basis differentials, etc. Some models are used to predict nodal LMP prices, while others operate on a zonal basis.

The models can provide a wealth of raw output data at the market and individual generator level. From these outputs, the project finance analyst can project and simulate the needed revenue and other forecasts to compensate for the inherent uncertainties over market drivers.

The data input and cost requirements for detailed fundamentals-based approaches are high, but an advantage of this approach is that it allows greater understanding of the drivers of future prices and project revenues. Therefore, the modeling process helps not only to forecast prices, but also to understand future market risks and develop power price risk management strategies.

Risk management strategies for merchant power price risks

Many residual merchant power risks cannot be practically hedged at the initial stages of a renewable energy project, especially longer-dated risks (hedging generation well out into the future). In this case, project sponsors need to design a prudent strategy for hedging these risks over the life of a renewable energy project. The remainder of this section provides a brief summary of a few approaches that may be used.

Stack-and-roll hedging

A common hedging problem in power markets is that there is little to no liquidity for long tenor hedges. In this case, one strategy is to roll the hedges, replacing hedge contracts as they roll off with new hedges. While this helps provide revenue stability over each hedge contract term, it does not, of course, provide long-term price stability, as the pricing may change over the term of each hedging contract.

Hedging in correlated markets

As discussed previously, longer-term hedges for power are constrained by liquidity considerations, but the market for long-term natural gas hedges is relatively deeper. If power market prices are expected to remain correlated with natural gas prices, it may be possible to hedge long-term price risks with natural gas futures or other derivatives. For example, in a market where natural gas-fired generation was expected to be marginal for the long-term, then an appropriately sized gas hedge (e.g. gas futures, swaps, or collars) can mitigate power price risk, even for a project (such as a wind farm) that uses no natural gas. Needless to say, using gas futures or forwards will generally provide quite a "dirty" hedge as power and gas prices can never be expected to track completely.

Project development and valuation

11

As previously detailed, large-scale renewable energy projects are complex to develop, build, and operate, and each phase of the process comes with its own operational and financial risks for investors. In this chapter, we explore the life cycle of a renewable energy project and its changing risk profiles, and discuss basic valuation techniques for renewable energy companies and individual projects in both the operations and development phases.

The renewable energy project lifecycle

Fig. 11.1 below illustrates a typical lifecycle for a renewable energy project, which is divided into three basic phases: development, construction, and operations. At the end of the operating life of a project, there may also be a requirement to dismantle the project, but these costs for solar and wind projects are relatively small.

In the development phase, the project developer or sponsor seeks to secure all of the elements necessary for a successful project. This, for example, will require securing a site (typically under an option for a lease, as discussed below), applying for and receiving environmental and other necessary permits, interconnection to the grid, and securing offtake arrangements, such as a PPA, for the output of the project. Preliminary engineering and project design work will also occur in this phase, as will efforts to secure financing, usually toward the end of the development phase. Development of a large-scale project can take years, especially if siting, permitting, and interconnection issues are significant.

At the end of the development process, if all the steps are completed successfully and financing can be secured, a Notice to Proceed (NTP) may be issued by the project developer or sponsor. The NTP is a formal notification from the project sponsor to the EPC contractor stating the date that the EPC contractor can begin work, subject to the conditions of the contract. Typically, the NTP coincides with the financial close for a project.

Final equipment procurement (e.g. delivery of the wind turbines or solar panels), site preparation, construction and equipment installation, and connection of the turbines or solar modules to the grid all occur during the construction phase. Compared to large thermal power plants, the construction phase for utility-scale wind and solar projects is often relatively short, ranging from 9 to 18 months.

The end of the construction phase is marked by the Commercial Operation Date (COD). The COD is the date on which an independent engineer certifies that the renewable facility has been built to the specifications of the EPC contract and completed all required performance tests. Sometimes, the transmission operator and

Renewable Energy Finance. https://doi.org/10.1016/B978-0-12-816441-9.00011-9

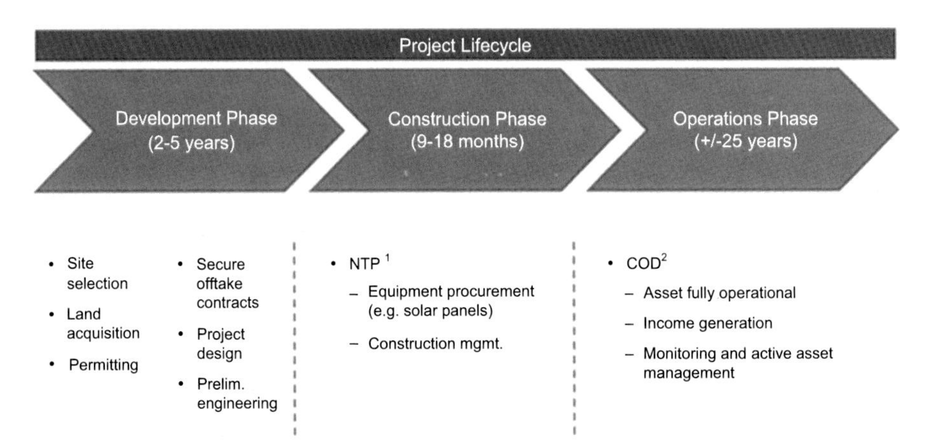

Fig. 11.1 Lifecycle for a renewable energy project.[1]

PPA offtaker may have their own testing and/or certification requirements as per the interconnection agreement and PPA, respectively. After the COD date, the project is typically deemed ready for commercial operation.

The COD triggers the start of the operations phase of the project, which can last for 25 years or more. Often, a project company will engage an O&M contractor to run the project and ensure that performance requirements are met. As with other critical project vendors, lenders and other investors are sensitive to their exposure to the experience and outcomes of the O&M contractor. Project finance documentation will often require investor approval before any new O&M contractor can be appointed by the project company or sponsor.

The project development phase

The development phase of a renewable energy project typically begins with analyzing the renewable resource and establishing control over the land necessary to site the project.

For a wind project, this involves erecting anemometers at the potential site to measure wind speeds at various heights. The data is then compiled and correlated with wind speed data from nearby weather stations that are maintained by government or quasi-government agencies, which usually have a longer data series. The correlation parameters are then used to estimate the wind resource through the useful life of the project using extrapolation techniques. The correlate-and-extrapolate methodology is subject to biases and errors. Therefore, the longer the wind measurement period, the better the quality of the estimates. The anemometers are set up at various heights to assess energy generation at various hub heights, as, generally speaking, higher hub height means a stronger wind resource. However, the question of whether the incremental wind generation justifies the incremental cost of a higher tower and additional

[1] 2016 AWEA Annual Report.

support infrastructure can only be answered through further analysis. There may also be constraints imposed by aviation traffic in the area or other siting authorities. In the United States, for example, wind developers need to get a permit from the Federal Aviation Administration (FAA) to ensure the wind farms do not pose any impediments to air traffic at nearby airports. Permits for structures higher than 500 feet usually require extensive review from the FAA and endure a longer public comment and approval process.

The resource analysis for a solar project is relatively straightforward and often can be done with a desktop review, as solar irradiation data has been studied over a long period and is much more stable than wind speed. As a result, solar irradiation can be predicted with sufficient accuracy without on-site data collection. Moreover, new software programs can precisely estimate the energy generation at a given solar site using data on solar panels and inverters from recognized manufacturers.

Given the large amount of land required for a renewable energy project, the land control process can be arduous, especially for wind projects. Generally, developers sign an option to lease available land for a nominal fee. The option covers the development period, and, if exercised, allows the developer to lease the land for a period of time that extends beyond the useful life of the project at an agreed upon royalty rate. This can be a fixed rate per acre of the land subject to the lease, or may be linked to the revenues earned by the project. The rate, in some cases, may change over the life of the lease in predefined increments. Over time, landowners have increasingly shown greater acceptance for hosting wind farms to generate supplemental revenue and have preferred a contract with lease rates linked to the revenues from the project.

Once the land is secured and resource analysis is underway, the developer may start the process of securing the necessary permits for project construction and operation. As discussed in Chapter 5, these may be numerous, time-consuming, and expensive.

The capital expenditure required to undertake the initial development activities is usually not significant. However, the developer is exposed to significant risks during the development phase, stemming from the fact that there are multiple risks that could undermine the viability of a project, for economic or other reasons. As a result, developers often create a portfolio of projects across a region to diversify the individual project risks.

Major capital outlays for new projects begin once the developer starts to secure interconnection agreements and offtake contracts, such as PPAs. As discussed in Chapter 5, the interconnection agreements allow a project to connect to the transmission grid, provided that its operation does not jeopardize the stability and operation of the transmission grid. As part of the interconnection process, the system operator or transmission owner may identify transmission upgrades or new facilities needed to accommodate the project; the project sponsor is often required to pay for these upgrades and facilities which may have considerable costs. Therefore, at the time of execution of the interconnection agreement, the transmission operator may stipulate that the project owner post a cash deposit or an LC in an amount equal to the cost of the upgrades, as well as defining a reimbursement schedule as the upgrades come online. The cash deposit or LC posting requirement is designed to ensure that the transmission operator can recoup the costs incurred if the project does not materialize as expected or is delayed beyond a reasonable amount of time.

Offtake counterparties may impose their own credit support requirements. As discussed in Chapter 5, off-taker(s) will not wish to assume the risk that the project does not come into service as expected and that it will be forced to enter into a new contract at a price higher than stipulated in the original contract. In order to cover the risk, the off-taker may ask for a cash deposit or an LC to cover the credit risks associated with the project. Such credit support generally needs to be posted within a few weeks of executing the PPA. In case of a commodities hedge or a proxy revenue swap, the credit support requirements need to be satisfied at the execution of the contract. Therefore, developers generally delay execution of the commodities hedges or proxy revenue swaps until the financial close, so that they can tap into the construction financing available to post the LCs. However, since the hedge prices in the commodities hedges vary with market conditions, the developers usually take the risk that the market conditions may materially impact the economics of the project.

The credit support requirements for interconnection agreements and off-take contracts can vary widely depending on the nature of the project, the entities involved, the term of the contracts, etc. Over time, these requirements have generally increased as transmission and system operators and contract counterparties have increasingly focused on the credit quality of the sponsors of renewable energy projects.

As illustrated in Fig. 11.2, risks tend to decline significantly over time while total capital outlay increases as the project matures through the successive development milestones. Accordingly, the projects under development may bring vastly different valuation based on where the project falls along the development spectrum.

The construction phase

By the time a renewable energy project reaches the construction phase, all development and permitting activities have been completed, various construction and equipment contracts have been executed, and financial close has been achieved. This assures the sponsor and other parties that the project has adequate financial resources.

The risks during the construction phase are primarily related to timely construction and commissioning of the project. Fortunately, the construction timeline for most renewable energy projects is relatively short – ranging from a few months to a couple of years – depending on the project site, and other characteristics. Construction of wind projects tends to be more logistically complex than solar projects, because solar projects are close to the ground and use smaller equipment. Construction of projects using certain technologies, such as biomass or geothermal, may be more complex and involve longer construction times.

The construction phase is not risk free. For example, based on historical data on project finance loans for infrastructure projects (which includes power projects) over the period 1983–2012, Fig. 11.3 shows that project finance loans for infrastructure projects have a relatively high default rate during the initial years, primarily due to construction risks. The following table and graph illustrate the evolution of the default rates, with the age of infrastructure projects based on historical data.

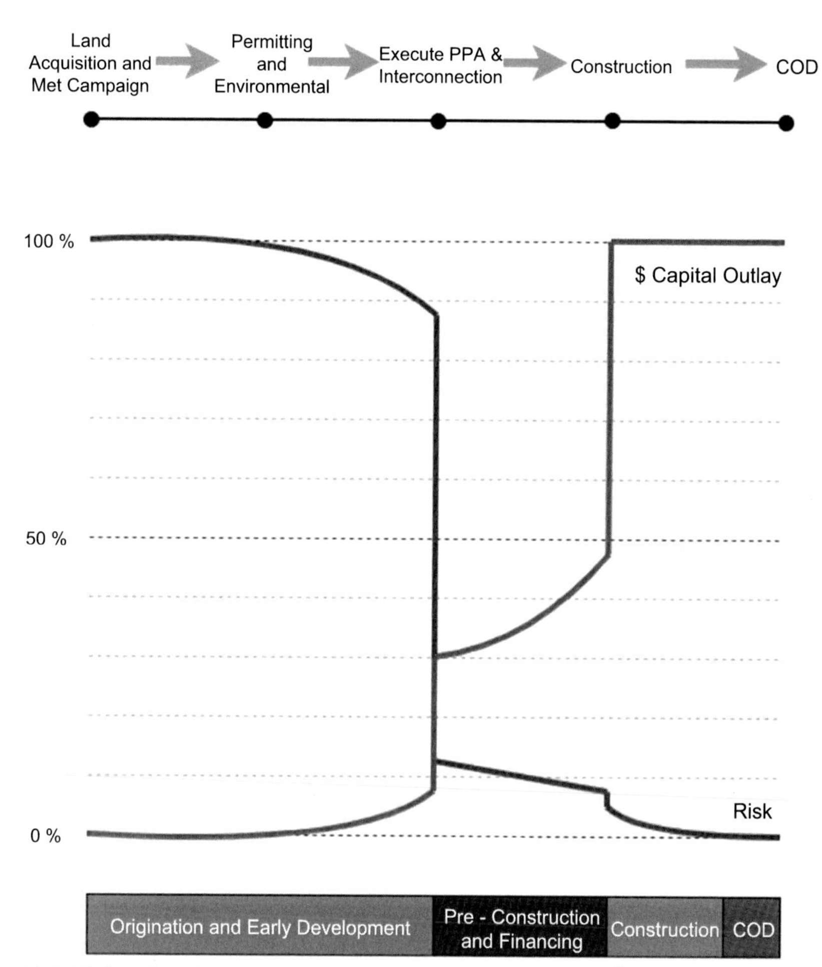

Fig. 11.2 Risk and capital outlay profiles for a development project.[2]

The primary conclusion from this historical loan analysis of defaults and recovery rates was that default risk decreases considerably as projects gained operational experience. For some projects, default rates were even lower in their later operational years than for debt securities rated as "A" grade by Moody's, even though at inception, such projects were rated at "Baa2" level.

Other relevant conclusions from the Moody's analysis included:

[2] "Investing in Development: Understanding the Risk-Reward Profile" Pattern Energy White Paper, 2017. (2017). *Investing in development understanding the risk-reward profile*. Pattern Energy Group Inc. Retrieved from https://investors.patternenergy.com/static-files/c0464707-fc27-4f3b-9a19-a7282249963cb

Marginal annual default rates

Year	Study Data (Basel)	Study Data (Moody's)	Moody's A	Moody's Baa	Moody's Ba
	Marginal Annual Default Rate %				
1	1.29%	0.95%	0.06%	0.18%	0.91%
2	1.21%	0.89%	0.12%	0.28%	1.68%
3	0.96%	0.70%	0.19%	0.32%	2.04%
4	0.74%	0.50%	0.20%	0.38%	2.19%
5	0.55%	0.37%	0.24%	0.39%	1.97%
6	0.39%	0.26%	0.27%	0.40%	1.81%
7	0.25%	0.16%	0.28%	0.38%	1.64%
8	0.18%	0.11%	0.29%	0.38%	1.57%
9	0.12%	0.06%	0.30%	0.39%	1.55%
10	0.08%	0.04%	0.29%	0.43%	1.57%

Incidence of projects originated, active and defaulted by year

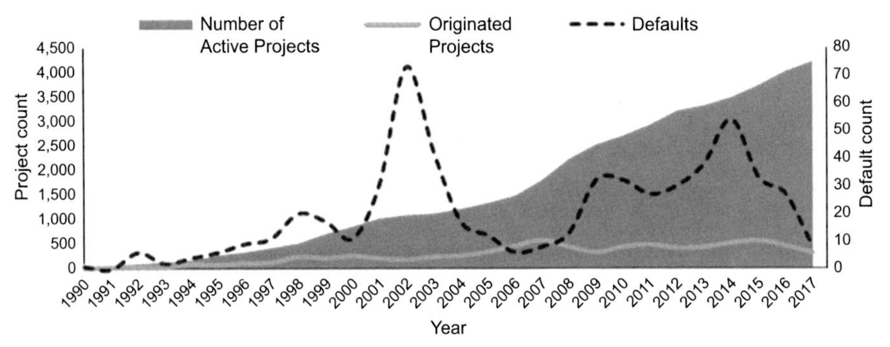

Fig. 11.3 Moody's analysis of project finance loans.[3]

- 2002 was a difficult year for project finance loans, but the deep recession beginning in 2008 only had a relatively modest impact on default rates.
- Recovery rates in the event of default are estimated by Moody's to be approximately 80% overall (and approximately 82%–83% after construction).

Higher default rates and lower recovery rates during the construction phase makes sense, as projects are more vulnerable during the construction than the operation phase. For example, there may be logistical issues in procurement and delivery of required equipment. Weather risks are also important during a major construction project. Reliability analyses for large engineering systems typically demonstrate that the failure rate is higher during the initial operation period and toward the end of the useful project life. For power projects, which often involve complex electromechanical systems, the failure rate is demonstrably higher during the first few months or years of operation. As the projects gain operational experience, the failure rate usually reverts to that of a steady state. Finally, a project defaulting during the construction phase is likely to see lower recovery for creditors than one in the operating phase.

[3] Moody's Investor Service. (March 5, 2014). *Default and recovery rates for project finance bank loans, 1983–2012.*

A default in the project construction phase is often related to construction delays or cost overruns. Unless some other investor is willing to put in more capital, the recovery prospects are usually expected to be low. An incomplete project under construction is not appetizing to new investors, especially if the original sponsor has chosen to walk away from the project instead of investing more equity capital. Moreover, if construction does not restart quickly, further delays may harm the project economics even more.

The operations phase

Once a project reaches its COD and enters full service, operational risks are generally relatively lower and dominated by resource risks (wind speeds or hours of sunshine, for example), assuming the underlying technology is reliable.

Valuation of renewable energy companies

Valuation of renewable energy companies is fraught with multiple challenges. Renewable generation companies' portfolios of operating projects generally use non-recourse financing and are owned through joint ventures and/or co-investments. This creates additional complexities in modeling cash flows and using typical valuation metrics.

In addition, many renewables companies have a significant development business and set of development projects underway, especially since the renewables sector is still going through a significant growth phase. Valuing these development projects is not trivial, since these projects are inherently risky (many may never come to fruition and hence have rather "binary" type risks) and there is not a standard framework for valuing these projects. To the outside observer, there is often limited visibility on the development pipeline, making it difficult to estimate of the probability of success. Finally, it is often difficult to develop market comparables for individual projects or portfolios due to the widely varying characteristics of projects and portfolios across companies.

Renewable energy generation companies are commonly valued as a portfolio of operating projects, and a development business which reflects the set of projects under development. Both can be understood in terms of cash flow analysis, but development projects, of course, have more uncertain cash flows until a project enters service.

Fig. 11.4 below illustrates the cash flow profile for a typical utility-scale renewable energy project. In this simplified diagram, all the cash outlays are assumed to have been incurred at a specific time interval. Accordingly, all development capital is assumed to be expended at the inception of the project at $T = 0$, whereas all of the construction capital is assumed to be drawn at $T = C$ when the construction begins. Furthermore, the cash flows during the operating phase are assumed to be equal for simplicity.

The modeling of these cash flows and discounting at appropriate rates is central to valuing both existing and potential projects, although the latter is more difficult.

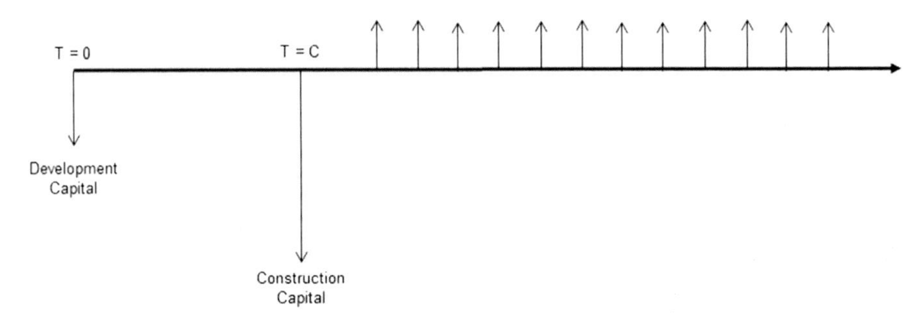

Fig. 11.4 Cash flow profile for a typical renewable energy project.

Valuation of an existing renewable project

The valuation of a project in its construction or operations phase is relatively straightforward using ordinary discounted cash flow (DCF) techniques. DCF is not the only valuation technique, of course, and other approaches may also be used in other contexts. However, DCF analysis is widely used in commercial project finance and will be used in the examples here. Under a DCF approach, the cost of equity is typically established using the Capital Asset Pricing Model (CAPM) or related techniques. This section provides a very brief introduction to the CAPM formula. Summaries of CAPM and related concepts can be found in many standard valuation and finance textbooks.[4] CAPM is not the only equity valuation method, but is widely used and reflects important concepts in modern financial theory about how investors view their portfolios.

The weighted average cost of capital (WACC) is calculated from the cost of equity and debt, and the capital structure of the business, while accounting for tax effects. In this simplified presentation, only common equity and a single tranche of debt are included.

$$\mathrm{WACC} = \mathrm{R_D} \times (1 - \mathrm{T_C}) \times \mathrm{D/V} + \mathrm{R_E} \times \mathrm{E/V}$$

$\mathrm{R_D}$ = Cost of Debt
$\mathrm{R_E}$ = Return on Equity
$\mathrm{T_c}$ = Tax Rate
D = Market Value of Debt,
E = Market Value of Equity,
V = D + E

Under the CAPM approach, the return on equity for a security is related to the systematic (undiversifiable) risk of a project, since idiosyncratic or unsystematic risks can be removed or mitigated by investors through diversification of their portfolios.

[4] Brealey, R., S. Myers and F. Allen (2007), *Principles of Corporate Finance* (9th. Ed.) McGraw-Hill.

The beta measures the volatility of returns of a specific security in comparison to that of the entire market.[5]

$$R_E = R_f + \beta \times R_P = R_f + \beta \times (R_m - R_f)$$

R_f = Risk-free Rate
R_m = Expected return for a market portfolio
β = Relative risk measure

In applying the CAPM, one important consideration is picking the right beta. Betas, in general, should reflect the systematic risks associated with a company or project, as idiosyncratic risks (e.g. how much the wind blows in a certain location) can be diversified away. In real-life applications size premia, country risk and other adjustments may also be made.

In practice, the appropriate beta for a specific company or project often cannot be determined directly, as there is no traded security reflecting these risks. In this case, investment analysts often pick a set of traded securities to represent reasonably the same underlying systematic risks in order to establish a beta. Since observed "regression" betas are influenced by the leverage of individual firms, the measured betas must be unlevered in order to calculate an equity beta for each comparable company.

For specific renewable energy companies and projects, there may be few directly comparable public companies, and therefore coming up with an appropriate proxy group can be difficult.

Once the WACC has been established using the CAPM and the cost of debt, valuation of a specific existing renewable project is usually fairly straightforward. Revenues for future periods are estimated using estimated outputs and future prices (from PPAs, other hedged revenues, or merchant price revenues – or some mixture of these for different periods, as appropriate). Project operating expenses can also be estimated.

Useful life estimates have extended as the renewable energy sector has gained more experience. Wind projects can have a useful life stretching as far as 30 years and solar projects can operate for as long as 35 years. Usually, the residual value of the projects is assumed to be zero because the scrap value of the equipment often pays only for the removal and disposal of the equipment.

Once future cash flows have been estimated, they can be discounted back to the valuation date, giving a net present value (NPV) to the project.

Project valuation example

The following simple example illustrates valuation of a solar energy project, including construction costs.

[5] In mathematical terms, the beta is the covariance of security returns and market portfolio returns over the variance in market returns over a period. Betas are often estimated using regression analysis

A renewable energy developer has a 150 MW solar project available for sale. The project is scheduled to achieve commercial operation a year from now, and is expected to operate at a capacity factor of 20%. The project has a useful life of 25 years and all energy will be sold under a PPA for the full term of useful life upon commercial operation. The PPA pays a rate of $75/MWh (no escalation). The operating expenses are expected to range $25/kW per year (again, with no escalation). The project will cost $200 million to build (fully loaded cost).

For simplicity of the example, assume that the full cost of the project can be depreciated over the 25-year useful term using the straight-line convention, and there are no tax credits, tax equity, or other complexities. Also assume a marginal corporate tax rate of 35%, no working capital at the project level, and that the project company will require no capital expenditure over its operating life.

In terms of the cost of capital, this example assumes that the risk-free rate is 2%, the equity risk premium is 6%, and the appropriate beta is 0.75. The cost of debt is 4%. The project will be 50% debt and 50% equity funded.

The expected return on equity can be calculated using the CAPM formula:

$$\text{Cost of Equity} = \text{Rf} + \beta \times \text{Rp} = 2.0\% + 0.75 \times 6\% = 6.5\%$$

The weighted average cost of capital is therefore:

$$\text{WACC} = (1 - 35\%) \times 50\% \times 4\% + 50\% \times 6.5\% = 4.55\%$$

The first step is to calculate the expected energy generation per year.

$$\text{Energy per year} = 150 \text{ MW} \times 20\% \times 8760 \text{ h/year} = 262{,}800 \text{ MWh}$$

Revenue can then be calculated from the PPA price.

$$\text{Revenue} = 262{,}800 \text{ MWh} \times \$75/\text{MWh} = \$19{,}710{,}000/\text{year}$$

Operating expenses (Opex) can be calculated from the fixed $/kW value provided.

$$\text{Opex} = \$25/\text{kW-year} \times 150 \text{ MW} \times 1{,}000 \text{ kW/MW} = \$3{,}750{,}000/\text{year}$$

Consequently, the EBITDA of the project is:

$$\text{EBITDA} = \$19{,}710{,}000 - \$3{,}750{,}000 = \$15{,}960{,}000/\text{year}$$

Finally, we know that the capital cost in year zero is $200,000,000. Thus, given the simplified assumption of straight-line depreciation over 25 years:

$$\text{Depreciation} = \$200{,}000{,}000/25 = \$8{,}000{,}000/\text{year}$$

Earnings before interest and tax (EBIT) is equal to EBITDA − depreciation, so

$$\text{EBIT} = \$15{,}960{,}000 - \$8{,}000{,}000 = \$7{,}960{,}000/\text{year}$$

Finally, free cash flow (FCF) in a year is equal to:

$$FCF = EBIT\ (1 - Tax\ Rate) + Depreciation + Change\ in\ Working\ Capital$$

In this case, the change in working capital is assumed to be zero in each period. The free cash flow in each year can therefore be calculated as:

$$FCF = \$7{,}960{,}000 \times (1-35\%) + \$8{,}000{,}000 + 0 = \$13{,}174{,}000/year$$

In this simple example, the free cash flow in each year is constant, since revenues and expenses are assumed to be constant in every year. This allows the present value of future operating cash flows to be calculated simply as the present value of 25 equally spaced yearly cash flows of $13,174,000 each. Using the WACC as the discount rate, this provides a present value of $194,345,391.

The net present value must also account for the initial $200 million construction cost of the project. So, the net present value of the project is *negative* $5,654,609.

The breakeven PPA price

The breakeven PPA price is defined as the fixed price of energy (in all periods) that equates the net present value of the project to zero. The breakeven PPA price is thus equivalent to the Levelized Cost of Electricity (LCOE) for the project.[6]

Mathematically, this is solving backwards to find the breakeven free cash flows in each period such that the present value at the WACC discount rate is equal to the cost of construction:

Present value (4.55%, 25 years, Breakeven FCF) = $200 mm.

A simple calculator or spreadsheet analysis shows that for these assumptions the breakeven FCF = $13,557,306/year.

Once we have the Breakeven Free Cash Flow that gives us the stipulated present value of $200 mm, we need to work backwards to calculate the breakeven revenues.

First, given the Breakeven FCF, we can calculate the Breakeven EBIT as follows:

$$Breakeven\ EBIT = (Breakeven\ FCF - Depreciation)/(1 - Tax\ Rate) = \$8{,}549{,}702/year$$

With the Breakeven EBIT, we can calculate the Breakeven Revenues as the adjustment to the original revenues needed as follows:

[6] The levelized cost of electricity (LCOE) is defined as the per-unit cost of generation over the life of the project. This measure is used to compare different generation technologies with unequal useful lives, project sizes, capital costs, return requirements, etc. Please refer to Appendix B for the detailed description of LCOE and the calculation methodology.

Breakeven Revenues = Original Revenues + (Breakeven EBIT − Original EBIT) = \$19,710,000 + (\$8,549,702 − \$7,960,000) = \$20,299,702/year

Breakeven PPA Price = \$20,299,702/262,800 MWh = \$77.244/MWh

Valuation of a renewable energy project under development[7]

In order to understand the valuation of the project under development, consider another simple example for a renewable energy project. Assume that the construction cost will be \$60 million, which would generate a forecast present value of operating cash flows of \$100 million when the project is operational, implying a net present value of \$40 million (at the time of construction). For the purpose of this simple valuation exercise, assume that the project will be a guaranteed success. Further assume that the project is anticipated to take one year for development work before the project can will be construction ready, as shown in Fig. 11.5. Furthermore, assume a cost of capital of 8%. What is the value of the project at present (T = 0)?

At year 0, the value of the project when it is commercially operational is only a forecast. Its present value is found by discounting the PV of cash flows at year 1 by the cost of capital of 8%. Therefore,

$$PV_0 = PV_1/(1 + Rc)$$

$$PV_0 = \$100 \text{ million}/(1 + 0.08) = \$92.59 \text{ million}$$

Assume that the \$60 million construction cost is known ahead of time and incurred all at once. It is important to note that the construction cost is assumed to be fixed and

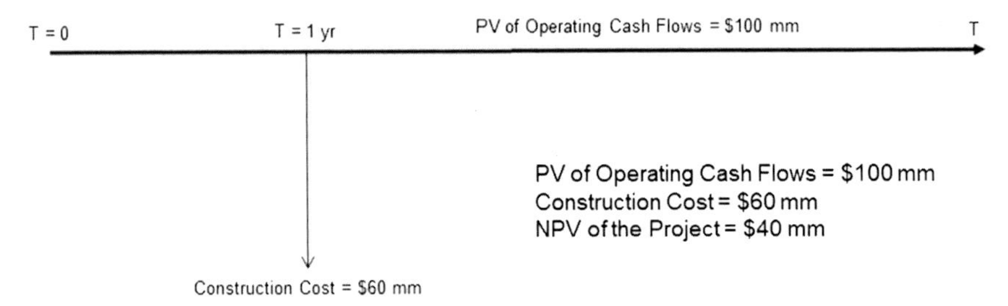

Fig. 11.5 Simple example of a development project.

[7] The analytical framework used in this section for valuing a development project is based on the structure used and described in Stewart Myers and Laxmi Shyam-Sunder for a pharmaceutical development project, "Measuring Pharmaceutical Industry Risk and the Cost of Capital", In R. Helms (Ed.), *Competitive strategies in the pharmaceutical industry*. AEI Press, 1996.

determinable, irrespective of the outcome of the development project. Consequently, the construction cost is treated as a fixed liability and accordingly discounted at the risk-free rate of 2%:

$$PV_{construction} = 60/(1 + 0.02) = \$58.82 \text{ million}$$

Therefore, the net present value of the project one year before commercial operation begins is

$$NPV_0 = PV_0 - PV_{Construction}$$

$$NPV_0 = \$92.59 \text{ million} - \$58.82 \text{ million} = \$33.77 \text{ million}$$

Please note that in arriving at a NPV_0, we implicitly discounted the project, which has a net present value of \$40 million at $T = 1$ year to \$33.77 million at $T = 0$. This implies a discount factor of 18.45% because

$$\$33.77 \text{ million} = \$40 \text{ million}/(1 + 0.1845)$$

Thus, the cost of capital applicable for the 1-year development period is 18.45% and not 8% as expected. Since we already have the inputs to the cost of capital, we can compute the appropriate β for the development period for this stylized example as follows:

$$\text{Risk Free Rate} = R_f = 2.00\%$$

$$\text{Equity Risk Premium} = R_P = R_m - R_f = 8\%$$

$$\text{Cost of Capital} = R_C = R_f + \beta_{Development} \times R_P = 2.0\% + 0.75 \times 8\% = 8\%$$

Therefore,

$$\beta_{Development} = 2.06$$

Here is another way of looking at the risk/reward of the project before and after the construction of the project. At $T = 0$, we have a project with a risk symbolized by $\beta_{Development}$ and present value of \$33.77 million. The project has a construction liability of \$58.82 million in present value terms at $T = 0$. For the purpose of this example, again assume the success of the project is a certainty. Therefore, $\beta_{Construction} = 0$. Upon construction of the project, expect its present value to be \$100 mm, at which $T = 0$ is \$92.59 mm and β_{Asset} of 0.75. Therefore, the risk of the project, when viewed at $T = 0$, is expected to be higher, in much the same way that the risk of a levered firm is expected to be higher even when the risk of the firm's assets is held constant. In other words, the betas of the project construction and the beta of the development must add up to the beta of the asset when commercial operation commences. Specifically,

$$\beta_{\text{Development}} \times PV_{\text{Development}} + \beta_{\text{Construction}} \times PV_{\text{Construction}} = \beta_{\text{Asset}} \times PV_{\text{Asset}}$$

$$\beta_{\text{Development}} \times \$33.77 \text{ million} + 0 \times \$58.82 \text{ million} = 0.75 \times \$92.59 \text{ million}$$

$$\beta_{\text{Development}} = 2.06$$

It should be re-emphasized that the $\beta_{\text{Development}}$ is materially higher than the β of the project, not because of the market or development risks the project faces before the construction begins, but rather because the project faces a future fixed liability (in terms of construction costs) before the project before the project can reach commercial operation. However, in the real world, renewable energy projects under development face a host of uncertainties until the project becomes commercially operational due to the permitting and licensing process. PPAs are also subject to regulatory approvals and ratemaking processes. Accordingly, assume that the commercial success of the project is no longer a certainty. Instead, due to regulatory uncertainty, the probability of success is now 80%, creating a payoff depicted below. Over the course of a 1-year development period, a successful development effort would yield an asset with net present value of \$40 mm, whereas a failed development effort, perhaps due to the lack of an important permitting approval, may lead to an asset with a value of zero (Fig. 11.6).

In this case, the project's net present value would fall to 80% of the value we had calculated before, or 80% × \$40 million/1.1845 = \$27.02 million. It should be noted that the cost of capital in this case is still 18.45%, as calculated before, because the risk of the failure is diversifiable. Therefore, the development risks considered in this example have the effect of reducing the net present value of the project alone.

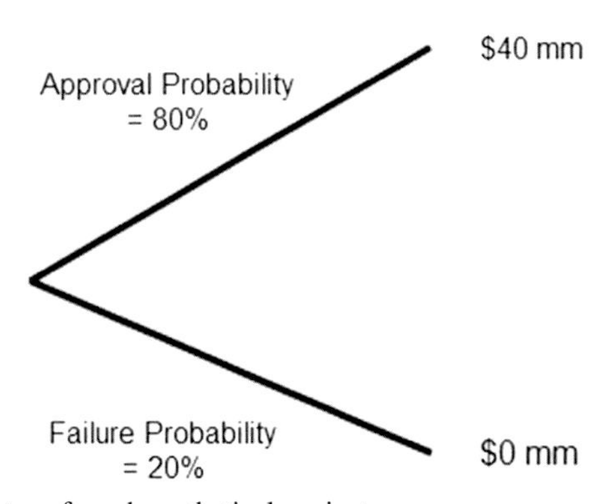

Fig. 11.6 Payoff structure for a hypothetical project.

The previous simplified examples could be extended to value a portfolio of development projects, often referred to as the development pipeline of a project developer. Each project would need to be valued independently, considering the development work as well as the risk of failure. Thereafter, the aggregate value of the portfolio would signify the valuation of the development pipeline. As discussed before, in a truly diversified portfolio of development projects, individual project failure risks may be diversified away in determining the cost of capital, but must be reflected in modeled cash flows. There may also be some residual risks of future cash flows which are systematic, and hence should be reflected in the valuation as illustrated previously.

Energy storage financing: opportunities and challenges

12

Energy storage is the capture and saving of energy produced to be used at a future point in time. The electrical grid needs constant adjustment to maintain balance between power supply and demand, and energy storage can be a critical component of maintaining that balance. The primary function of energy storage is to maintain a flexible and reliable grid by removing power from the grid when it is oversupplied and returning power to the grid when it is undersupplied.

Types of energy storage

There are several methods of storage. Pumped hydro storage has existed for decades. This method consists of pumping water behind hydroelectric dams, absorbing excess energy from the grid to carry out the pumping, and returning the energy later through the release of the water through the turbines, thereby generating electricity. Pumped hydro storage makes up the largest capacity of storage globally, with about 170 GW of pumped storage hydropower deployed across the world.

An other conventional means of energy storage is compressed air energy storage (CAES) and flywheels.

CAES works in a similar way as pumped hydro storage; during times of excess electricity supply, air is pressurized and pumped into a cavern that was created and designed specifically for energy storage. When additional electricity is demanded, the stored air is heated and passed through a steam turbine, thereby generating electricity.

Flywheels, on the other hand, store mechanical energy. These are cylindrical rotors that spin at speeds as high as 50,000 rpm. When there is excess supply of electricity, the excess is absorbed by the flywheels by increasing the speed of the rotors. Conversely, when electricity is needed, rotor speed is reduced so that electricity is generated and fed to the electric grid.

These conventional energy storage mechanisms are primarily used for short-term balancing of energy supply and demand on the grid. With advances in solid-state electronics and other technology, another source of energy storage that holds significant potential is battery storage. As we will explore later, battery storage is more versatile than conventional energy storage systems because, in addition to balancing the energy supply/demand, the battery storage can provide other ancillary services. Moreover, battery storage generally has a faster response time than traditional generation and conventional storage technologies.

Due to the versatility of battery storage, more than 3.3 GW of battery storage is deployed globally, with the US and China leading the charge. 750 MW of storage

Renewable Energy Finance. https://doi.org/10.1016/B978-0-12-816441-9.00012-X

have been installed in the US, with a record 311 MW of deployment in 2018. During the first quarter of 2019, an additional 149 MW of grid storage capacity was added to the grid. Battery storage capacity in the US is expected to double over the course of 2019 to over $1 billion in overall value, triple in 2020, and rise to over $4 billion in value by 2024.[1] China, which announced a 10-year plan for grid-scale storage development in November 2017, closely followed the US with 300 MW of storage deployment in 2018. Together, the US and China are projected to account for 54% of global storage deployments through 2024. The next tier of storage leaders is comprised of Japan, Australia, and South Korea, with Germany, Canada, India, and the UK following thereafter. Storage deployment in these countries is motivated by a number of factors, including a significant capacity value increase, REC's from renewables-plus-storage in South Korea, competitive energy markets driving residential solar-plus-storage in Australia, and duration-length tied incentives in the UK Battery storage markets across the globe are only at the beginning of their promising life cycle;[2] consequently, we will devote the rest of the chapter to battery storage.

There are four types of battery storage currently used in power applications. Lithium-Ion (Li-Ion) batteries are the most widely known and deployed, due to their proven technology with long run times, high efficiency, and good cycle life. Sodium-metal batteries are a distant second in terms of deployment, with significantly fewer installations than Li-Ion. Flow batteries and lead-acid batteries are the remaining two technologies and they have a much lower penetration. The Clean Energy States Alliance (CESA) estimates that more than 78% of the battery installations in the US use Li-Ion technology. Each of these technologies comes with its own set of benefits and challenges. The technology for the flow battery is still evolving, but once fully commercialized flow batteries could become the most competitive for large-scale power applications.

Value of storage

As mentioned earlier, the principal use of batteries is to balance electricity supply and demand on the grid. However, batteries can also be used to provide effective ancillary services:

- *Demand Response or Frequency Regulation:* In order to work reliably, the grid needs to operate within a specified frequency band. As a result, the grid operator needs access to resources that can respond quickly and robustly to changes in demand. Traditionally, utilities

[1] John, J. S. (2019, April 10). *Global energy storage to hit 158 gigawatt-hours by 2024, led by US and China*. Retrieved from https://www.greentechmedia.com/articles/read/global-energy-storage-to-hit-158-gigawatt-hours-by-2024-with-u-s-and-china#gs.zt174j.

[2] Wood Mackenzie data from the report titled "Global Energy Storage Outlook 2019" reported in the GTM article titled "Global Energy Storage to Hit 158 Gigawatt-Hours by 2024, Led by US and China" by Jeff St. John published on April 10, 2019. John, J. S. (2019, April 10). *Global energy storage to hit 158 gigawatt-hours by 2024, led by US and China*. Retrieved from https://www.greentechmedia.com/articles/read/global-energy-storage-to-hit-158-gigawatt-hours-by-2024-with-u-s-and-china#gs.zt174j.

have relied on spinning reserves, which are generators that are already on line and generating electricity at a capacity lower than their nominal capacity. Batteries are particularly good at providing this service because they can discharge quickly and are much faster than ramping up than traditional energy generation sources. As a result, batteries can provide much better frequency regulation than traditional generators.

- *Peak Shaving:* Batteries can be charged during times of low demand and discharged during a time of peak demand, thereby reducing the peak demand a system needs to fulfill. This technique, known as peak shaving, can be useful for both utilities and customers. The utility can use peak shaving to defer the construction of expensive power plants.[3] On the other hand, when deployed on the customer side for an industrial or commercial application, the customer can use peak shaving to reduce demand charges as these charges are typically based on peak demand.
- *Renewable Energy Intermittency*: Renewable energy resources are dependent on external factors (e.g. wind speeds) and hence somewhat unpredictable and intermittent. However, when coupled with renewable energy resources, batteries can smooth out intermittent generation to make such resources dispatchable. For example, when coupled with a utility-scale solar project, a battery can charge up during the day, when the solar power system produces energy, then discharge overnight, when the solar system is idle. As a result, the solar-plus-storage plant can be dispatched more like a conventional generation source.
- *Microgrids*: Battery storage is particularly important for isolated communities, some of which may use microgrids, as access to centralized power generation can be limited; battery storage can mitigate the unpredictability associated with power generation in these areas.
- *Price Arbitrage:* Since power prices vary substantially, it can be profitable to charge energy storage devices (such as batteries) at low prices periods (e.g., overnight) and discharge during high price periods.

Battery business models and financing strategies

Different business models may be employed depending on the batter application:

- *Capacity Payment for Demand Service* – In the basic business model, the battery storage project is paid a periodic capacity payment for being "available". The project can then earn additional revenues when called into action to provide actual demand service. The business model can be expanded to cover other ancillary services, such as frequency regulation.
- *Merchant Energy Storage* – In this business model, the battery is charged during the off-peak hours by buying electricity when prices are low, then discharged by selling into the market during peak times when electricity prices are high.
- *Storage Coupled with Renewable Energy Resource* – In this business model, battery storage is integrated with a solar or wind project, making the system partially or completely dispatchable, similar to a traditional power plant. Such a system should draw better pricing through a PPA (once contracted) or in a merchant market (relative to a stand-alone renewable energy project) since some electricity sales can be "shifted" to the peak hours from the

[3] A utility responsible for serving a given area needs to have enough generating capacity (with sufficient reserve margin to account for contingencies) to serve the peak load. A peak shaving technique could reduce the peak demand the utility needs to serve. Therefore, if implemented successfully across the serving area, with peak shaving techniques the utility can potentially defer the construction of new power plants to future periods.

off-peak hours. Storage coupled to renewable projects may also help lower the interconnection costs of new projects.
- *Behind-the-Meter Distributed Energy Resources* – As discussed before, on a stand-alone basis, a battery can be used for peak shaving applications to reduce the peak demand and, by extension, the demand charges. Alternatively, a battery can be combined with a solar project to "shift" the excess electricity exports to the peak hours, when time-of-the-day pricing is presumably higher.

Financing strategies vary depending on which business model is adopted. As discussed in Chapter 4, financial institutions usually have little appetite to finance a merchant power project. Financing a storage project on a merchant basis is even more difficult, due to the challenge in forecasting the patterns of peak and off-peak prices.

In the capacity payment business model, a utility is bound by a long-term contract to provide a stream of availability-based payments and a demand charge that is based on the prevailing merchant energy prices at the time of the storage dispatch. The utility contracts for capacity-based payment to the storage project which essentially acts like a traditional PPA and which may be financed using traditional project finance structures. In the absence of such long-term contracts, commodities trading houses may be able to provide hedges that grant the hedge counterparty a right to dispatch the project to arbitrage the on-peak/off-peak prices in return for a stream of cash flow (akin to a capacity payment). The market for such hedges is still far from mature.

A storage project coupled with battery storage works like a dispatchable power plant and can tap into either the traditional PPAs or commodities hedges to enable financing.

Tax incentives for storage projects in the US

The US federal government provides valuable depreciation and tax credits to encourage deployment of energy storage, such as battery systems. The eligibility for the benefits depends on the use of the energy and whether it is integrated with a renewable energy source. For example, stand-alone energy storage (i.e. storage with no tie-in to a renewable energy resource) qualifies for a 7-year MACRS depreciation. Energy storage that is charged by a renewable energy source can qualify for a portion of an investment tax credit, up to 30% of the cost of the system. When the battery is charged by renewable energy for more than 75% of the time annually, it qualifies for the credit in an amount equivalent to the following:

Annual time charged by renewable energy (>75%) × 30% ITC = Amount of ITC (as %)

This concept is depicted in Fig. 12.1 below. For example, if battery storage is charged by an adjoined solar power system for 90% of the year, the battery would qualify for 90% × 30% = 27% of the ITC. It should be noted that if the batter storage project is charged by renewable energy resources for less than 75% of the time, the batter storage project does not qualify for any ITC.

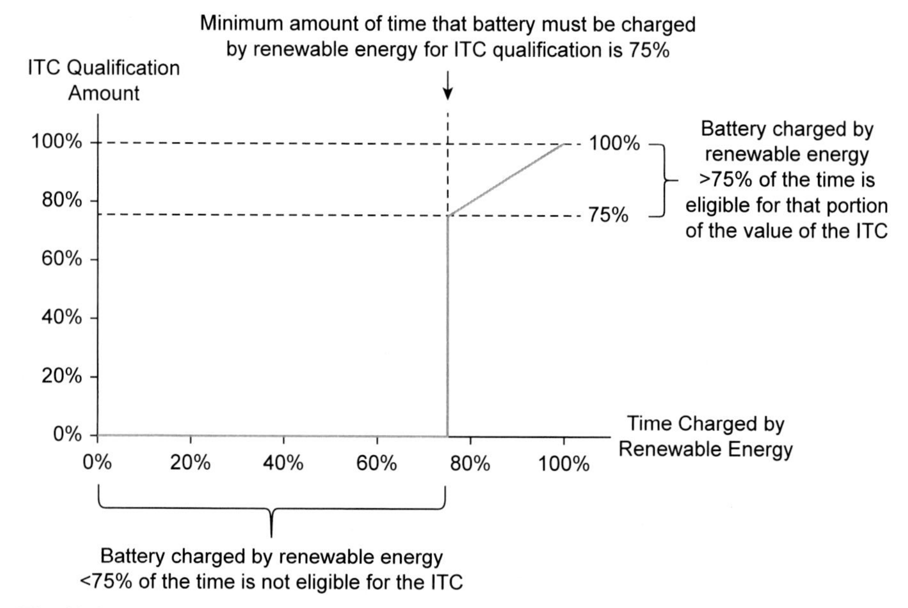

Fig. 12.1 Depiction of the amount of ITC that a battery can qualify for based on the amount of time (annually) the battery is charged by renewable energy.

In addition to the tax credits, a battery storage charged by a renewable energy resource for more than 75% of the time annually also qualifies for the 5-year MACRS (rather than a 7-year MACRS).

The ITC rules for energy storage follow the same rules and considerations discussed in Chapter 6 for solar and wind projects. Specifically, the depreciation basis for the energy storage claiming an ITC is reduced by 50% of the ITC claimed. Furthermore, the ITC claimed for battery storage is subject to the same recapture rules applicable for solar and wind projects.

The ITC and depreciation eligibility guidelines described above apply to energy storage systems installed at the same time as the renewable energy system. In its paper, titled "Federal Tax Incentives for Energy Storage Systems,"[4] NREL assumes that energy storage added to an existing renewable energy system would qualify for the same benefit as a new renewable-plus-storage system; this is based on a precedent set by a 2012 private-letter ruling that allowed a wind farm owner to add energy storage to an existing wind farm and claim the tax benefit.[5] NREL further argues that the solar power system and energy storage would need to be in close proximity and under common ownership (e.g. the same taxpayer).

The ability to claim the ITC and 5-year MACRS depreciation provides a significant boost to the economics of the battery storage system when combined with a renewable

[4] *Federal tax incentives for energy storage systems*. National Renewable Energy Laboratory. Retrieved from https://www.nrel.gov/docs/fy18osti/70384.pdf.

[5] IRS Private Letter Ruling PLR 201208035, February 24, 2012.

energy source. Consequently, there has been a significant number of new solar-plus-storage projects. Such projects are much more attractive to utilities than stand-alone solar energy projects due to the additional ancillary services such combined projects can provide. As battery costs continue to decline, the penetration of solar-plus-storage systems is likely to increase substantially.

Challenges to the adoption of battery systems

The battery storage market is heavily reliant on incentives and policy mechanisms. In the US, the California and PJM systems currently dominate in terms of cumulative installed capacity, but capacity in other states is starting to expand as state policies incentivize more battery storage. For example, large-scale storage projects have come online in Hawaii, Colorado, Texas, and Minnesota, with more expected across the country. State-level policy updates in Massachusetts, New York, and New Hampshire are expected to open the East Coast storage market even further. The market outlook for storage has also been expanded through utility plans. In early 2019, the Hawaiian Electric Company requested approval from regulators for a large group of storage PPA's, which will add 1,048 MW to their electric grid. Not long after, Puerto Rico published its new utility plan, which called for 440−900 MW of storage over the next 4 years, and Arizona Public Service announced it planned to install 850 MW storage by 2025.

On a broader policy level, the battery storage market in the US received a strong impetus in February of 2018 with the issuance of FERC Order 841. This Order removed barriers for storage to participate in the capacity, energy, and ancillary service markets by requiring Regional Transmission Operators and Independent System Operators to establish rules around the participation and compensation of storage in these markets. RTO's and ISO's filed their plans to comply with the Order in late 2018 and final implementation of the Order is expected by the end of 2019.

FERC Order 841 may prove an important step towards widespread adoption of battery storage systems. Because batteries provide ancillary services and peak shaving applications, they can be a fundamental building block for a transmission grid, especially a grid with high renewable energy penetration. In fact, with their compact size and short construction times, battery storage systems can sometimes be used in lieu of building new transmission lines. This is especially relevant in the US, given the difficulties in acquiring necessary right of way easements and permits for building transmission lines. However, this is dependent on reliable pricing signals for market participants that empower developers to build battery systems in optimal locations to provide support for reliable operation of the electrical grid. Such pricing signals can only come from regional transmission operators, Independent System Operators, and Public Utility Commissions, who are tasked with long-term transmission system planning.

Furthermore, a capacity market mechanism is crucial to the financing and deployment of the batteries in key areas. To explain further, consider the ERCOT market, which is the only market in the US that does not have a capacity market. The ERCOT claims that the energy market alone should provide adequate signals to the market participants to boost the reserve margins. However, the ERCOT's rampant wind farm

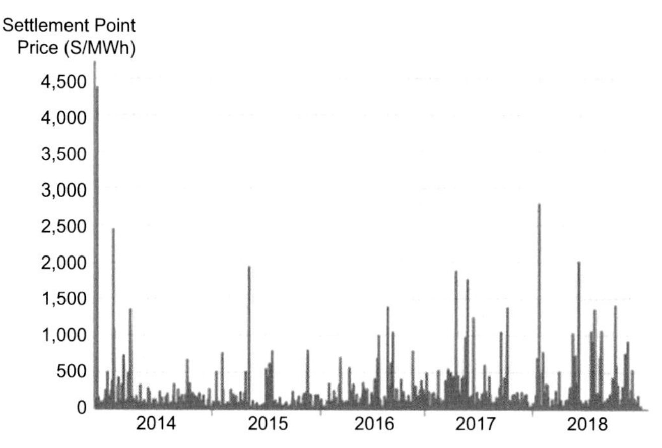

Fig. 12.2 15-min ERCOT settlement point prices in $/MWh over 5 years.

development in the western part of Texas is hobbled by the lack of transmission support. If there was a market mechanism to encourage wind developers to install batteries to shift generation to on-peak hours, some transmission bottlenecks could be avoided. Figs. 12.2 and 12.3 below illustrate the historical ERCOT market prices and annual planned reserve margins. Even a marginal amount of battery storage capacity on the transmission grid could have mitigated a significant amount of price volatility.

In addition to policy incentives, cost reduction is crucial to the expansion of the battery storage market. The cost of batteries has generally declined over the years, but the process of converting batteries into energy storage involves more than just the physical battery. There is a suite of systems integration that includes packaging, thermal management systems, power and control electronics, and conversion systems, all of which

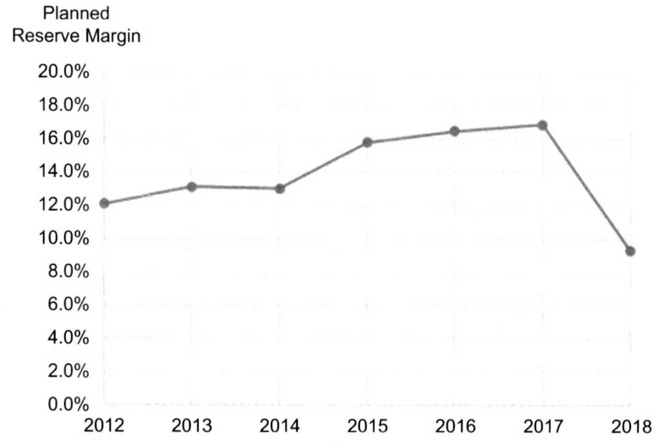

Fig. 12.3 ERCOT-planned reserve margins from 2012 to 2018.

require costly engineering work. In addition to falling costs of battery components, the costs of associated systems must come down as well.

Despite the large potential that battery storage holds for the renewable energy market, particularly with declining costs and increased storage-focused policy, there are also concerns around the physical materials that comprise different batteries. The primary concern is the recyclability of these materials, and how batteries should be disposed of at the end of their life cycle. Another major concern is the sourcing of battery materials, as the raw materials often must be imported. These challenges may grow with the continuing rapid development of the energy storage industry.

Renewable energy finance in the international context

13

Private investment in renewable energy, made using project finance structures or not, is substantially shaped by regional government policies.

This chapter explores how government policies have shaped investment in large-scale renewable energy projects in four case study countries: Chile, India, Germany, and China. These short case studies provide insight into how policy in the real world stimulates or hinders large-scale renewable energy private investment.

Recent global experience in private renewables finance

As shown in Fig. 13.1 below, total annual investment in renewable energy assets has almost doubled since 2008.[1] While these amounts are large in absolute terms, they are small in comparison to the amounts needed to decarbonize the global energy sector. Despite shifts in investment toward renewables in recent years, global carbon emissions were expected to grow in 2018 and 2019, significantly driven by high growth in energy demand.[2] A significant decarbonization, even of

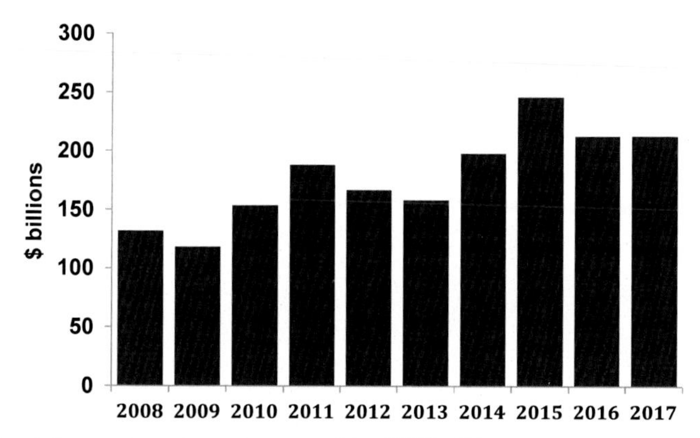

Fig. 13.1 Asset finance investment in renewables 2008–17.
Source: Data from IRENA.

[1] IRENA, & CPI. (2018). *Global landscape of renewable energy finance, 2018.* Abu Dhabi: International Renewable Energy Agency.

[2] Jackson, R. B., Le Quéré, C., Andrew, R., Canadell, J. G., Korsbakken, J. I., Liu, Z., Peters, G. P., & Zheng, B. (2018). Global energy growth is outpacing decarbonization. *Environmental Research Letters, 13.* http://doi.org/10.1088/1748-9326/aaf303.

Renewable Energy Finance. https://doi.org/10.1016/[illegible]

only the global electric power sector, would require much larger levels of annual investment.

Approximately 90% of investment in renewable energy assets comes from the private sector, dominated by renewable developers, corporate banks, and other financial institutions. Households, which include the personal investment funds of high net worth individuals, also play a major role.[3]

Trends in investment by region and type

As illustrated in Fig. 13.2, total investment in renewables (which includes R&D and other large asset financings) increased to over $250 billion per year by 2017, with China replacing Europe as the predominant region for renewable investment in dollar terms in recent years.

In earlier years, before most capital markets had extensive experience with large renewable projects and related project finance techniques, projects were typically financed on a balance sheet from corporate funds. Now, project finance makes up almost half of global private investment in renewable energy assets – close to $100 billion each year.[4]

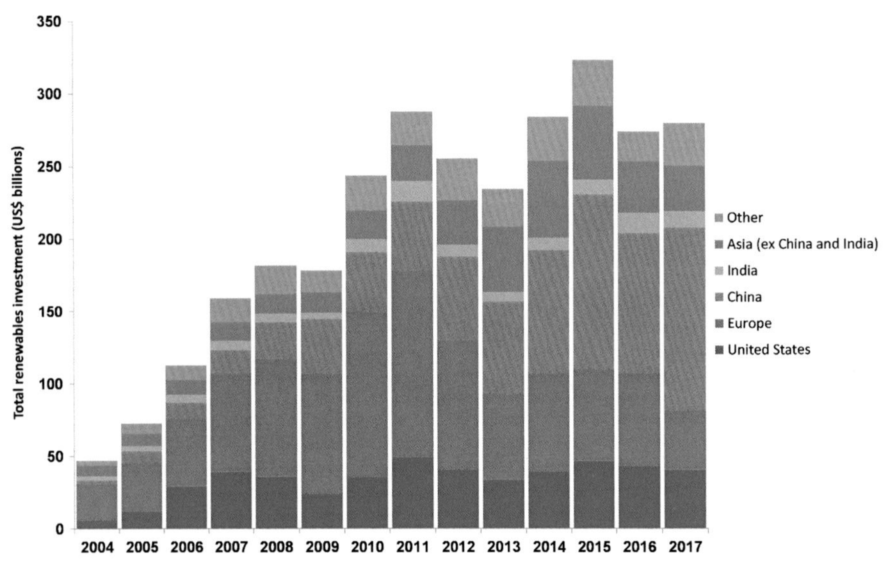

Fig. 13.2 Total investment by region 2004–17.
Source: Data from IRENA.

[3] See footnote 1.
[4] Frankfurt School – UNEP Centre/BNEF. Global trends in renewable energy investment 2018. Available from www.fs-unep-centre.org.

Project finance bonds, as discussed in Chapter 3, remain a relatively minor vehicle for private investment in the sector, more frequently used in refinancing activities.

Future needs for private renewables investment

While private investment in the renewable energy sector has increased substantially over the last decade, it is still well below the levels many experts believe are needed to stabilize carbon concentrations in the atmosphere.

The International Energy Agency (IEA) has published a "450 Scenario –" a conceptual future pathway for the energy and process industry-related sectors designed to limit global temperature increases to less than 2 °C with a 50% probability.[5] While the IEA 450 Scenario has been subject to criticism, it provides a basis for comparing the estimated capital investment needed for renewable energy and other sectors to limit the potential impacts of climate change.

The IEA 450 Scenario shows that investment in clean energy, energy efficiency and other measures would have to increase substantially over current levels to meet the above goal. A 2017 report from a Stanford group suggests that annual renewable energy investment to 2040 would need to be over $500 billion per annum (approximately double current investment levels), as part of a $2.3 trillion annual investment in global clean energy.[6] The total investment needed globally to limit expected temperature changes, even to 2 °C, could be in the tens of trillions of dollars.

The Stanford analysis suggests several major challenges to mobilizing the scale of private investment needed – through project finance or other means. First, the sheer scale of investment needed is huge. The amounts required make up a significant portion of the world's total annual investible capital.[7]

Second, renewable energy project development remains relatively risky, which limits the pool of capital available. As has been discussed throughout this book, a major aspect of project finance is putting contractual, regulatory, and financing structures in place to manage the risks of large capital-intensive renewable energy projects. Accessing the private capital needed will require better alignment between renewable project risks and the conservative risk profile of institutional investors, who control most major private capital flows.[8] Lowering technical barriers to new project development (such as transmission access for renewable power projects) will also be required.

Third, there is a strong need for inter-regional capital flows to support clean energy investment. Most capital is held in the developed world, while much of the need in the

[5] See IEA description available at https://www.iea.org/weo/energyandclimatechange/.

[6] Reicher, D., Brown, J., Fedor, D., Carl, J., Seiger, A., Ball, J., & Shrimali, G. (October 27, 2017). *Derisking decarbonization: Making green energy investments blue chip*. Stanford Law School. Retrieved from https://law.stanford.edu/publications/derisking-decarbonization-making-green-energy-investments-blue-chip/.

[7] See footnote 6 (2017).

[8] See footnote 7.

next decades will be in emerging market countries. To be successful, investment funds sourced in the developed world will need to flow to developing countries, with consequent need for managing policy, political, currency, and other risks.

In the remainder of this chapter, we explore how policies in four countries have affected investment risks and private capital flows into the renewable energy sector.

Chile

Latin America has seen rapid growth in renewable energy investment in recent years. In particular, Chile has been a major destination for foreign investment. In 2015, for example, Chile attracted more than half of the total investments in renewables within the Latin American and Caribbean region.[9]

Growth in the Chilean renewables sector has been aided by the general macroeconomic performance of the country – Chile has seen a substantial period of sustained macroeconomic performance over this period underpinned by a strong natural resources sector, good wind and solar resources, and a relatively stable legal, regulatory, and policy environment.

The Chilean electricity system

The Chilean power system and market design are both shaped by the country's unusual geography, which extends over 4,000 km entirely along a north-south axis. With a very low population density outside of the central region, Chile has relied on the operation of four separate transmission systems throughout its history. The *Sistema Interconnectado Central* (SIC) covered much of Chile's population, while the *Sistema Interconnectado del Norte Grande* (SING) served the northern regions where much of the load came from mining activities. In 2017, the SIC and SING systems were combined to form the *Sistema Eléctrico National* (SEN), which serves more than 97% of Chile's population. There are two additional isolated grids – the *Sistema Eléctrico de Aysén* (SEA) and the *Sistema Eléctrico de Magallanes* (SEM). Together, these two isolated systems have less than 1% of Chile's installed capacity (Fig. 13.3).

Chile is rich in mineral resources but has limited domestic energy resources. Historically, most of its electric generation has been fueled by coal, natural gas, and petroleum (often imported) to complement the country's large hydroelectric generation plants. In recent years, there has been substantial development of new renewable resources such as wind, biomass, solar, and mini-hydro (included in the category of *Energia Renovable no Convencionales* (ERNC) in the Chilean regulatory system), and these sources now make up a rapidly growing share of total installed generating capacity.[10]

[9] Norton Rose Fulbright. (2017). *Renewable energy in Latin America.*

[10] Comisión Nacional de Energía (2018), *Anuario Estadístico de Energía.* Retrieved from https://www.cne.cl/wp-content/uploads/2019/04/Anuario-CNE-2018.pdf.

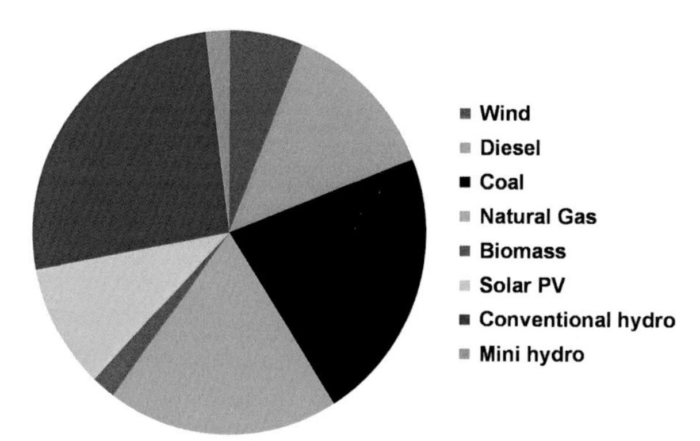

Fig. 13.3 Installed capacity in the SEN (2019). Data from CNE.

Chile has substantial wind resources, especially in the southern part of the country, and one of the best solar resources in the world in the northern Atacama Desert region. This region is also home to much of Chile's mining industry and industrial electric load. Several major mining groups have built substantial solar PV capacity at their operations.

The Chilean electricity market

Chile was one of the first countries in the world to restructure its electricity industry toward a competitive model for the generation sector, starting as far back as 1982.[11] Chile, which operates a generation pool with a centralized dispatch and system operator, has seen the basic principles of its market design emulated by several other Latin American countries.

Chile has both spot and forward contract markets for electricity. Unlike the UK, US, and many other countries, the Chilean spot market is dispatched against regulated variable costs in strict merit order, which matches supply and demand. The Chilean spot market also includes a capacity component, an annual payment that is allocated proportionally to the actual capacity of each plant to cover the peak demand. The sum of marginal energy cost (based on the most expensive unit dispatched) and the capacity charge equals the marginal cost of generation in the spot market.[12]

There are also forward contract markets, in which generators sell power forward to distributors (who serve regulated customers) and large customers that are free to

[11] Rudnick, H. & Zolezzi, J. M. (2001). Electric sector deregulation and restructuring in Latin America: Lessons to be learnt and possible ways forward. *Generation, Transmission and Distribution, IEE Proceedings, 148*, 180–184. 10.1049/ip-gtd:20010230.

[12] Bustos-Salvagno, J., & Fernando Fuentes, H. (2017). *Electricity interconnection in Chile: Prices versus costs*.

purchase in the open wholesale market. Large customers and distributors are required to purchase supply to meet their demand from generators.

Electricity prices in Chile have historically been relatively high at both the household and industrial level.[13] In 2005, under Law 20,018, the Chilean government instituted a tender mechanism for long-term supply.[14] The tender process is run by the *Comisión Nacional de Enérgia* (CNE), which coordinates future demand projections with the distribution companies. The CNE modified the tender process in 2015 to create tenders for different blocks of products (e.g., year-round, quarterly, and day/night differentiated contracts) and to increase the length of the resulting PPA contracts up to 20 years.

From a financing perspective, these contracts have several critical features. To allow time to site and construct the required generation facilities, the tenders are organized at least five years before the start of the supply contract so that new entrants can bid. New project sponsors may also postpone or cancel their projects if these are delayed by factors outside their control. However, in early 2017, CNE introduced new rules to increase penalties for project cancellations to ensure delivery of needed resources.[15]

Renewable support policies in Chile

In 2008, Chile established a target to generate at least 10% of electricity from non-conventional renewable sources (ERNC in Spanish) by 2024. In 2013, the country increased the target to 20% of generation by 2025. In practice, qualified ERNC generation greatly exceeded these obligations over the last few years, as shown in Fig. 13.4.

Investment in new renewables projects has continued despite falling prices for electricity in the tenders run by CNE in recent years, as shown in Fig. 13.5. Partially due to the rapidly falling costs of renewables (especially PV solar) in Chile, contract prices have decreased substantially in recent years, which should, over time, lower retail prices for Chilean customers.

Over 50% of the new generating projects built over the period 2015 to 2018 were renewable projects, according to CNE. Another 1,500 MW of new renewable generation is expected from 2018 to 2020.[16]

Why has Chile succeeded?

Chile's relative success in developing and financing new renewable generation as compared to many of its peers is due to several factors that emphasize long-term investment and stability. First, the Chilean government established a privatization and

[13] *Energy policies beyond IEA countries: Chile (2018)*. S.l.: Organisation for Economic Co-operation and Development.

[14] Norton Rose Fulbright. (2017). *Renewable energy in Latin America.*

[15] See footnote 13.

[16] See footnote 10.

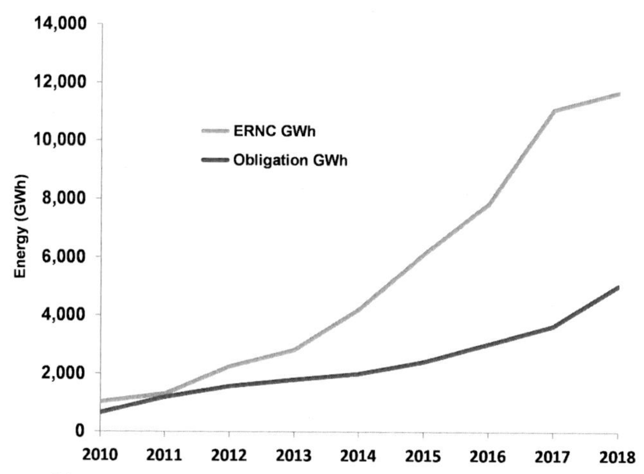

Fig. 13.4 Renewable generation versus obligated quantity.
Source: Data from CNE.

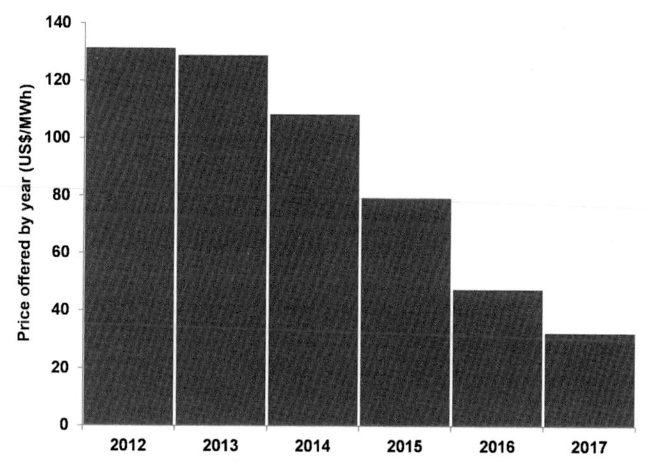

Fig. 13.5 Contract prices offered in auctions by year (US$/MWh).
Source: Data from CNE.

market-based reform program that boasted a privately-owned utility sector, a feature that attracted significant private foreign investment. Second, as discussed earlier, the Chilean electricity market structure has generally been stable and effective enough that other Latin American countries adopted it as a model. Notably, the primary PPA counterparties (local distribution companies) for new projects have creditworthy profiles. Third, the Chilean tender system for new PPAs has generally supported new renewable construction, with long-term indexed contracts and attractive pricing levels (compared to the costs of new entry for solar or wind projects). Finally, as mentioned previously, Chile's geography is well suited to renewable generation; much of the new solar development has been in the Atacama Desert in northern Chile, a region with a world-class solar resource that can be exploited to meet the substantial energy demands from the mining operations located there.

Chile has also attracted substantial capital investment in the energy sector for new projects. According to the OECD, this included almost $7 billion in investment up to early 2016, with almost half of that directed to the solar PV sector. This investment was split 52% debt and 48% equity, with about two-thirds of the equity coming from international investors. Chilean lenders were principal providers of private debt, with some public debt finance coming from the German development bank KfW and other agencies.[17]

The recent wave of solar projects in Chile illustrates how new PV projects are financed in the country with domestic and international capital. The 143 MW Sonnedix project raised $99 million in a non-recourse project financing, with a group of Chilean lenders acting as lead arranger and senior lender.[18] From an international investor perspective, Spanish developer Solarpack raised $90 million (out of a total cost of $114 million) for a 123 MW Atacama PV project, which has 19-year PPAs with a set of regional distribution companies for the period 2021 to 2040, with remaining output sold on the spot market. KfW-IPEX, a German export finance lender, provided the loan financing.[19]

While the renewables targets set out in 2005 (and strengthened in 2013) were an important starting point, the PPA tendering process has been the primary means for securing new capacity – much of it well beyond the ERNC requirements imposed on distributors and large customers. Chile once had relatively high historical wholesale and retail prices, but with rapidly falling costs and a favorable financing climate, new renewable projects have recently undercut conventional resources in the competition to secure new long-term contracts.

India

India has ambitious goals for its renewable energy sector. The government of India announced plans to install 175 GW (175,000 MW) of renewable energy projects by 2022 and 275 GW by 2027.[20] If successful, this would imply that

[17] García, J. (2016). *Renewable energy financing: the case of Chile*. Working document prepared for the Research Collaborative on Tracking Private Climate Finance. Available at: www.oecd.org/env/researchcollaborative.

[18] *Sonnedix expands Chile presence with financial close on USD 99 million project financing for 171MWp solar PV plant in the Atacama desert on 29 May, 2019*. Retrieved from https://sonnedix.com/news/sonnedix-expands-chile-presence-with-financial-close-on-usd-99-million-project-financing-for-171mwp-solar-pv-plant-in-the-atacama-desert/.

[19] See footnote 18.

[20] Prayas Energy Group. (October 2016). *India's journey towards 175 GW renewables by 2022*. Available at http://www.indiaenvironmentportal.org.in/files/file/Indias%20Journey%20towards%20renewable%20energy.pdf.

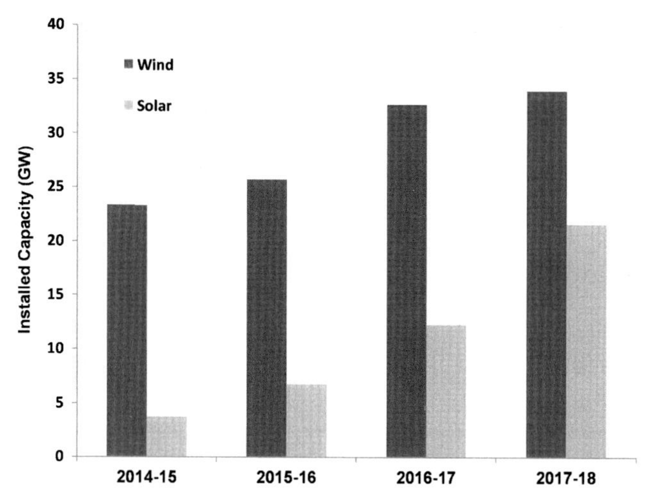

Fig. 13.6 Wind and solar capacity in India.
Source: Data from CEA.

almost a quarter of generation would come from new renewable resources by 2027 – an extraordinary transition for a country somewhat lacking in domestic energy resources, and long dominated by inefficient coal generation. In comparison, the total installed capacity in the country in 2019 was approximately 356 GW, of which almost half was coal-fired.[21] Despite the potential for massive economic growth, meeting energy demand in the coming decades while lowering emissions poses a major policy challenge for India due to the country's low per capita consumption and significant populations still without conventional grid-connected electricity.

Until 2016, new capacity additions in India were dominated by conventional thermal generation, primarily coal.[22] Since then, renewable generation has dominated with fairly rapid growth in wind and solar capacity, as shown in Fig. 13.6. Projected additions by 2022 are expected to be mostly solar and wind.

Most renewable additions in India in recent years have been utility-scale solar, as shown in Fig. 13.7.[23] While the wind resources are abundant in some regions of India, wind farms require substantial quantities of consolidated land that has sometimes proven difficult to assemble in the country.

India has relatively weak transmission and distribution infrastructure, so rooftop and off-grid solar, in theory, might be attractive. As discussed below, however, customer credit concerns have drastically slowed uptake of these solar options.

[21] *Policies and publications*. Retrieved from https://powermin.nic.in/en/content/power-sector-glance-all-india.

[22] Government of India. (2018). *Annual Report 2017–18*. Ministry of Power, Government of India. Retrieved from http://www.cea.nic.in/reports/annual/annualreports/annual_report-2018.pdf.

[23] Bridge to India. (2019). *India RE outlook 2019*. Bridge to India Energy Private Limited.

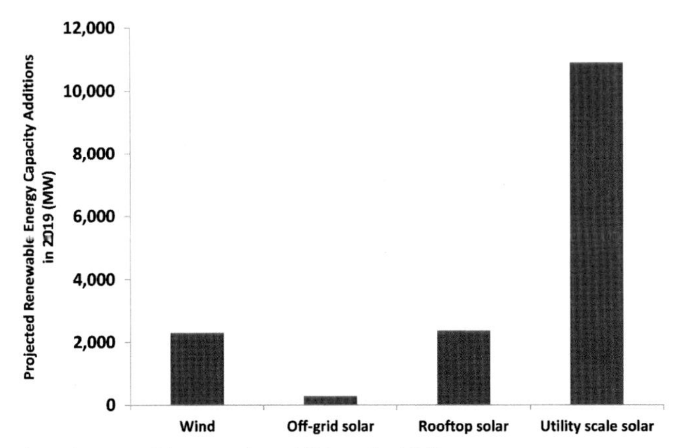

Fig. 13.7 Projected renewable capacity additions in 2019.
Data from Bridge to India.

Renewable policy and support mechanisms in India

India has renewable support mechanisms active at both the national and individual state levels. At the national level, this includes a complex mix of direct production subsidies, accelerated depreciation and tax breaks, and other measures.

State governments control much of the Indian power sector and many distribution companies ("DISCOMs") are state controlled. State DISCOMs for some years had renewable obligations and traded REC-type systems, but these were unsuccessful. They have since been largely replaced with tender mechanisms, in which developers bid to supply renewable energy under long-term PPAs to state or national offtake counterparties.

Two national government enterprises, the National Thermal Power Corporation (NTPC) and the Solar Energy Corporation of India (SECI) are the major national level off-takers for new renewable projects. SECI, under the Ministry of New and Renewable Energy (MNRE), is charged with implementing the National Solar Mission for developing solar power in the country.

SECI and NTPC have substantially better credit profiles than most or all of the state DISCOMs. As shown in Fig. 13.8 below, SECI and NTPC have been the designated off-takers for much of the new utility-scale solar projects underway in India.[24] Much of the remaining capacity, developed with offtake arrangements with state-level DISCOMs, is concentrated in a small number of states – usually those with better credit profiles.

Recent SECI tenders highlight the key aspects of the PPA and tender which affect the risks of project development.[25] For example, the tender requires that

[24] See footnote 23.

[25] Solar Energy Corporation of India Limited. (June 28, 2019). Request for selection (RfS) document for selection of solar power developers for setting up of 1200 MW ISTS-connected solar power projects in India under global competitive bidding (ISTS-V). Available from www.seci.co.in.

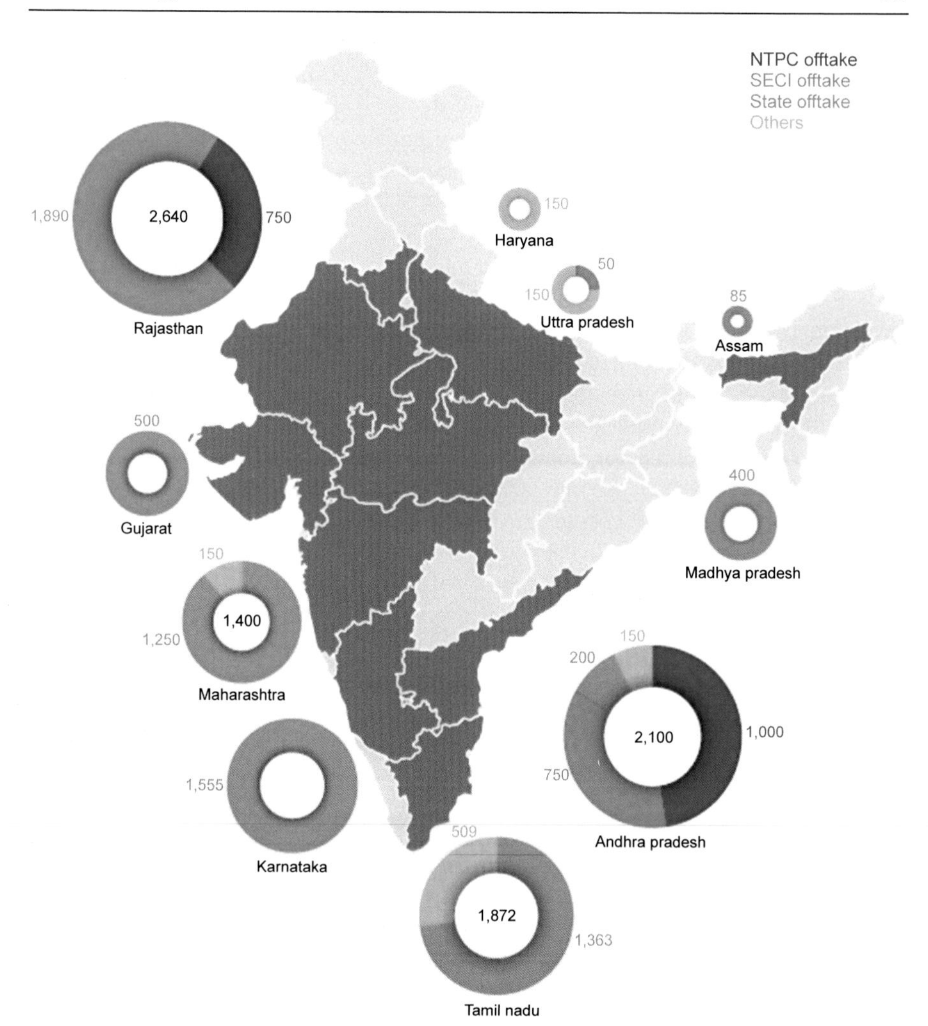

Fig. 13.8 Projected 2019 utility-scale solar capacity additions by state and offtake entity. *Source:* Bridge to India. Used by permission.

the PV technology be commercially available and operational to minimize completion risks, and that the developer is responsible for building any facilities necessary to connect the project to the grid. The PPA is for a period of 25 years, with a maximum tariff rate provided for in the tender, a maximum capacity of 300 MW, and a minimum capacity of 50 MW. Projects can be located anywhere in India, but must be integrated into the national grid system. Transmission risks on sellers are minimized as sales are made at the interconnection and delivery point of the project.

Financing challenges for India's renewable energy plans

To meet India's short-term targets for renewable energy will require vast amounts of new capital. For example, the Climate Policy Initiative (CPI) estimated in late 2016 that meeting the target of 175 GW of new renewable generation by 2022 would require $189 billion in additional investment, including $57 billion in equity and $132 billion in debt.[26]

The CPI analysis identified a number of major barriers to institutional investment in the Indian renewables sector for both foreign and domestic institutional investors. For foreign investors, the greatest risks were offtake entity credit and payment risks, lack of transmission infrastructure, currency risks, regulatory and policy risks, and unfavorable return expectations. For domestic institutional investors, the largest single risk factor was unfamiliarity with the sector.[27]

Much of the offtake risk for renewables in India is associated with the financial position of the DISCOMs, whose rates are set by state governments and which, in many cases, do not cover the full cost of providing electricity, especially given the level of subsidies common in India in the agricultural and residential sectors. For example, for fiscal year 2014–15, the average revenue to DISCOMs (including subsidies from governments) was 4.62 rupees/kWh, while the average cost of service for these companies was estimated to be 5.20 rupees/kWh.[28]

DISCOM operational and financial performance is evaluated periodically by rating agencies and summarized by the Ministry of Power. In July 2018, only 7 of the 41 ranked state power distribution utilities were graded as having "high" or "very high" operational financial performance capabilities, with the top performing utilities concentrated in a few Indian states.[29] There have been several attempts to reform the finances of the distribution sector over the years, including the UDAY program of 2015, in which state governments were to take over some debts of DISCOMs. However, many DISCOMs clearly remain in a non-creditworthy condition, with attendant risks for renewable investors.

The current move toward tenders for renewable capacity, backed by PPAs with SECI and NTPC as counterparties, has addressed some of these risks. SECI and NTPC are perceived to have stronger counterparty credit risk profiles and payment histories than many state DISCOMs, and the 25-year PPAs minimize recontracting and price risks. Since sales are made at the point of interconnection, the offtaker bares

[26] Sen, V., Sharma, K., & Shrimali, G. (2016). *Reaching India's renewable energy targets: The role of Institutional Investors*. Climate Policy Initiative. Retrieved from https://climatepolicyinitiative.org/publication/reaching-indias-renewable-energy-targets-role-institutional-investors/.

[27] See footnote 26.

[28] Shakri Sustainable Energy Foundation (2018). *Analysis of financial health of DISCOMs and its link with end-use efficiency implementation*. Shakti Sustainable Energy Foundation. Retrieved from https://shaktifoundation.in/wp-content/uploads/2018/10/Analysis-of-Financial-Health-of-DISCOMs-16Oct18.pdf.

[29] Ministry of Power, Government of India (2018). *State distribution utilities sixth annual integrated rating*. Ministry of Power, Government of India. Retrieved from http://pfcindia.com/DocumentRepository/ckfinder/files/GoI_Initiatives/Annual_Integrated_Ratings_of_State_DISCOMs/6th_rating_booklet.pdf.

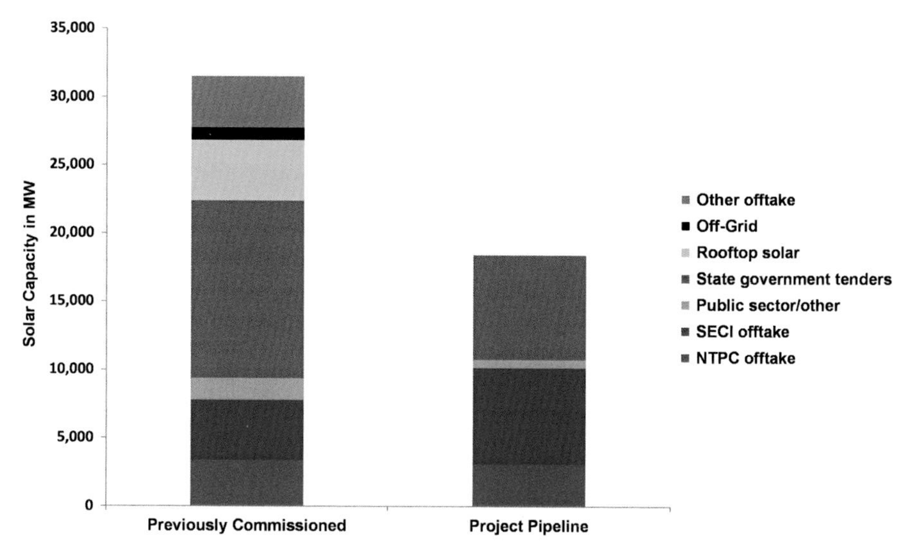

Fig. 13.9 Commissioned and project pipeline solar capacity as of March 2019. *Source:* Data from Bridge to India.

most transmission risks, although there can be some sharing of curtailment risks in the PPAs.

It is therefore unsurprising that many new projects in the development pipeline, as shown in Fig. 13.9, are expected to have SECI and NTPC as the preferred PPA counterparties.[30] However, the total credit capacity of these state enterprises is limited; more fundamental reform is needed for the PPA counterparty credit risk to be addressed systematically.

Recent developments have highlighted both potentially positive and negative forces with respect to DISCOM counterparty credit risk. In June 2019, the central Ministry of Power issued an order requiring DISCOMs to maintain adequate LCs to support their new obligations under PPAs.[31] According to India Ratings and Research (Ind-Ra), an affiliate of Fitch, this could require DISCOMs to create approximately 300 billion rupees worth of LCs, and still would not resolve all past delayed payments. However, the potential for central government to restrict dispatch of power to DISCOMs not meeting the LC requirement could strengthen DISCOM compliance.[32]

On the negative side, there is also pressure from some DISCOMs to review and renegotiate already signed PPAs. Andhra Pradesh, a leading state for renewables in

[30] Data from Bridge to India. (2019). *India solar compass 2019: Q1.*

[31] Ministry of Power, Government of India, Press Release dated June 28, 2019.

[32] Market wire. (July 2, 2019). *Market wire: Enforcing payment security mechanism - A bold move, challenging to implement, benefits uncertain.* Retrieved from https://www.indiaratings.co.in/PressRelease?pressReleaseID=37476&title=Market-Wire:-Enforcing-Payment-Security-Mechanism—A-Bold Move, Challenging To Implement, Benefits Uncertain

India, has recently appointed a committee to try and renegotiate executed PPAs, with consequent potential risks for these renewable investors.[33] Another negative for the sector is that projects which were awarded PPAs under the tender system but did not complete them under tender requirements have been substantially delayed, canceled, or hit with fines.

Until 2016, wind projects in India could rely on a FIT that was set by a central government agency. However, the government decided to instead institute an auction mechanism to allot the PPAs. The regulatory uncertainty has not only stalled the construction of new wind projects, but also created financial distress for local wind turbine manufacturer Suzlon – again, an outcome opposite of what the central government is trying to achieve through its domestic content policies.

The Indian auction mechanism has its own perils. The implementation of the National Solar Mission was hobbled by delays in reaching commercial operation for the projects that received allocation of the PPAs due to lack of technological expertise, delays in land acquisition, and difficulties in financial closing. However, as a result of the delays in commercial operation, NTPC canceled 14 of the PPAs granted and triggered the guarantees. The parties approached the courts to settle the disputes. The situation has not necessarily improved, despite the growth of new solar projects in India: some projects were canceled and others hit with fines for not completing projects under tender requirements. Most of the recent delays were due to challenges with land acquisition.

India represents one of the brightest spots in terms of the potential for harnessing renewable energy resources. However, until certain thorny issues such as credit quality of DISCOMs, regulatory uncertainty, political risks, and antiquated laws for land acquisition are resolved, widespread deployment of renewable energy may be significantly impeded.

Germany

Germany's *Energiewende* ("Energy Transition") refers to a long-term shift from coal, nuclear, and other conventional power sources to renewable-generated electricity (primarily wind, solar, and biomass). The Renewable Energy Act (EEG in its German acronym), first introduced in 2000, has been Germany's main legislative tool to develop renewable energy.

In 2002, a law was adopted to phase out nuclear energy by approximately 2022. Going beyond individual targets, the German government in 2010 introduced an overarching energy concept to provide policy pathways for the country's energy future. It created ambitious targets for renewable energy, efficiency, and carbon emissions reductions. In the process, the German federal government also decided

[33] India Ratings & Research. (July 4, 2019). *Renegotiation of power sale contracts in Andhra Pradesh exerts added cashflow pressure and impairs investor confidence.* Retrieved from https://www.indiaratings.co.in/PressRelease?pressReleaseID=37507&title=Renegotiation-of-Power-Sale-Contracts-in-Andhra-Pradesh-Exerts-Added-Cashflow-Pressure-and-Impairs-Investor-Confidence.

to delay the nuclear phase-out until 2032. This was upended by the Fukushima Daiichi nuclear accident in Japan in March 2011; Germany then decided to readopt its 2022 nuclear closure plan, starting with the immediate closure of the eight oldest nuclear plants.

German renewable policy

Germany has a long history of government policies designed to promote renewable energy. The 1990 Electricity Feed Law (StrEG) was important for the development of the renewable sector in the 1990s, especially for wind.[34] Public utilities were obligated to buy renewable-generated electricity (somewhat similar to PURPA in the United States) on a yearly fixed-price basis. Wind producers received 90% of the average retail electricity rate, and other renewables received 65–80% of the average retail rate, depending on plant size. This subsidy mechanism created rapid growth in renewables output.

In 2000, the EEG came into effect, which created the FIT system. Under the FIT program, a long-term (generally 20 years) fixed price was set for every kWh produced from the project. These prices were generally above market prices and FIT costs were recovered from surcharges on energy customers. Certain large industrial customers were exempted from some or all EEG costs and other energy related levies, implying a higher rate on other customers.

The FIT system was designed to provide certainty to investors by effectively guaranteeing fixed prices and priority transmission access. Different prices were paid to different types of renewable energy projects to stimulate new technologies coming to market. Moreover, remuneration levels for wind power projects were adjusted by quality of the wind resource, with low-wind sites receiving higher payments. Thus, installations were spread more equally around the country and windfall profits at high-wind sites were reduced.

The EEG over the years has been changed multiple times. FIT prices have been changed to reflect falling costs but other more structural changes have also occurred. The 2012 amendments to the EEG reduced FIT rates paid for some types of renewable generation, increased degression rates (an automatic reduction over time in FIT rates conditional on installation levels), and encouraged the use of a FIP mechanism to incentivize renewable projects to sell their energy through spot markets. Under the (sliding) FIP system, the subsidy payment made to the renewable generator was set as the difference between the FIT price and a reference price, defined as the difference of the average of wholesale spot prices in the previous month and a defined "management premium".[35] This adjustment of the payment depending on power price levels erased most power price exposure and continued to provide project developers with

[34] Runci, P. J. (2005). *Renewable Energy Policy in Germany: An Overview and Assessment.* Joint Global Change Research Institute. Retrieved from http://www.globalchange.umd.edu/data/publications/PNWD-3526.pdf.

[35] Deutsche Bank Group. (2012). The German feed -in tariff: Recent policy changes. Deutsche Bank Group. Retrieved from https://www.db.com/cr/en/docs/German_FIT_Update_2012.pdf.

high levels of revenue certainty. However, it also rendered them responsible for balancing costs and, thus, incentivizing good output forecasts.[36,37]

The year 2014 brought further changes to German renewable policy, making the FIP system mandatory for large projects (while eliminating the "management premium"), requiring direct marketing of the electricity produced by projects. The 2014 amendments to the EEG also set up a system of tenders for the future.

The 2017 EEG rules set up a reverse auction tendering system for most renewable energy projects, with defined quantities to be acquired. This marked the change from administratively-set remuneration levels to levels identified in the tenders. However, the basic remuneration mechanism, the sliding FIP system, remained in place. Tenders were to be technology specific, evaluated only on price, with the bidder to be paid its bid market premium if its bid was accepted. The 2017 EEG rules also opened some portion of the tenders to cross-border EU participants.[38]

Development of renewables in Germany

The implementation of the EEG rules created a rapid increase in generation of electricity from renewable resources such as wind and solar, as shown in Fig. 13.10.

Costs of wind and solar installation in Germany, as in other countries, have declined significantly over the last two decades. These declining costs have also been reflected in the large drop in the remuneration rates paid to solar PV generation over the decade from 2007 to 2017, as shown in Fig. 13.11. These have dropped by approximately 85% since 2000.

Data from more recent onshore wind and solar PV procurement auctions under the EEG 2017 amendments, as shown in Fig. 13.12, show recent accepted bid prices have leveled out, with solar prices lower than wind on a cents per kWh basis. Some of the initial auction results were probably impacted by peculiarities in the auction design. Somewhat higher later auction prices are likely due to difficulties in securing permission or new projects, which has lowered competition, rather than rising underlying costs.

Some in Germany have called for additional measures to meet more stringent renewables targets for the country, beyond the current auction framework. New wind installations have slowed, at least partially due to issues with construction permits and local opposition to new wind farms. This has led to lower participation in wind tenders.[39]

[36] May, N. G. (2017). The impact of wind power support schemes on technology choices. *Energy Economics, 65*, 343–354. http://doi.org/10.2139/ssrn.2616715.

[37] Klobasa, M., Winkler, J., Sensfuß, F., & Ragwitz, M. (2013). Market integration of renewable electricity generation — The german market premium model. *Energy and Environment, 24*(1–2), 127–146. http://doi.org/10.1260/0958-305x.24.1-2.127.

[38] BMWi. *2017 German Renewable Energy Law (EEG 2017) and cross-border renewable energy tenders.* Available from www.bmwi.de.

[39] Wettengel, J. (February 15, 2019). *German grid agency worries about low participation in wind tenders.* Retrieved from https://www.cleanenergywire.org/news/german-grid-agency-worries-about-low-participation-wind-tenders.

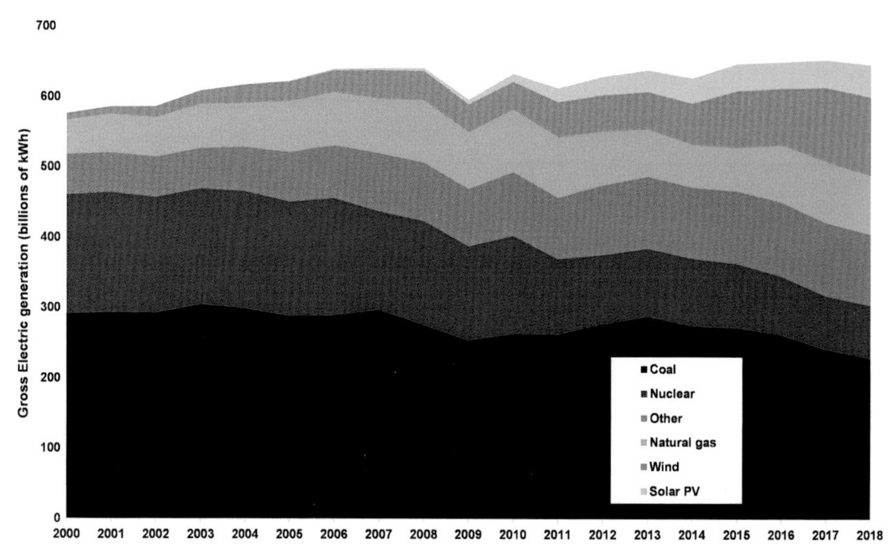

Fig. 13.10 German electricity production by source (2000–18).
Data from Statisches Bundesant and BDEW.

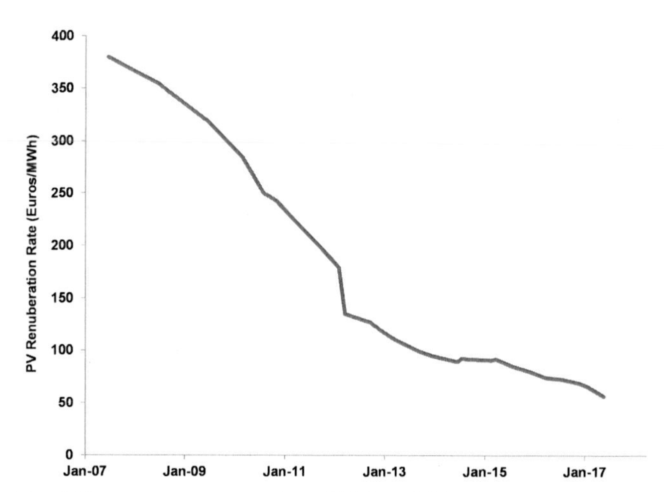

Fig. 13.11 Solar PV remuneration rates from 2007 to 2017.
Data from DIW Berlin.

Effectiveness of German policies

The EEG and Germany's other renewable policies have been effective in growing renewables and the FIT and FIP programs have allowed substantial private investment to take place, by transferring all or most price risks for many years onto consumers through cost recovery by TSOs under the FIT program and later by the regulator

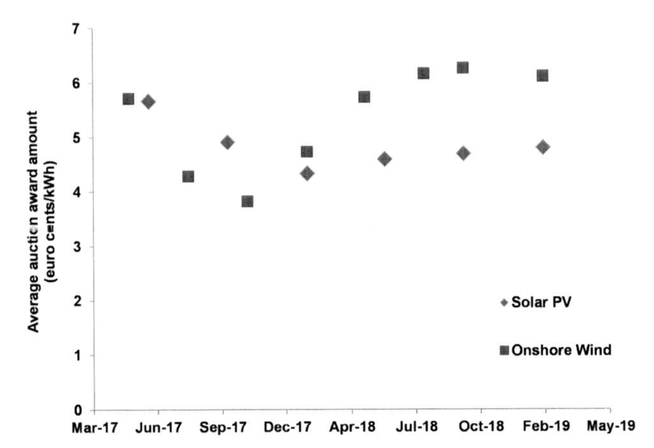

Fig. 13.12 Wind and solar average auction award prices.
Data from Bundenetzagentur.

through electricity levies. German renewables investment has also benefitted from a strong domestic legal framework for investment, preferential transmission access policies, and substantial availability of domestic capital.

Despite Germany's achievements in growing wind and solar power, the cost of these policies for consumers has been controversial in some quarters, due to the magnitude of the associated electricity price surcharges required to support renewable energy payments. The EEG surcharge compensates for the difference between the payments made to renewable suppliers (originally by TSOs and later through levies) and the sales revenues received, plus some other items.[40] In 2018 almost half of industrial consumption was exempted from the EEG surcharge, which increased costs on residential consumers.

In 2018, renewables surcharges were expected to cost consumers over 25 billion Euros.[41] Renewable surcharges on a per kWh basis have started to decline in recent years, as shown in Fig. 13.13.[42]

Germany has the highest cost for renewable energy support per kWh of gross electric generation of any European Union country, as shown in Fig. 13.14.[43] While German costs for wind and solar power projects are relatively low in comparison to other European countries, especially when accounting for Germany's relatively poor

[40] Fraunhofer ISE (2019). *Recent facts about photovoltaics in Germany.* Fraunhofer ISE. Retrieved from https://www.ise.fraunhofer.de/content/dam/ise/en/documents/publications/studies/recent-facts-about-photovoltaics-in-germany.pdf.

[41] Appunn, K. (October 15, 2018). *Renewables surcharge set to fall by six percent in 2019.* Retrieved from https://www.cleanenergywire.org/news/renewables-support-set-fall-six-percent-2019.

[42] Data available from https://www.netztransparenz.de/EEG/EEG-Umlagen-Uebersicht.

[43] CEER (2018). *Status review of renewable support schemes in Europe for 2016 and 2017.* Council of European Energy Regulators. Retrieved from https://www.ceer.eu/documents/104400/-/-/80ff3127-8328-52c3-4d01-0acbdb2d3bed.

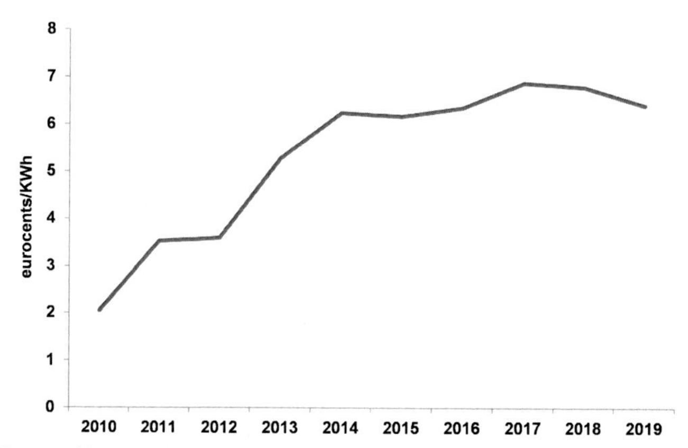

Fig. 13.13 Renewables surcharge in Germany – 2010 to 2019.
Data from Netztransparenz.de.

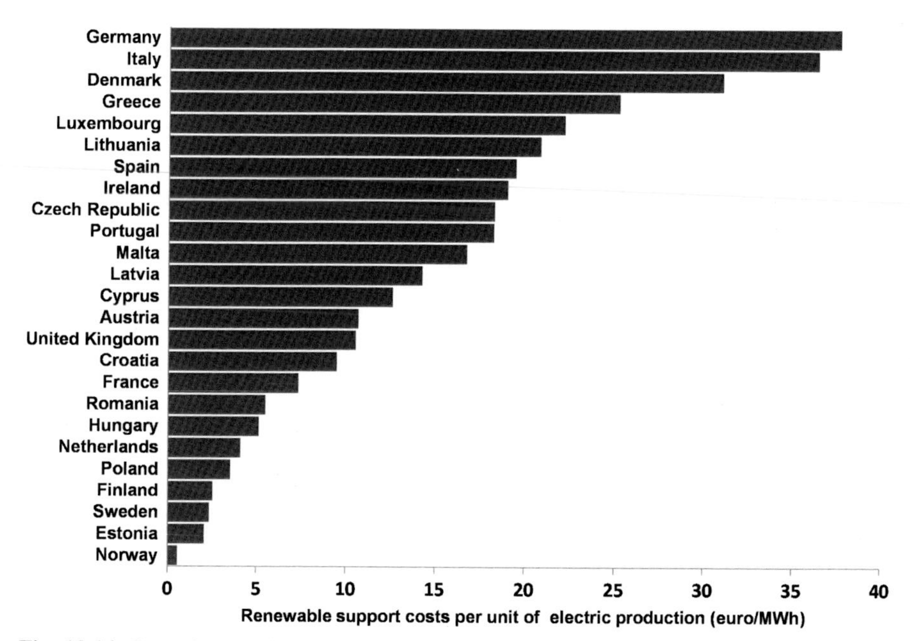

Fig. 13.14 Cost of renewable support payments per unit of gross electric generation in 2016 (euro/MWh).
Data from CEER.

wind and solar resources, Germany particularly pushed early solar power deployment. These early solar developments had considerably higher costs than wind power, or current renewable energy projects.

Some critics have claimed that high power costs in Germany, which are at least partially attributable to renewable support mechanisms, have decreased the competitiveness of German businesses, especially for smaller firms.[44]

German power sector policies have also been criticized by some as not being very effective at actually lowering carbon emissions. While the large scale of investment in new wind and solar in Germany has certainly made an impact, German carbon emissions have remained stubbornly high, due to continuing high emissions from coal-fired generation and the transportation sector.[45] Economic analysis suggests that the avoided costs per ton of carbon emissions from German renewables are very high. A 2014 study by Marcatonini and Ellerman, for example, found that for 2006–10 German wind subsidies translated to an implicit carbon price of €57 per ton of CO_2 while solar PV subsidies translated to an implicit carbon price of €532/ton CO_2.[46] The latter in particular is well above any expected carbon price, suggesting solar PV in this period was not likely to have been a very cost-effective policy for avoiding carbon emissions. A more recent study (using data from 2010 to 2015) suggests implied abatement costs of €105–270/ton CO_2 from wind subsidies in Germany and much higher implied abatement costs for solar.[47]

Despite the cost impacts to consumers, growth of renewable energy and the *Energiewende* retain broad support in Germany.[48] Beyond the high level of environmental and energy consciousness common in Germany, some of the support for the *Energiewende* may come from the community investment model typical in the German renewables sector. More than half of German energy projects have been built using funding from energy cooperatives, which typically have a "one member, one vote" governance and often invest in projects in the local community.[49] Unlike in many countries, where large renewable projects have often been built by large corporate developers, the German model has in general emphasized smaller, local projects. The community investment model has likely been a major driver of public acceptance for renewable

[44] Wilkes, W., & Parkin, B. (September 24, 2018). *Germany's economic backbone suffers from soaring power prices - BNN Bloomberg*. Retrieved from https://www.bnnbloomberg.ca/germany-s-economic-backbone-suffers-from-soaring-power-prices-1.1141809.

[45] Hockenos, P. (December 13, 2018). *Carbon crossroads: Can Germany revive its stalled energy transition?* Retrieved from https://e360.yale.edu/features/carbon-crossroads-can-germany-revive-its-stalled-energy-transition.

[46] Marcantonini, C., & Ellerman, A. D. (2016). The implicit carbon price of renewable energy incentives in Germany. *The Energy Journal, 37*(1). http://doi.org/10.5547/01956574.37.1.cmar.

[47] Abrell, J., Kosch, M., & Rausch, S. (2019). Carbon abatement with renewables: Evaluating wind and solar subsidies in Germany and Spain. *SSRN Electronic Journal*. http://doi.org/10.2139/ssrn.3313637.

[48] Amelang, S., Wehrmann, B., & Wettengel, J. (February 14, 2019). *Polls reveal citizens' support for Energiewende*. Retrieved from https://www.cleanenergywire.org/factsheets/polls-reveal-citizens-support-energiewende.

[49] Wettengel, J. (October 25, 2018). *Citizens' participation in the Energiewende*. Retrieved from https://www.cleanenergywire.org/factsheets/citizens-participation-energiewende.

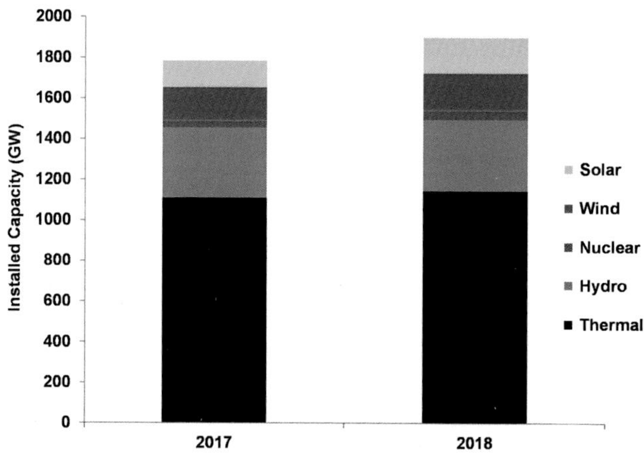

Fig. 13.15 Installed capacity by type in China – 2017 and 2018.
Data from CEC.

energy, despite the likely higher project and transaction costs associated with smaller project sizes.

China

China has experienced immense growth in electricity demand and generating capacity over the past few decades. China is now the largest generator of electricity in the world by a substantial margin. In the past, coal-fired generation has dominated, and much of economic transformation of China has been powered by coal. In recent years, as shown in Fig. 13.15, the share of hydropower and other renewables (primarily wind and solar) have increased rapidly.[50]

China continues to plan for rapid growth in renewable energy and slowing coal generation to meet its targets under the Paris Accord. China has also seen alternative sources of energy as an attractive means of meeting other environmental goals, such as reductions in particulate pollution which pose major health issues in many Chinese population centers.

Renewable support mechanisms in China

China has traditionally relied on a mixture of renewable energy support mechanisms to support its policy goals. A FIT for wind power was introduced in China in 2003, which required regional power grid operators to purchase energy from registered wind energy suppliers at prices fixed by the National Development and Reform Commission

[50] Data from China Energy Council. https://chinaenergyportal.org/en/2018-electricity-other-energy-statistics/.

(NRDC). The FIT prices were reformed on a regional basis in 2009, and a solar FIT program was introduced in 2011.[51] China has also used RPS type mechanisms.[52]

In addition to these mechanisms, China has also relied substantially on direct subsidies to support renewable energy, especially in the solar sector. For example, under the "Golden Sun" program, the government would subsidize a substantial portion of the costs of solar PV projects. By 2015, total subsidies and other support mechanisms were estimated to cost $12 billion per year.

Financing mechanisms for Chinese renewables

The Chinese experience stands in contrast to the previous three country examples of renewable energy finance. The rapid growth in Chinese renewable energy capacity has been supported by the availability of low-cost financing from state-owned banks. Large developers and wind and solar companies have been able to access billions of dollars of credit from Chinese state banks at relatively low-cost rates to investment in renewable projects.[53] Relatively few installations are project financed.

Renewables integration challenges

While China has invested billions into the renewable energy sector, the operational outcomes have often been lacking. Fig. 13.16 below shows percentage of wind generation in China that was rejected (curtailed) due to transmission constraints for the period 2010–15.[54] These curtailment rates are very high compared to most countries and given the huge scale of the Chinese wind industry reflect a large loss of total wind power output.

Chinese renewable policies have created poor locational incentives for investment and many projects have been built in areas with substantial transmission constraints. Several northern regions have continued to see high levels of wind curtailment. Zhung et al. state that in 2016 alone China saw more than 49,000 GWh of wind generation curtailed, and that solar PV generation had a curtailment rate of approximately 15%

[51] Yan, Q. Y., Zhang, Q., Yang, L., & Wang, X. (2016). Overall review of feed-in tariff and renewable portfolio standard policy: A perspective of China. *IOP Conference Series: Earth and Environmental Science* (Vol. 40) (p. 012076). http://doi.org/10.1088/1755-1315/40/1/012076.

[52] Lo, K. (2014). A critical review of China's rapidly developing renewable energy and energy efficiency policies. *Renewable and Sustainable Energy Reviews, 29*, 508–516. http://doi.org/10.1016/j.rser.2013.09.006.

[53] Bridle, R., & Kitson, L. (2014). *Public finance for renewable energy in China: Building on international experience*. International Institute for Sustainable Development. Retrieved from https://www.iisd.org/sites/default/files/publications/public_finance_renewable_energy_china.pdf.

[54] Zhang, Y., Tang, N., Niu, Y., & Du, X. (2016). Wind energy rejection in China: Current status, reasons and perspectives. *Renewable and Sustainable Energy Reviews, 66*, 322–344. http://doi.org/10.1016/j.rser.2016.08.008.

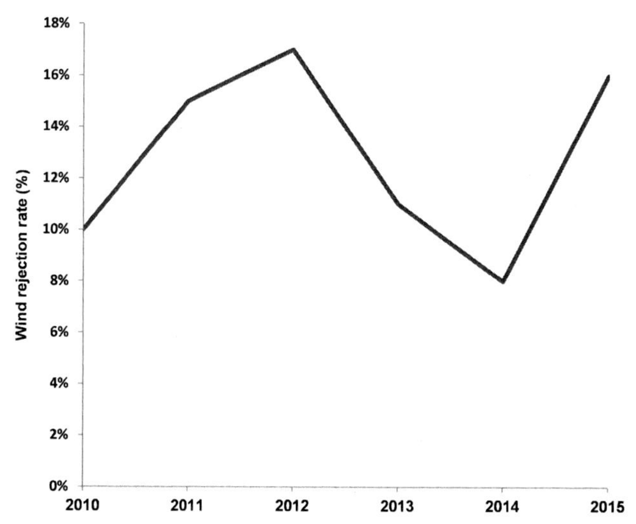

Fig. 13.16 Wind curtailment in China.
Data from Zhang et al. (2016).

over the period 2013–16.[55] As demand has slowed, mismatching renewable supply and energy demand have also become more of a problem.

The curtailment issues in China have been made even worse by the relatively poor performance of regional power grid governance and operations in economic terms. As was described in Chapter 8, absent binding transmission constraints, to maximize efficiency generators should be dispatched in "merit order", using lowest cost generation first (e.g. wind and solar) and then moving on to higher marginal cost fossil-fired generation as needed. In China, however, dispatch is often controlled by provincial agencies. These often use an "equal shares" approach of non-economic dispatch, basing total generation from a generator or class in a period to a planning target. While this non-economic dispatch conflicts with the priority dispatch of renewables required under China's 2009 Renewable Energy Law, it may allow provincial authorities to meet various local objectives with respect to generator output.

Lessons from the international experience

As discussed in Chapter 2, renewable energy projects are capital intensive and involve large sunk costs. In most cases, historically at least, renewable energy support mechanisms have been required to improve the return and risk profile of renewable energy projects. This has typically involved paying renewable generation a higher price than

[55] Zhang, S., Andrews-Speed, P., & Li, S. (2018). To what extent will Chinas ongoing electricity market reforms assist the integration of renewable energy? *Energy Policy, 114*, 165–172. http://doi.org/10.1016/j.enpol.2017.12.002.

conventional market prices and establishing means to stabilize these renewable output prices to reduce investment risks.

The four international case studies presented in this chapter, plus the experience of the United States and other countries, show that it is possible to mobilize private financial investment on a large scale – hundreds of billions of dollars per year – with appropriate policies, although the experience also shows that the details matter a great deal with respect to effectiveness.

Policy and financing costs

Different renewable energy support mechanisms create different types of risks for renewable energy sponsors and investors. Long-term fixed price PPAs and FITs fix the price for renewable project generation and hence create lower financial risks. A FIP system may create some additional risks. The US experience suggests that RPS-type mechanisms are primarily useful for stimulating utilities to sign PPAs to hedge long-term price risks, as REC prices themselves tend to be highly volatile.

The different risks to renewable energy investors can affect the cost of capital, which is a significant driver of total renewable project costs. May and Neuhoff analyzed the impacts of different renewable support mechanisms on the weighted average cost of capital (WACC) for 23 European Union countries and found that the use of a tradable green certificate of TREC system increased the WACC for renewable projects by 1.2%, which may be important given the capital intensity of most projects.[56] Sliding FIPs did not impact the cost of capital compared to fixed FIT mechanisms, but this may have been dependent on design specifics in the programs examined. These authors also found that REC and tradable green certificate systems create additional financial risks for off-takers, as the regulatory and market risks associated with such schemes often cannot be completely hedged.

Polzin, Egli, Steffen and Schmidt reached broadly similar conclusions with respect to policy choice, but emphasize that specific policy designs are critical.[57] Long-term fixed price mechanisms, such as FIT programs, have a high potential for stimulating renewable energy investment, but with the tradeoff that total policy cost may be hard to predict in advance. These authors also note that FIT programs, which are often differentiated by source (e.g., solar projects receiving a different tariff than wind projects), appear effective at introducing new technologies, but likely with higher overall costs.

The German case study provides an excellent example of a regulatory and support policy environment which has supported both substantial early investment in renewable energy and low financing costs. For example, a 2016 study suggested that German

[56] May, N. G., & Neuhoff, K. (2017). Financing power: Impacts of energy policies in changing regulatory environments. *SSRN Electronic Journal*. http://doi.org/10.2139/ssrn.3046516.

[57] Polzin, F., Egli, F., Steffen, B., & Schmidt, T. S. (2019). How do policies mobilize private finance for renewable energy? A systematic review with an investor perspective. *Applied Energy, 236*, 1249–1268. http://doi.org/10.1016/j.apenergy.2018.11.098.

renewable policies supported an 80%/20% debt/equity structure for onshore wind projects, with a WACC of approximately 3.5%–4.5%.[58] Germany also likely benefitted from well-developed domestic capital markets, a strong legal framework for support payments (through TSO charges or later imposed electricity levies), and possibly from lower return threshold requirements by community-based investors.

Chile has done a good job at attracting private capital at reasonable returns in recent years, aided by an overall favorable investment climate, bankable PPAs with creditworthy utilities, and a predictable regulatory regime.[59] India, on the other hand, has a relatively high WACC for renewables investment, but has also seen very competitive bidding for new projects and low construction costs as an offset to high financing costs.[60] Reliable estimates for the WACC for renewable energy projects in China are hard to find, especially since most funding comes from state-controlled banks with limited transparency. Renewable energy projects in China appear to pay a higher cost of capital than conventional power developments, and face additional hurdles in accessing the substantial capital needed.[61]

Renewables and market integration

As described in Chapters 8 and 9, renewable energy projects need to be integrated into electricity markets, transmission grids, and the larger structure of the power industry to be. The four countries examined in this section have taken different approaches to integrating renewable energy projects into their power systems.

Germany minimized project risk through emphasis on large FIT and FIP projects and by providing priority right of interconnection to the transmission grid for renewable projects. The zonal transmission pricing system used in Germany and other EU countries under market coupling is relatively inefficient, compared to LMP systems. However, under German TSO rules, renewable energy generators are compensated at a high-level during curtailment periods, so projects have had less financial exposure than in many other systems.[62] To combat rising congestion costs, the federal government is planning to build large transmission projects to accommodate these higher levels of renewable generation demand in the future.

As previously noted, Chile previously adopted a market-based restructuring for its electricity sector, one of the first countries in the world to do so. The resulting industry structure was stable for some years and the credit profiles of

[58] Diacore (2016). *The impact of risks in renewable energy investments and the role of smart policies.* Diacore. Retrieved from http://diacore.eu/images/files2/WP3-FinalReport/diacore-2016-impact-of-risk-in-res-investments.pdf.

[59] Bloomberg New Energy Finance. (November 27, 2018). Emerging markets outlooks 2018, *Climatescope.*

[60] See footnote 59.

[61] Hong, M., & Wang, Y. (September 26, 2018). *To supercharge Chinese renewables, fix their financing.* Retrieved from https://www.wri.org/blog/2018/03/supercharge-chinese-renewables-fix-their-financing.

[62] Joos, M., & Staffell, I. (March 7, 2018). *Short-term integration costs of variable renewable energy: Wind curtailment and balancing in Britain and Germany.* Retrieved from https://www.sciencedirect.com/science/article/pii/S1364032118300091.

the distribution utilities in Chile as counterparties aided renewables development. With the share of renewable generation expected to grow, especially from solar power in the northern Atacama region, transmission constraints became a major issue, with curtailment of 16% of renewable generation in 2017.[63] A new transmission line linking two of the major grids has reduced congestion more recently.

India has faced a more difficult path in developing the renewable energy sector, with weaker institutions in the power industry, especially the DISCOMs. The lack of creditworthiness of distribution entities limits many new projects to offtakes with SECI, NTPC and a small number of better credit profile DISCOMs. India also faces substantial challenges in coordinating regional and inter-regional transmission operations and expansion to accommodate high penetration of renewable energy, as envisioned in recent policies.

China's experience highlights the many problems of rapid growth in renewable energy supply. The scale of Chinese investment has been tremendous, but industry and market institutions are relatively weak and much renewable generation has been built in areas where the power cannot be delivered though the grid. While reform efforts are underway, the lack of coherent coordination and dispatch mechanisms has also led to the inefficient use of renewable generation in service, protecting the market shares of coal-fired generation.

[63] Rypl, N. (October 12, 2018). *Chile's new coal fleet challenged by renewables and air pollution.* Bloomberg New Energy Finance.

Appendix A: glossary of terms and energy units

Energy

The basic SI unit for energy is a Joule (J), named after the English physicist James Prescott Joule. Energy can also be expressed in terms of calories:[64]

$$1 \text{ calorie} = 4.184 \text{ J}$$

A calorie is an amount of energy required to raise the temperature of 1 g of water by 1 °C. Accordingly, the amount of energy contained in 1 J is small, such as lifting a 1 kg weight (2.2 pounds) up one meter. Therefore, most of the time, energy is expressed in units of gigajoules (GJ), which equals a billion joules.

Being an SI unit, the Joule is widely used across the world. In the US, British Thermal Unit (BTU) is a prevalent unit for measuring the energy content of fuels. One BTU of energy equals 1054.3503 J.

Power versus energy

Power is defined as the rate of doing work (in order to generate energy). The basic unit of measuring power is a Watt, named after the English scientist James Watt. One Watt (W) equals 1 J/s. Since W is a small unit, for electricity, kW and MW are commonly used.

$$1 \text{ kW} = 1{,}000 \text{ W}$$

$$1 \text{ MW} = 1{,}000 \text{ kW}$$

$$1 \text{ GW} = 1{,}000 \text{ MW}$$

Given the relationship between work and energy, the energy can be found by multiplying power by time.

$$\text{Energy} = \text{Power} \times \text{Time}$$

Example

A power plant generates 10 MW of power each hour for 5 h. Therefore, the energy produced would equal 10 MW × 5 h = 50 MWh.

[64] Please note that calorie should not be confused with Calorie. A Calorie, often referred to as Kcal to avoid confusion, equals 1000 calories.

Power plant efficiency metrics

The capacity of a power plant is defined as the maximum amount of power a plant can produce. Therefore, a 500 MW power plant is capable of producing 500 MW at full capacity.

Capacity Factor is often used to express how much a power plant generates in a year. The Capacity Factor is defined as the ratio of the amount of energy generated over a given period to the amount of energy the plant could have generated had it run at full capacity.

Example

A power plant with a 100 MW of capacity produces 350,000 MWh in a year. What is the capacity factor for the project?

Total amount of energy produced = 350,000 MWh

Total amount of energy that the project could generate at full capacity = 100 MW × 1 year × 365 days × 24 h = 100 MW × 8,760 hours[65] = 876,000 MWh

Therefore, the Capacity Factor is the ratio of the two quantities, namely

Capacity Factor = 350,000 MWh/876,000 MWh = 39.95%

The capacity factor varies for generation sources depending on the technology. Some hydro, geothermal nuclear and other plants are used to serve base load power, and are run most of the time. Therefore the capacity factor for such plants is sometimes very high – close to or higher than 90%. Wind and solar power projects depend on the availability of the resource. Therefore, the capacity factor for the two technologies is much lower – ranging from 25% to 55% for wind power projects and 12%–25% for solar power plants.

Sometimes, solar power plant efficiency is measured in terms of Energy Yield. Energy Yield is defined as the ratio of energy generated in a given period to the capacity of the power plant and is expressed as Kwh/KW or MWh/MW.

Example

A 100 MW solar power project in California can generate 219,000 MWh over a year. What is the Energy Yield for the project?

Energy Yield = 219,000 MWh/100 MW = 2,190 MWh/MW or 2,190 kWh/kW

Since energy equals power multiplied by the number of hours, the Energy Yield is essentially a measure of operating time.

[65] While the number of days in a year varies depending on the leap year, 8760 hours in a year is a widespread measure and easy to remember.

Availability Factor is another metric used to measure efficiency of a power plant. Availability Factor is defined as a ratio of the number of hours that a power plant was "available" without any outages during a given operating period over the number of hours in the same period.

Example

A developer owning a wind farm has a planned outage of 120 h for maintenance over a one-year period. The plant suffered a precautionary unscheduled outage due to the passage of a storm (from which the plant escaped without damage) of 72 h. What is the Availability Factor for the wind farm?

Total number of hours in the year = 8,760

The plant was available during the year = 8,760 − 120 − 72 = 8,568 h

Therefore, the Availability Factor for the plant would be 8,568/8,760 = 97.81%.

Appendix B: Levelized Cost of Electricity

Levelized Cost of Electricity (LCOE) is an economic measure used to compare the lifetime costs of generating electricity across various generation technologies. The lifetime costs for generation can be categorized into the following groups:

- Capital Costs: up-front costs to construct a power plant
- Operation and Maintenance (O&M) Costs: costs incurred to run a power plant. These costs can be sub-categorized into fixed and variable costs. Fixed O&M costs are incurred regardless of the plant generating electricity; they are comprised of personnel salaries, security costs, insurance, etc. Variable O&M costs are directly linked to the generation of the power project. Fuel costs for conventional plants also vary with output.
- Disposition Costs: costs typically incurred at the end of the useful life. The disposition costs for certain generation technologies, such as nuclear power plants, can be huge. In most of the instances, the disposition costs for the solar and generation projects are assumed to be zero because the scrap value of the equipment generally should cover the cost of removal.

As shown in the examples below, financing costs are internalized in the LCOE calculation. The LCOE methodology also considers various tax benefits, including depreciation that may provide a tax shield.

LCOE is a useful tool as it allows comparison of various generation technologies with different capital costs, O&M costs, useful life, etc. LCOE can be viewed from an economic perspective as an "average" electricity price that must be earned by a specific generation source to break even. LCOEs are used as a relative scale to compare various technologies rather than an absolute measure informing investment decisions. Actual system planning must also consider reliability issues (such as availability at periods of peak demand) as well as other factors.

Accordingly, LCOE is primarily used by policymakers for long-term planning, as well as devising incentive mechanisms. Developers and independent power producers may use the metric as a broad planning tool to compare the attractiveness of various generation technologies. Finally, investors are interested in LCOEs to understand long-term economic trends, especially for renewables, for which the decrease in cost has greatly improved their competitiveness.

To understand the LCOE concepts, consider a simple example of a 100 MW wind farm with the following parameters:

Total Capital Cost = \$1,400/KW
Fixed O&M Cost = \$45/KW-year
Capacity Factor = 40%
Useful Life = 30 years
Discount Rate = 6%

Using the calculations below, we can calculate the Total Capital Cost and Fixed O&M Cost as follows:

Total Capital Cost (I_0) = \$1,400/KW × 100 MW × 1,000 KW/MW = \$140 million
Fixed O&M Cost (M) = \$45/KW-year × 100 MW × 1,000 kW/MW = \$4.5 million/year

Table 13.1 LCOE estimates by Lazard.

		LCOE
Alternative Energy	Solar PV - Rooftop Residential	\$160−\$267
	Solar PV - Rooftop C&I	\$81−\$170
	Solar PV - Community	\$73−\$145
	Solar PV - Crystalline Utility Scale	\$40−\$46
	Solar PV - Thin Film Utility Scale	\$36−\$44
	Solar Thermal Tower With Storage	\$98−\$181
	Fuel Cell	\$103−\$152
	Geothermal	\$71−\$111
	Wind	\$29−\$56
Conventional	Gas Peaking	\$152−\$206
	Nuclear	\$112−\$189
	Coal	\$60−\$143
	Gas Combined Cycle	\$41−\$74

For this simplistic model, the cost structure looks as follows:
The wind farm's annual electric generation can be calculated as follows:

$$\text{Annual electricity generation } (\boldsymbol{E}) = 100\text{ MW} \times 8{,}760\text{ h/year} \times 40\% = 350{,}400\text{ MWh}$$

In order to calculate the LCOE, we need to equate the present value of the lifetime costs with the present value of the lifetime energy generation.

In other words, assuming that all capital expenditure is incurred at the beginning ($t = 0$) and the project starts generating electricity overnight, we get

$$\sum_{t=1}^{n} \frac{E_t \times \text{LCOE}}{(1+r)^t} = I_0 + \sum_{t=1}^{n} \frac{M_t + F_t}{(1+r)^t}$$

Simplifying the equation further, we get

$$\text{LCOE} = \frac{I_0 + \sum_{t=1}^{n} \frac{M_t + F_t}{(1+r)^t}}{\sum_{t=1}^{n} \frac{E_t}{(1+r)^t}}$$

Since the PV of an annuity can be calculated as

Onshore Wind - Low Wind Case

In USD 000, unless otherwise noted

Assumptions

Capacity	150	CapEx ($/kW)	1,550
Capacity Factor %	38.0%	Other Costs ($/kW)	-
Fixed O&M ($/kW-yr)	36.50	Total CapEx ($/kW)	1,550
Variable O&M	-	Total CapEx ($mm)	232.50
O&M Escalation Rate	2.25%		
Debt %	60.0%	MACRS Schedule	
Cost of Debt	8.0%	1	20.0%
Equity %	40.0%	2	32.0%
Cost of Equity	12.0%	3	19.2%
Combined Tax Rate	40.0%	4	11.5%
Economic Useful Life = Debt Term	20	5	11.5%
MACRS Depreciation (Years)	5	6	5.8%
LCOE	**60.21**		
Conversion to 000	1,000		

Pro Forma

Period	0	1	2	3	4	5	6	7	8	9	10
Capacity		150	150	150	150	150	150	150	150	150	150
Capacity Factor		38.0%	38.0%	38.0%	38.0%	38.0%	38.0%	38.0%	38.0%	38.0%	38.0%
Total Generation (MWh)		499,320	499,320	499,320	499,320	499,320	499,320	499,320	499,320	499,320	499,320
Levelized Energy Cost (LCOE)		$60.21	$60.21	$60.21	$60.21	$60.21	$60.21	$60.21	$60.21	$60.21	$60.21
Total Revenues		30,066	30,066	30,066	30,066	30,066	30,066	30,066	30,066	30,066	30,066
Operating Expense											
Fuel Cost		-	-	-	-	-	-	-	-	-	-
O&M		5,475	5,598	5,724	5,853	5,985	6,119	6,257	6,398	6,542	6,689
Total Operating Costs		5,475	5,598	5,724	5,853	5,985	6,119	6,257	6,398	6,542	6,689
EBITDA		24,591	24,468	24,342	24,213	24,081	23,946	23,809	23,668	23,524	23,377
Debt Outstanding BoP		139,500	136,452	133,159	129,604	125,764	121,616	117,137	112,300	107,075	101,433
Interest Expense		11,160	10,916	10,653	10,368	10,061	9,729	9,371	8,984	8,566	8,115
Amortization		3,048	3,292	3,556	3,840	4,147	4,479	4,837	5,224	5,642	6,094
Debt Outstanding EoP	139,500	136,452	133,159	129,604	125,764	121,616	117,137	112,300	107,075	101,433	95,339
Levelized Debt Service		14,208	14,208	14,208	14,208	14,208	14,208	14,208	14,208	14,208	14,208
EBITDA		24,591	24,468	24,342	24,213	24,081	23,946	23,809	23,668	23,524	23,377
Less: Depreciation		46,500	74,400	44,640	26,784	26,784	-	-	-	-	-
Less: Interest Expense		11,160	10,916	10,653	10,368	10,061	9,729	9,371	8,984	8,566	8,115
Taxable Income		(33,069)	(60,849)	(30,951)	(12,940)	(12,764)	14,217	14,438	14,684	14,958	15,262
Tax Liability (Benefit)		(13,228)	(24,339)	(12,380)	(5,176)	(5,106)	5,687	5,775	5,874	5,983	6,105
After-Tax Net Equity Cash Flow	(93,000)	23,610	34,599	22,514	15,180	14,978	4,051	3,825	3,586	3,332	3,064

Period	11	12	13	14	15	16	17	18	19	20
Capacity	150	150	150	150	150	150	150	150	150	150
Capacity Factor	38.0%	38.0%	38.0%	38.0%	38.0%	38.0%	38.0%	38.0%	38.0%	38.0%
Total Generation (MWh)	499,320	499,320	499,320	499,320	499,320	499,320	499,320	499,320	499,320	499,320
Levelized Energy Cost (LCOE)	$60.21	$60.21	$60.21	$60.21	$60.21	$60.21	$60.21	$60.21	$60.21	$60.21
Total Revenues	30,066	30,066	30,066	30,066	30,066	30,066	30,066	30,066	30,066	30,066
Operating Expense										
Fuel Cost	-	-	-	-	-	-	-	-	-	-
O&M	6,839	6,993	7,151	7,312	7,476	7,644	7,316	7,992	8,172	8,356
Total Operating Costs	6,839	6,993	7,151	7,312	7,476	7,644	7,316	7,992	8,172	8,356
EBITDA	23,226	23,072	22,915	22,754	22,590	22,421	22,249	22,074	21,894	21,710
Debt Outstanding BoP	95,339	88,758	81,650	73,974	65,684	56,730	47,[illegible]60	36,616	25,337	13,156
Interest Expense	7,627	7,101	6,532	5,918	5,255	4,538	3,[illegible]65	2,929	2,027	1,052
Amortization	6,581	7,108	7,676	8,290	8,954	9,670	10,[illegible]44	11,279	12,181	13,156
Debt Outstanding EoP	88,758	81,650	73,974	65,684	56,730	47,060	36,616	25,337	13,156	(0)
Levelized Debt Service	14,208	14,208	14,208	14,208	14,208	14,208	14,208	14,208	14,208	14,208
EBITDA	23,226	23,072	22,915	22,754	22,590	22,421	22,249	22,074	21,894	21,710
Less: Depreciation	-	-	-	-	-	-	-	-	-	-
Less: Interest Expense	7,627	7,101	6,532	5,918	5,255	4,538	3,735	2,929	2,027	1,052
Taxable Income	15,599	15,972	16,383	16,836	17,335	17,883	18,435	19,144	19,867	20,657
Tax Liability (Benefit)	6,240	6,389	6,553	6,735	6,934	7,153	7,3[illegible]4	7,658	7,947	8,263
After-Tax Net Equity Cash Flow	2,778	2,475	2,153	1,811	1,447	1,060	6[illegible]7	208	(261)	(761)

After-Tax IRR to Equity	**12.0%**

$$PV = \frac{C\left[1 - (1+r)^{-n}\right]}{r}$$

Integrating the present value formula into the LCOE formula gives us the following:

$$\text{LCOE} = \frac{I_0 + \dfrac{M \times \left[1 - (1+r)^{-n}\right]}{r}}{\dfrac{E \times \left[1 - (1+r)^{-n}\right]}{r}}$$

Incorporating the inputs from 100 MW wind farm example:

$$\text{LCOE} = \frac{\$140\text{ MM} + 13.76 \times \$4.5\text{ MM}}{350,400\text{ MWh} \times 13.76} = \$41.87/\text{MWh}$$

The National Renewable Energy Laboratory (NREL) maintains a LCOE calculator on its website.[66]

Each year, the investment bank Lazard publishes LCOE estimates for various generation technologies. Table 13.1 provides the results of the survey published in November 2018.[67]

The following figure provides a spreadsheet example of the LCOE calculation using the methodology and assumptions described in Lazard's report for a 100 MW onshore wind project. The spreadsheet snapshot provides the assumptions underlying the LCOE calculations as well.

For the purpose of the analysis, the useful life of the wind farm and the debt tenor is assumed to equal 20 years. The LCOE in the bold cell is calculated using the Goal Seek function in Microsoft Excel such that the equity IRR equals 12%.

[66] https://www.nrel.gov/analysis/tech-lcoe.html.

[67] The full report and the underlying assumptions can be found at https://www.lazard.com/media/450784/lazards-levelized-cost-of-energy-version-120-vfinal.pdf.

Glossary

- advance rate (Chapter 3) - the amount of an asset (in a project finance transaction) that is financed by debt. Alternatively, it is the percentage of an asset value that a lender is willing to extend as a loan.
- back-leverage (Chapter 6) - a type of project finance loan which is secured by the sponsors equity interests in the project company. The loan is serviced solely through the cash flow distributions allocable to the sponsor's equity interests.
- blocker corporation (Chapter 6) - a type of corporate legal structure used by tax exempt investors (e.g., pension funds, sovereign wealth funds) to invest in project companies that may qualify for tax credits. Blocker corporations are taxable and the use of blocker corporations enable the project companies to use tax credits. The project companies may be prohibited from claiming tax credits if the tax-exempt investors invest directly without a blocker corporation.
- capacity factor (Chapter 4) - the ratio of the actual electrical energy output to the maximum possible energy output over a specified amount of time.
- CFADS (Chapter 4) – Available cash that the project finance entity can use for payment of debt interest payments and mandatory amortization for a given payment period.
- commercial operation date (COD) (Chapter 11) - the date on which an independent engineer certifies that the renewable facility has been built to the specifications of the EPC contract and completed all required performance tests. The COD triggers the start of the operating phase of the project, which can last for 25 years or more.
- conditions precedent (Chapter 4) - a condition that must be fulfilled before a contracting party can be called upon to perform its obligation.
- contingent liability (Chapter 3) - a liability that may potentially occur, depending on the outcome of a future uncertain event.
- cross-currency swap (Chapter 5) - an agreement between two counterparties to exchange payments (usually debt payments - interest and principal) denominated in two different currencies. The contracts are highly customizable and can include floating or fixed interest rates.
- deadweight loss (Chapter 2) - the measure of the loss of total economic welfare caused by pricing production at the marginal private cost rather than the marginal social cost.
- degression rate (Chapter 13) - a planned reduction over time in FIT rates to account for falling costs.
- distributed generation (Chapter 1) - electricity generated from sources near the point of use (as opposed to centralized generation sources from power plants).
- DSCR (Chapter 4) - DSCR is defined as the ratio of Cash Flow Available for Debt Service (CFADS), to mandatory debt service, which is the sum of the interest and mandatory amortization.
- feed-in premium (FIP) (Chapter 2) - a policy mechanism whereby a renewable energy producer receives a premium to the prevailing market prices for electricity.
- feed-in tariff (FIT) (Chapter 2) - a policy mechanism whereby renewable energy producers are offered payment for electricity fed into the supply grid, that is usually higher than prevailing market prices for electricity.
- FTR (Chapter 9) - FTRs, or financial transmission rights, are financial instruments, which merely hedge congestion price differences between two locations. An FTR is defined by a source node and a sink node, which then determines the locational congestion price components as these vary by node.

- impedance (Chapter 8) - the measure of the resistance of a circuit to an alternating electric current.
- Independent System Operator (ISO) (Chapter 5) - a standalone (and usually non-profit) organization that administers the operation of the transmission system.
- inside basis (Chapter 6) - the partner's share of the tax basis of partnership assets.
- interconnection agreement (Chapter 5) - a legal agreement between a project company and a grid operator giving the project company the right to feed electricity from a given power project into the operator's grid.
- interest rate swap (Chapter 5) - a contract wherein a stream of future floating interest payments is exchanged for one with fixed payments.
- investment grade (Chapter 5) - the credit quality or credit profile of a firm. To be rated investment grade, a firm must have a credit rating of Baa3 or BBB- or higher from Moody's or Standard and Poor's respectively (or equivalent from other rating agencies).
- ITC (Chapter 2) – An Investment Tax Credit is a federal policy mechanism that provides a tax credit of up to 30% of the installed costs for certain technologies including solar and small wind turbines. It is a one-off tax credit, paid when the facility goes into service.
- LMP (Chapter 8) – LMP, or Locational Marginal Pricing, is the cost to buy and sell power at different points in the wholesale electricity market. ISO's usually have day-ahead and real time LMP's. Day – ahead LMP's represent prices in day – ahead markets which let participants buy and sell electricity a day before the operating period to avoid real-time volatility.
- marginal private cost (Chapter 2) - the cost of producing an additional unit of a good experienced by a household or a firm. This cost does not include social or environmental costs that may arise from the production of the good.
- marginal social cost (Chapter 2) - the total cost to a household or firm of producing an additional unit of a good. This includes the marginal private cost along with social and environmental costs.
- mini-perm (Chapter 4) - a debt repayment structure in which a loan has a defined amortization structure until maturity date and any remaining balance is then made as a balloon payment at maturity.
- negative carry (Chapter 3) - a condition where the cost of holding a security or asset is greater than the income earned from it.
- Notice to Proceed (NTP) (Chapter 10) - a letter from the owner of the project company to the EPC contractor to begin construction. NTP is usually also the date of financial close when construction of the project begins
- off-take agreement (Chapter 2) - a contract (often long-term) between an energy producer and buyer to purchase or sell a portion of the producer's generated energy. This agreement helps the project company to obtain a guaranteed source of revenue, which in turn helps to secure lower-cost financing.
- outside basis (Chapter 6) - the tax basis of each investor's interest in the partnership.
- peak shaving (Chapter 12) - a technique wherein energy storage systems are charged during low demand and discharged during periods of high demand to meet and reduce the peak demand a system needs to fulfill.
- probabilistic maximum loss (PML) (Chapter 5) - the largest possible loss with a certain percentage of surety. For example, a 99% PML of $100 would imply that the maximum possible loss with 99% certainty is $100.
- PTC (Chapter 2) – A production tax credit is an inflation-adjusted corporate tax credit for each kilowatt-hour (kWh) of electricity produced from qualifying facilities. The PTC is a tax credit in addition to any other revenues received for power sold. The tax credit lasts for the first 10 years of facility operation.

- quasi-merchant financing (Chapter 4) - a project finance structure wherein a portion of the project output is contracted while the remainder bears merchant price risk. In this type of structure, debt may be financed in a way where the first lien debt coincides with the contract term and the second lien debt bears all the refinancing risk and is priced accordingly.
- REC (Chapter 2) – A REC is an attribute correlated with one unit of renewable energy generation (such as one MWh of wind power) in a location. A utility which generates or purchases more renewable energy than it needs to meet its RPS requirement can sell the associated RECs to another utility that is short against its requirement.
- recourse (Chapter 3) - A recourse financing provides the lender the right to make claims against the assets of the entity that provides the recourse.
- RPS (Chapter 2) – A renewable portfolio standard requires utilities to supply a certain percentage of their power generation from renewable sources such as wind and solar. An RPS usually establishes incremental targets which increase over time. For example, a state could require utilities to source 20% of their supply from renewable sources by 2020, and 25% by 2022.
- S-curve (aka power curve) (Chapter 4) – a transformation function applicable to a wind turbine that converts a given wind speed into the generation output.
- tax basis (Chapter 6) - the value of an asset eligible for receiving investment tax credits. This may be the cost of the asset or the fair market value, or may include value of the parts of an asset that qualify for ITC.
- time-tranching (Chapter 4) - a form of debt structuring wherein the loan is broken down into different pieces of varying maturities. The cash flows are prioritized to the different tranches sequentially. All cash flows remaining after the interest payment for all the tranches, are directed toward the shortest maturity bond until it is fully amortized, then shift to the next shortest maturity bond, and so on.
- tradable permit system (Chapter 2) - a system in which the maximum possible emission level is determined by the government and each firm/participant is allocated a permit for a unit of production. The sum of all permits is equal to the maximum emission. These permits can then be traded between firms.
- transmission basis (or transmission basis differential) (Chapter 4) - the difference between the price of electricity at two different nodes and/or hubs. It can also be thought of as the price of transporting electricity from the source to the point of delivery.
- variable interest entity (VIE) (Chapter 3) - a legal corporate structure where an investor may have controlling interest despite not having a majority of voting rights. VIEs are often formed as Special Purpose Vehicles (SPVs) to finance an asset without putting the whole company at risk.

Appendix C: Sample term sheets

1. Construction loan term sheet for a utility-scale wind project
2. Tax equity term sheet for a utility-scale solar project
3. Back leverage term sheet for a portfolio of residential solar projects
4. Sample confirm for a commodities hedge for a utility-scale wind project[1]

[1] Please visit http:www.isda.org for the latest forms of Master Agreement, Credit Support Annex, and schedules therefore prepared by the International Swap Dealers Association (ISDA).

1. Construction loan term sheet for a utility-scale wind project

Confidential

[__] MEGAWATT WIND FARM

$[__] Construction Credit Facility

Indicative summary of terms and conditions

Please be advised that the summary of material terms and conditions set forth in this indicative term sheet is for discussion purposes only and subject to change. This indicative term sheet does not constitute an offer or commitment by or on behalf of any potential participant to arrange, syndicate, underwrite, purchase, provide, place or otherwise participate in any capacity in the proposed facility, offering or transaction described herein or in any other financing, nor does it constitute an agreement by or on behalf of any potential participant to prepare, negotiate, execute or deliver any such proposed facility, offering or transaction described herein or any other financing, and is without legal effect whatsoever. This indicative term sheet and any other information provided in connection herewith must be held in strict confidence and may not be disclosed to any other person (other than to your affiliates and advisors on a confidential and need to know basis), or referred to publicly, unless with the prior written consent of each potential participant. By accepting this indicative term sheet, the addressee hereof agrees to be bound by the foregoing agreements, restrictions and limitations.

General

Project:	A [__] megawatt ([__] turbine) wind farm to be constructed in [__] (the "Project"). The Project will interconnect directly into a [__] kV transmission line, which is approximately [__] miles from the Project. The Project has secured two power purchase agreements ("PPAs"), one with [__] for [__] megawatts over a term of [__] years and another with [__] for [__] megawatts over a term of [__] years.
Credit Facility:	The loan facility (the "Credit Facility") will comprise: a senior secured Project Company construction loan facility of up to $[__] (the "Construction Loan Facility"), which amount shall be based

Continued

	on the lesser of a (i) [100]% advance rate on the purchase price commitment pursuant to a Purchase and Sale Agreement (the "PSA") and (ii) 90% of Project Costs.
Purchase and Sale Agreement:	Sponsor has entered into a definitive agreement with a strategic buyer for the sale of a [__]% interest in the Project at COD. The buyer or its parent is investment grade and will be disclosed during diligence. The purchase price is expected to be well in excess of project costs.
Project Costs:	The aggregate cost of the Project, inclusive of development costs, site acquisition and easement costs, construction costs, commissioning costs, start-up costs, financing fees, interest during construction, closing costs, a contingency allowance, reserves and all other costs related to the Project. Project costs are expected to be approximately $[__].
Financing Documents:	The terms and conditions applicable to the Credit Facility, including financial terms, representations and warranties, covenants, conditions, events of default and remedies typical for this type of construction loan financing, will be contained in one credit agreement (the "Financing Document"). The Financing Document will contemplate and require the execution and delivery of other customary credit documentation, including, but not limited to, collateral documents and legal opinions. The Financing Document is expected to be based on financing documentation typical for this type of construction loan financing.
Use of Proceeds:	The proceeds of the Construction Loan Facility will be used by the Project Company to finance: **(i)** the development and construction of the Project (including the purchase of major equipment to be included in the Project), payments for land rights related to the Project, costs for certain interconnection facilities to be used by the Project and other Project costs;

	(ii) the payment of transaction expenses and third party fees related to the development and construction and financing of the Project; **(iii)** the payment of transaction expenses and third party fees in respect of the Credit Facility; and **(iv)** the payment of interest and fees on the Credit Facility that accrue during construction.
Project Company:	[__], a Delaware limited liability company (the "Project Company") and the owner of the Project.
Holdco:	[__], a Delaware limited liability company ("Holdco") and 100% owner of the Project Company.
Borrower:	the Project Company, the ("Borrower").
Sponsor:	[__]
Lead Arranger and Bookrunner:	[__]
Administrative Agent:	[__]
Collateral Agent:	[__]
Depositary Bank:	[___]
Lenders:	The Arrangers and other financial institutions selected by the Arrangers in consultation with the Sponsor (the "Lenders").
Closing:	The date of the execution and delivery of the Financing Document and the related credit documentation (the "Financial Closing Date") is expected to occur on or before [__].
Sponsor Advances:	Advances by Sponsor and/or any other affiliate of Sponsor will be made to Project Company such that the amount of any and all such advances equal at least $[__] (the "Sponsor Advances") on or prior to the Financial Closing Date. The Sponsor Advances as of the Financial Closing Date and each borrowing date thereafter shall in aggregate be no less than [__]% of anticipated Project Costs.
Date Certain:	[__]

[Intentionally omitted]

[Intentionally omitted]

Construction Loan Facility

Commitment:	$[__] for the Project Company Construction Loan Facility (the "Construction Loan Commitment") which amount shall be based on the lesser of (i) [100]% advance rate on the price pursuant to the PSA and (ii) 90% of Project Costs.
Availability:	Loans under the Construction Loan Facility ("Construction Loan") will be available from time to time from the Financial Closing Date until the earliest of: **(i)** the full utilization of the Construction Loan Commitment; and **(ii)** the date of acceleration of the Credit Facility upon the occurrence of an event of default under the Credit Facility.
Use of Proceeds:	As described in General section.
Term:	Commencing on the Financial Closing Date and maturing on the "Construction Loan Maturity Date", defined as the earliest of: **(i)** the occurrence of the "Funding Date" under the PSA; **(ii)** the Date Certain; and **(iii)** the date of acceleration of the Credit Facility upon the occurrence of an event of default under the Credit Facility.
Interest Rate:	As described in General terms and conditions section.
Commitment Fee:	[__]% *per annum* on the average daily unused portion of the Construction Loan Commitment, payable quarterly in arrears, which shall start accruing on the Financial Closing Date.

General terms and conditions

Interest Rate for Credit Facilities:	1, 2, 3 or 6 month LIBOR, plus the Applicable Margin (as defined below). With respect to the Construction Loans the term "Applicable Margin" for any period shall mean [__]%.

	The Financing Document shall also provide that loans may be funded at the base rate with an applicable margin equal to 1.00% less than the Applicable Margin for LIBOR loans.
Interest Payments:	Accrued interest will be payable at the end of each Interest Period, but no less frequently than quarterly. Interest to be calculated on the basis of (i) a 360-day year and the actual number of days elapsed days for LIBOR loans and (ii) a 365/366-day year and the actual number of days elapsed for base rate loans.
Default Rate Margin:	2.00% *per annum* above the Applicable Margin described above, payable on all outstanding obligations under the Credit Facility upon the occurrence and during the continuance of an event of default under the Credit Facility.
Arrangement/Upfront Fee:	Paid to the Lead Arranger and Bookrunner on the Financial Closing Date.
Type of Loan Construction Loan	**Upfront Fees** [__]%
Administrative Agent Fee:	One time payment of $[__] to be payable at the first Borrowing in respect of the Credit Facility.
Pro Rata Funding:	All borrowings under the Credit Facility will be funded *pro rata* from each of the Lenders
No Reborrowing:	Once repaid or prepaid, in full or in part, loans provided under the Credit Facility may not be re-borrowed.
Mandatory Prepayments:	The Financing Document will contain customary mandatory prepayment provisions typical for this type of financing transaction.
Voluntary Prepayment:	Voluntary prepayments in a minimum amount of $500,000 and in increments of $100,000. All voluntary prepayments shall be applied *pro rata* to scheduled payments
Application of Prepayments:	Prepayments of the Credit Facility may be made without premium or penalty, subject to payment of customary breakage expenses, if any. All prepayments shall be applied among the financing parties according to their *pro rata* portion of the aggregate principal amount of loans outstanding at the time of the applicable prepayment.

Continued

Withholding Taxes:	Subject to usual and customary exceptions to be included in the Financing Document, all payments will be made free and clear of, and without deduction or withholding for, any taxes, such that each Lender will receive an amount equal to the sum it would have received had no deductions or withholdings been made
Illegality; Change of Circumstances; Capital Adequacy; Increased Costs:	The Financing Document will include customary provisions protecting the Lenders in the event of unavailability of funding, illegality, increased costs, capital adequacy charges and funding losses, and withholding tax gross-up. The provisions regarding the increased costs of Lenders' funds shall be consistent with what is typical for this type of transaction.
Non-Recourse:	Other than as described under the headings "Sponsor Advances", the Lenders will have recourse solely to the Collateral (as defined below).
Collateral:	The collateral and security provisions shall be consistent with what is typical for this type of transaction. The obligations under the Credit Facility will be secured by a first priority security interest in, among other things, the following collateral (the "Collateral"): **(i)** the real property rights and all inventory, machinery and equipment comprising the Project; **(ii)** the accounts receivable, bank accounts and general intangibles of the Project Company; **(iii)** all agreements entered into by the Project Company; **(iv)** all governmental approvals for the Project, to the extent assignable as collateral; **(v)** all insurance and condemnation proceeds receivable by the Project Company; **(vi)** all membership interests in the Project Company; **(vii)** the PSA and the rights of the borrower entity party to or third party beneficiaries thereunder (including a consent to collateral assignment entered into with the PSA counterparty and any related guarantor, if applicable); and **(viii)** all other real or personal property of the Project Company (including cash and financial instruments). Upon the Construction Loan Maturity Date and full repayment of the Credit Facility, the security interests granted by the Project Company will be released.

Conditions Precedent:	The Financing Document will contain customary conditions precedent to the Financial Closing Date, each borrowing, for financings of this type and consistent (as applicable) with precedent transactions among the parties, including, but not limited to, the following conditions, as customary for financings of this type, certain conditions precedent will be subject to a standard as to Material Adverse Effect (as defined below), subject to agreed other materiality and knowledge qualifiers, and consistent (as applicable) with precedent transactions among the parties: Conditions to Closing **(i)** All material Project contracts, the PSA, the PPAs, the construction agreements (including any BOP and turbine supply contracts), the O&M agreement, the interconnection agreement and any other material contract necessary for the construction and initial operation of the Project, and the Financing Document and other credit documentation associated with the Credit Facility shall have been executed and delivered, shall be in full force and effect and shall be in form and substance satisfactory to the Administrative Agent (including with respect to security interests in the Collateral). **(ii)** The Administrative Agent shall have received reports from the Independent Engineer and the Wind Resource Consultant (each as defined below) in form and substance satisfactory to the Lenders, covering the technical and economic feasibility of the Project (including, but not limited to, a review of transmission, engineering design, equipment selection, site characteristics, Project contracts, construction budget and schedule, status of permits and licenses, earthquake and flood risks, confirmation of the amount equity contributions made as of the Financial Closing Date, ability of the Project to meet regulatory and contractual requirements, completion of acceptance tests and the net production forecast for the Project, including the probabilities of exceedance of 50%, 75%, 90%, 95% and 99% on a one and 20 year basis). **(iii)** The Administrative Agent shall have received a report from the Transmission Consultant (as defined below) in form and substance satisfactory to the Lenders.

Continued

	(iv) The Administrative Agent shall have received a report from the Insurance Consultant (as defined below) in form and substance satisfactory to the Lenders, addressing the adequacy of the insurance coverage to be maintained (including coverage levels, deductibles, insurance carriers and form and scope of endorsements, including earthquake and flood coverage as necessary with customary exclusions and limitations appropriate for the Project) and all required construction period policies shall have been issued. **(v)** The Administrative Agent shall have received evidence reasonably requested by it in connection with its due diligence review of the real property rights associated with the Project in form and substance satisfactory to the Lenders. **(vi)** The Administrative Agent shall have received a report from the Environmental Consultant (as defined below) in form and substance satisfactory to the Lenders, addressing, without limitation, the presence or absence of hazardous materials on the Project site. **(vii)** All then required permits for the construction of the Project shall have been issued and shall be in full force and effect and in form and substance reasonably satisfactory to the Lenders. **(viii)** The Lenders shall have received such financial, organizational and other information concerning the Project Company, Holdco, the Sponsor, any other relevant borrower entities (including the affiliate of the Project Company that is party to the PSA, collectively, the "Credit Parties"), as well as the counterparty to the PSA (and any affiliate guarantor, if applicable) as it shall have reasonably requested in order to complete each Lender's respective "Know Your Customer" processes. **(ix)** The Lenders shall have received customary legal opinions (including a tax opinion) consistent with what is typical for this type of transaction, in form and substance reasonably satisfactory to the Lenders. **(x)** If the consultants' reports are not addressed directly to, and permit reliance by, the Administrative Agent, the Arrangers and the Lenders, the Administrative Agent shall have received customary reliance letters from the relevant consultants allowing such parties to rely on such reports.

	(xi) The Administrative Agent shall have received (a) the final Project budget and construction schedule and (b) the Project's base case projections (the "Financial Model), in each case, in form and substance acceptable to the Lenders. **(xii)** The representations and warranties (other than such that, by their terms, refer to specific dates) are true, complete and correct in all material respects. **(xiii)** Perfection of the security interests granted by the security documents and delivery of reasonably satisfactory consents to collateral assignment for agreed material Project contracts consistent with the precedent transactions among the parties. **(xiv)** Mortgage and title insurance for the Project satisfactory to the Lenders, to be in full force and effect. Real property survey shall be delivered. **(xv)** Customary resolutions, incumbency, corporate documents and officer's certificates from the Credit Parties and other relevant borrower entities. **(xvi)** The accounts of the Project Company shall be established (as described below). **(xvii)** Payment of all fees and expenses. **(xviii)** Other reasonable conditions that may arise as a result of the conduct of due diligence.
	Conditions to All Borrowings The Financing Document will contain customary conditions precedent to borrowings under the Construction Loan Facility including, but not limited to, the following: **(i)** For the first Borrowing, evidence of Sponsor Advances in an amount equal to the greater of $[__] or 10% or Project Costs. **(ii)** Delivery of a customary drawdown certificate by the Project Company and a confirmatory certificate of the Independent Engineer. **(iii)** Delivery of lien waivers from third party contractors (subject to thresholds to be agreed). **(iv)** Delivery of date down endorsement to the title insurance policy. **(v)** Delivery of a notice of borrowing. **(vi)** No default or event of default under the Credit Facility shall have occurred and be continuing. **(vii)** Representations and warranties are true, complete and correct in all material respects.

Continued

Material Adverse Effect:	(a) any change, event or effect that is, or could reasonably be expected to be, materially adverse to the status of the business, Assets, liabilities, results of operations or condition (financial or otherwise) of any borrower entity, the Project or any major project participant that is an Affiliate of Sponsor or any change, event or effect that is, or could reasonably be expected to be, materially adverse to (i) the ability of Project Company to achieve commercial operation on or prior to the Construction Loan Maturity Date, (ii) the ability of PSA buyer to meet its obligations under the PSA, (iii) the likelihood of the occurrence of the funding date under the PSA or (b) any change, event or effect that is, or could reasonably be expected to be, materially adverse to (i) the ability of the Project Company, any Credit Party or any other person to perform any material obligations under the operative documents and, to the extent material to the Project, any individual real property document to which it is a party, (ii) the ability of the Lenders to enforce any of the obligations or (iii) the validity, priority or perfection of the secured parties' security interests in and Liens on the Collateral pledged or granted by the Project Company or any Credit Party ("Material Adverse Effect").
Accounts:	Accounts of the Project Company will include a "Construction Account" and a "Construction Period Local Account", which shall be funded with the proceeds of the Construction Loans.
Representations:	Usual and customary.
Covenants:	The Financing Document will contain customary covenants for financings of this type, subject to agreed materiality qualifiers, including, but not limited to, the following: **(i)** Delivery of certain periodic and other information concerning the Project (including monthly progress reports during the construction period, unaudited quarterly financial statements and audited annual financial statements, in each case prepared in accordance with GAAP) and notices of material events. **(ii)** No additional indebtedness other than as permitted by the Financing Document. **(iii)** No liens on any assets or properties other than certain permitted liens to be specified in the Financing Document. **(iv)** No fundamental changes (e.g., mergers, acquisitions, consolidations, liquidations) or formation of new subsidiaries. **(v)** Limitations on affiliate transactions.

	(vi) Maintenance of existence, rights and records; access to properties. **(vii)** Procurement of insurance in accordance with requirements to be negotiated. **(viii)** No amendment, waiver, termination, modification of a Project Document (unless such amendment, waiver or termination is administrative in nature and could not be reasonably expected to have a material adverse effect).
Events of Default:	The Financing Document will contain customary events of default for financings of this type including, but not limited to, non-payment; breach of covenants; default under or termination or invalidity of any material Project contract or permit; prior to Construction Loan Maturity Date, breach of the PSA; invalidity of any Financing Document or other credit documentation or a lien on the Collateral; failure to make equity contributions as described under the heading "Sponsor Advances" above; loss events; abandonment of Project; judgments; ERISA events; material misrepresentations; bankruptcy of the Credit Parties (including, for the avoidance of doubt, the Sponsor) or, subject to certain replacement rights to be agreed among the parties, the construction contractors; cross-defaults; change of control (as described under the heading "Change of Control" below); and failure of the Project to achieve the "Funding Date" under the PSA by the Date Certain. The Financing Document will contain cure periods and materiality and other qualifiers relating to certain events of default customary for financings of this type.
Required Lenders:	Excluding loans held by a Lender that is an affiliate of the Project Company, the holders of more than 50% of the aggregate commitments and loans under the Credit Facility. Any "defaulting lender" shall be disregarded in determining Required Lenders at any time. Amendments and waivers under the credit documentation shall require the approval of the Required Lenders, each Lender directly affected thereby or 100% of the Lenders, as applicable.
Assignments:	Each Lender may assign its rights and obligations under the Financing Document to other financial institutions subject to (i) the applicable Borrower's consent (in the absence of a default or event of default under the Credit Facility) not to be unreasonably withheld, (ii) no assignment may be for less than $5,000,000 and (iii) no assignment may

Continued

	result in more than four (4) Lenders under the Credit Facility. Transfers by Lenders to affiliates will not require the Borrower's consent. Each assignee will become party to the Financing Document to the extent of the interest assigned, and the assignor shall be relieved of its obligations to the extent of the interest assigned. The Project Company shall not bear any increased costs resulting from an assignment.
Change of Control:	The following transfer restrictions will apply prior to the Construction Loan Maturity Date: **(i)** no transfer of the direct ownership interests in the Project Company or Holdco; and **(ii)** no transfer by the Sponsor of any indirect voting or indirect economic interests in the Project Company or Holdco other than with the prior written consent of the Required Lenders (which consent shall not be unreasonably withheld or delayed).
Lenders' Counsel:	[__]
Lenders' Local Counsel:	[__]
Borrower's Counsel:	[__]
Borrower's Local Counsel:	[__]
Wind Resource Consultant:	[__] (the "Wind Resource Consultant").
Independent Engineer:	[__] (the "Independent Engineer").
Transmission Consultant:	[__] (the "Transmission Consultant").
Environmental Consultant:	[__] (the "Environmental Consultant").
Insurance Consultant:	[__] (the "Insurance Consultant").
Expenses:	The Borrowers will pay the reasonable and documented costs of the Wind Resource Consultant, Independent Engineer, Transmission Consultant, Environmental Consultant and Insurance Consultant and the Lenders, including the legal costs incurred in connection with the negotiation, review, documentation, structuring, closing, administration and enforcement of the proposed financing (provided that, to the extent expenses exceed an amount to be agreed, the Borrowers shall be provided with notice of such excess amount not less than biweekly).
Governing Law:	State of New York except, where necessary, Texas or elsewhere, as may be relevant.

2. Tax equity term sheet for a utility-scale solar project

Terms and Conditions
[__] MW [__] Solar Project

The terms and conditions set forth in this Terms and Conditions (this "Term Sheet") are to be used as a basis for continued discussions relating to the potential tax equity investment transaction between [__] ("Investor") and [__] ("Sponsor") in respect of the [__] MW (ac) solar generation project (the "Project") to be constructed in [__] by Sponsor, with an expected commercial operation date in [__]. The equity investment will be made under an Investment Tax Credit flip structure in a limited liability company (the "Company") that will own and operate the project company (the "Transaction"). This Term Sheet does not constitute a commitment of Investor to prepare, negotiate, execute or deliver such a commitment. Investor's decision to participate in the Transaction is contingent on investor's completion of due diligence and obtaining internal approvals, including approval of its investment committee, and the execution of final documentation in a form and substance satisfactory to Investor, among other conditions. This Term Sheet is not to be released to or discussed with any third party without the prior written consent of Investor. Unless the context otherwise requires, capitalized terms used but not defined in this Term Sheet have the meanings given to them in the proposal letter to which this Term Sheet is attached.

Sponsor anticipates that the Project will qualify for, and elect to claim, the investment tax credit under Section 48 of the Internal Revenue Code.

General	
The Company	A limited liability company organized under the laws of the state of Delaware that owns 100% of the equity in [__], the limited liability company that owns the Project (the "Project Company").
The Managing Member	Sponsor or one of its affiliates will be the Managing Member of the Company.
Guarantor	[Sponsor]
The Tax Equity Investor	Investor or an affiliate thereof.
Manager and Operator	Sponsor or an affiliate of Sponsor will be the (1) manager/administrator of the Project pursuant to a Management Services Agreement ("MSA") and (2) the operator of the Project pursuant to an Operations and Maintenance Agreement (the "O&M Agreement").
Commitment Date	Investor requires that the equity capital contribution or purchase agreement (the "ECCA") be signed no later than [__].

Continued

Initial and Final Funding Dates	The outside date for the final funding of the Project, which will occur at the substantial completion of the Project, will be [__], 2017, however, Sponsor anticipates that the final funding date for the Project will occur by [___], 2017 (the "Final Funding Date"). Following the mechanical completion of the first block of the Project but prior to the subsequent testing, power generation, synchronization or back feeding, etc of the Project (each a "Subsequent Development Action"), there will be an initial funding by the Class A Member of the Project as specified below in "Class A Member Funding Amount" (the date of such funding being the "Initial Funding Date"). The Initial Funding Date and the Final Funding Date are referred to herein as "Funding Dates".
Deal Structure	
Class B Shares/Class B Member	Sponsor or an affiliate will purchase or retain 100% of the Class B Shares of the Company.
Class A Shares/Class A Member	The Tax Equity Investor will receive 100% of the Class A Shares of the Company upon the Initial Funding Date.
Target IRR/Flip Date	An Internal Rate of Return of [__]% (the "Initial Target IRR") expected to be realized in the Base Case Model no later than [__] years following the Final Funding Date (the "Target Flip Date"). The date upon which the Target IRR is achieved being the "Flip Date". The Class A Member's [5]% residual share is subject to adjustment to maintain an after-tax IRR at the twenty-fifth (25th) anniversary of the Final Funding Date of at least [100] basis points over the Target IRR (the "All-In Return"). The Base Case Model shall demonstrate a pre-tax return, treating the ITC as cash, of at least [2.00]% as of the twenty-fifth (25th) anniversary of the Final Funding Date and contemplates the receipt of the ITC on the remaining tax payment dates during the calendar year following the Final Funding Date.
Re-Pricing Metrics	(i) the occurrence of the Flip Date by the Target Flip Date in the Base Case Model, (ii) achievement of an All-In Return of at least 100 basis points over the Target IRR, (iii) the Base Case Model demonstrating a pre-tax return, treating the ITC as cash, of at least 2.00% as of the twenty-fifth (25th) anniversary of the Final Funding Date and (iv) a deficient restoration obligation no greater than the DRO Cap.

Cash Distributions and Tax Allocations[a]	*Phase I*: Beginning at the Final Funding Date until [__] (end of the ITC year), the Class A Member shall receive [__]% of the cash items and [__]% of the tax items, including the ITC. The Class B Member shall receive [__]% of the cash items and [__]% of all tax items. *Phase II*: Beginning after the expiration of Phase I until [__] ([ ] years after the Final Funding Date), the Class A Member shall receive [__]% of the cash and [__]% of the tax items and the Class B Members shall receive [__]% of the cash and 33[__]% of the tax items. *Phase III*: Beginning after the expiration of Phase II until the earlier to occur of (i) the Target Flip Date and (ii) the Flip Date, the Class A Member shall receive [__]% of the cash and [__]% of the tax items and the Class B Members shall receive [__]% of the cash and [__]% of the tax items. [*Phase IV*: For the period after the Target Flip Date until the Flip Date, the Class A Member will receive 100% of cash and 99% of the tax items and the Class B Member will receive 0% of cash and 1% of the tax items.] *Phase V*: For the period after the Flip Date, the Class A Member will receive [5]% of cash and tax items and the Class B Member will receive [95]% of cash and tax items Notwithstanding the forgoing, in the event of a Change in Tax Law subsequent to the Final Funding Date results in the Flip Date being projected to occur more than [] months following the Target Flip Date (the "Outside Target Flip Date"), the cash distribution percentages set forth in Phases I through III above shall be adjusted in amounts sufficient to cause the Base Case Model (updated to take into account such Change in Law and the corollary adjustent in cash distribution percentages) to demonstrate the achievement of the Flip Date no later than the Outside Target Flip Date. Notwithstanding the foregoing, cash associated with the sale of SRECs prior to the Flip Date (i.e., during Phases 1 – III described above) shall be distributed [100]% to the Class B Member and [0]% to the Class A Member, provided, that if the Target IRR has not been achieved by the Target Flip Date, then from the Target Flip Date until the Flip Date, cash associated with the sale of SRECs shall be distributed as set forth in Phase IV above. Cash distributions shall be made by the Managing Member no less frequently than monthly.

Continued

Post Flip DRO reimbursement	Notwithstanding the cash distributions and tax allocation set forth above, following the Flip Date, any income allocation in excess of the [5]% base allocation for deficit cure which is not offset by the carry forward loss limitation allowance, and which will then generate "excess" taxes, will be compensated by an additional cash distribution to Investor equal to those "excess" taxes (uncovered income times the applicable tax rate) (such amount the "Post-Flip DRO Cure"). Any Post-Flip DRO Cure to Investor will be in addition to the minimum [5]% cash distributions.
Class A Members Funding Amount	Based on the Terms and Conditions herewith and the Base Case Model [____].xls ("Base Case Model"), the Class A Member agrees to fund (i) [20]% of its aggregate anticipated funding in respect of the Project (the "Initial Class A Funding Amount") on the Initial Funding Date and (ii) the balance of its funding in respect of the Project on the Final Funding Date of such Project (the "Final Class A Funding Amount") (the sum of (i) and (ii) for the Project being the "Aggregate Class A Funding Amount"), which represents the aggregate expected Class A funding required for the Project. The Initial Class A Funding Amount and Final Class A Funding Amount shall be adjusted on the applicable Funding Date as per the section below entitled "Update to Class A Funding Amount" (such updated amount being the "Class A Funding Amount"), subject to a maximum Class A Funding Amount equal to $[__] million (the "Class A Funding Commitment").
Update to Class A Funding Amount	Immediately prior to the Initial Funding Date and Final Funding Date, the Class B Member shall rerun the Base Case Model to determine any adjustments to the applicable Initial Class A Funding Amount, Final Class A Funding Amount, and/or cash distributions in the LLC Agreement, as applicable, by updating the data and inputs used for the Base Case Model delivered at execution of the ECCA to take into account any change to the assumptions based on updated information from the independent engineer, the appraiser, the market and transmission consultant, the insurance consultant and the environmental consultant.

Change in Tax Law and Proposed Change in Tax Law	In the event of Change in Tax Law or a Proposed Change in Tax Law, the Base Case Model shall be rerun to incorporate such Change in Tax Law or Proposed Tax Law Change, as applicable, and the Initial Class A Funding Amount, Final Class A Funding Amount, and/or cash distributions in the LLC Agreement, as applicable, shall be adjusted so as to demonstrate achievement of the Re-pricing Metrics, it being understood that the aggregate of such payment plus the Aggregate Class A Funding Amount shall not exceed the Class A Funding Commitment. To the extent that adjustments are made as set forth in the foregoing paragraph to address a Proposed Tax Law Change and such Proposed Tax Law Change (or any further Proposed Tax Law Change that has the same or similar impact as such Proposed Tax Law Change) does not become a Change in Tax Law before the date that is ten (10) days (excluding Sundays) after adjournment *sine die* of the second session of the One Hundred and Fifteenth United States Congress (or such earlier date when the Members agree in writing that the Proposed Tax Law Change will not become a Change in Tax Law), then the Base Case Model from the Final Funding Date shall be rerun, without changing any inputs or parameters other than to remove the effect of assumptions made in connection with such Proposed Tax Law Change, to achieve the Re-pricing Metrics, and the Class A Member shall make an additional payment to the Class B Member, which shall be treated as a purchase price adjustment, it being understood that the aggregate of such payment plus the Aggregate Class A Funding Amount shall not exceed the Class A Funding Commitment, and the LLC Agreement will be amended as necessary to take into account any changes to allocations or distributions resulting from the rerun of the Base Case Model. "Change in Tax Law" shall be defined as (i) any change in or amendment to the Code or other applicable federal income tax statute; (ii) any issuance, promulgation or change in, or of, any proposed, temporary or final Treasury Regulations; (iii) any IRS guidance published in the Internal Revenue Bulletin and/or Cumulative Bulletin, notice, announcement, revenue ruling, revenue procedure, technical advice memorandum, examination directive, or similar authority published by the IRS of general applicability, that applies, advances or articulates a new or different interpretation or analysis of any provision of the Code, any other applicable federal tax statute or any temporary or final Treasury Regulation promulgated

Continued

	thereunder; or (iv) any change in the interpretation of any of the authorities described in clauses (i) through (iii) above by a decision of the U.S. Tax Court, the U.S. Court of Federal Claims, a U.S. District Court, a U.S. Court of Appeals or the U.S. Supreme Court. "Proposed Tax Law Change" shall be defined as (a) any federal income tax legislation that is passed by either house of Congress or reported by the House Ways and Means Committee, the Senate Finance Committee, the House Committee on Appropriations, the Senate Committee on Appropriations, the House Committee on Energy and Commerce, the Senate Committee on Energy and Natural Resources or any bill sponsored or co-sponsored by the chairman or ranking member of such committee being referred to such committee, (b) any proposed change in or amendment to the Code or any other applicable federal income tax statute included in currently proposed federal legislation from (1) the Executive Branch, (2) the Majority Leader of the United States Senate or (3) the Speaker of the United States House of Representatives, or (c) the issuance of proposed Treasury Regulations that, in each case, if it becomes law or, in the case of proposed Treasury Regulations, become final, would materially adversely affect the modeled ITC or other tax items, including the ability of the Class A Member to use the ITC and other tax items.
Fixed Tax Assumptions	For purposes of calculating whether the Flip Date has occurred, the following items will be fixed and not change: **(i)** the Company is a partnership for United States federal income Tax purposes; **(ii)** after the Initial Funding Date, the Class B Member and the Class A Member (the "Members") are the sole partners in the Company; **(iii)** the Company, after the Initial Funding Date, is for United States federal income Tax purposes the owner of the Project; and **(iv)** the allocations described in the limited liability company operating agreement for the Company (the "LLC Agreement") will be respected by the Internal Revenue Service either because they have "substantial economic effect" or are otherwise consistent with the Members' interests in the Company within the meaning of Section 704(b) of the Internal Revenue Code of 1986, as amended from time to time.

Capital Accounts	Capital accounts will be maintained in accordance with Section 704(b) of the Internal Revenue Code of 1986, as amended from time to time, and the Treasury Regulations promulgated thereunder. The Class A Member will have a Deficit Restoration Obligation Cap of [__]% (the "DRO Cap") of its investment.
Terms and Conditons	
Expenses and Fees	Sponsor will be responsible for documented third party fees, Investor's out-of-pocket expenses, and expenses of third party consultants, including the independent engineer, the appraiser, the market and transmission consultant, the insurance consultant and the environmental consultant. With respect to legal fees, Sponsor will pay the Class A Member's transaction and local counsel any amounts due upon the execution of the ECCA and upon each of the Initial Funding Date and Final Funding Date. Sponsor will review and approve a budget for Class A Member's local counsel, which will be the cap for payment by Sponsor to such local counsel.
Structuring Fee	Upon the execution of the ECCA, Sponsor will pay Investor, a non-refundable structuring fee equal to [__]% of the Class A Funding Commitment (the "Structuring Fee").
Conditions Precedents to each Funding Date	The conditions precedents to the obligation of the Class A Member to fund the Class A Member Funding Amount under the ECCA on each Funding Date shall be subject to the following conditions precedent: **1.** The Project shall be free and clear of liens, other than permitted liens as specified. **2.** Each of the Company and the Project Company has performed its obligations under the material contracts to which the Company or the Project Company is party to be performed prior to such date. **3.** All representations and warranties made by the Class B Member are true and correct in all material respects. **4.** All governmental approvals are validly issued, except for those not yet required to be obtained. **5.** Each material contract is executed, is in full force and effect and is acceptable to the Class A Member. **6.** No condemnation is threatened or pending. **7.** No proceeding has been instituted in writing by any governmental authority to prohibit consummation of the transaction. **8.** Notice of Funding 10 business days prior to funding date.

Continued

	9. An updated Base Case Model, reasonably satisfactory to the Class A Member has been received. **10.** Receipt of an appraisal satisfactory to the Class A Member. **11.** Satisfactory audited balance sheets of the Company and the Guarantor have been received. **12.** The Members have received all legal opinions specified in the ECCA, including corporate, regulatory, environmental, real estate and state and local tax, as well as an opinion from tax counsel to the Class A Member. **13.** Good standing certificates and incumbency certificates have been delivered for the Class B Member and its applicable affiliates. **14.** The cost segregation report has been received and incorporated into the funding model. **15.** Satisfactory first Annual Budget has been received. **16.** The Class A Member has received an estoppel from the EPC Contractor, counterparty to the Interconnection Agreement, the Off-Taker and the landowners. **17.** All consents, approvals, and filings required to consummate the transaction have been completed. **18.** For the Initial Funding Date, the Independent Engineer has certified that mechanical completion of the first block[b] has occurred and no Subsequent Development Action has occurred, for the Final Funding Date, the Independent Engineer has provided a certificate of substantial completion. **19.** Bring-downs of the reports and reliance letters from the insurance consultant, environmental consultants, independent price forecasting reports (for SRECs, merchant power prices, and merchant capacity prices), and the independent engineer. **20.** Insurance Certificates have been received by all Members. **21.** The Title Proforma has been received by all Members and the insured amount shall be no less than the fair market value of the Project(s). **22.** All amounts required to be paid to complete construction of the Project, or reserves to make such payments, have been paid, deposited or established. **23.** Satisfactory due diligence and receipt of related reports with respect to endangered species. **24.** Satisfactory due diligence with respect to the Off-Take Agreement. **25.** Satisfactory due diligence with respect to the Off-Taker for the appropriations, termination for convenience and set-off risks associated therewith.

	26. Each Member has received the flow of funds. **27.** All the Members have received the executed LLC Agreement. **28.** The Members have received the executed ASA and O&M Agreement. **29.** No adverse Change in Tax Law or Proposed Change in Tax Law, unless the impact of such laws have been accounted for in the model as otherwise set forth under "Change in Tax Law and Proposed Change in Tax Law" to the Class A Member's satisfaction; **30.** Use of major equipment (panels, inverters and trackers) from approved manufacturers with warranties satisfactory to the Class A Member. **31.** No occurrence of a material adverse effect on the Class B Member, the Guarantor or the major project participants. **32.** Such other conditions precedent as are identified in due diligence.
Representations and Warranties	The ECCA shall include customary representations and warranties.
Property Insurance	Sponsor will procure (i) casualty insurance for the Project in an amount no less than 100% of the total replacement costs of the Project, (ii) an endorsement to the casualty insurance to cover the economic impact of ITC recapture associated with such casualty and (iii) earthquake insurance. The Project's insurance coverage shall be provided for each of the named perils for the Project. If the Project suffers a casualty it shall be rebuilt with the insurance proceeds. Class A Member will be added as "Additional Insured".
Management and O&M Services	Pursuant to the Administrative Services Agreement ("ASA") and the O&M Agreement, Sponsor or an affiliate will manage and provide O&M, administrative and other services to the Project (for the avoidance of doubt, the O&M Agreement does not include any panel or other equipment maintenance to be provided directly by the suppliers). • The agreements will have initial terms of [10] years and will be automatically extended for five-year periods unless either the Company, Sponsor, or the appropriate Sponsor affiliate, elects not to renew. • The Company will pay Sponsor or its affiliate annual fees under the ASA and the O&M Agreement, plus

Continued

	reimbursement for costs and overhead, subject to the limitations of the Approved Budget per the LLC Agreement. • The annual base fee under the ASA and the O&M Agreement will be $[ ] and $[ ], respectively (both fees escalating each year by the CPI). • The agreements will contain customary limits on liability and indemnities.
Approved Budget	The Managing Member shall prepare and submit to the Members no later than [__] of each fiscal year, a budget and operations plan for the Company for the subsequent fiscal year, which sets forth the anticipated revenues and expenditures for such fiscal year as determined by the Managing Member (the applicable operating budget for such fiscal year being referred to as the "Operating Budget").
Transfers	
Transfers Generally	The LLC Agreement will include usual and customary transfer restrictions. Without the consent of the Class A Member, the Class B Member may transfer up to 49% of the Class B Membership Interest. Without the consent of the Class A Member, the Class B Member may transfer more than 49% of its Class B Membership Interest if the transferee meets the following requirements: **(i)** either (A) has, or is controlled by an affiliate that has, a rating not less than "BBB" for S&P or "Baa2" from Moody's or (B) has a tangible net worth of at least $1,000,000,000 if determined prior to the Flip Point or at least $250,000,000 if determined after the Flip Point; and **(ii)** prior to the Flip Date, (A) has, (A) for the three (3) preceding years, owned or operated (or had access to the expertise required in order to operate through committed management agreements) at least 500 MWs of solar generation assets or 1,000 MWs of renewable energy generation assets (of which at least 100 MWs were solar generation assets) or (B) after the Flip Date, for the preceding year, owned or operated (or had access to the expertise required in order to operate through committed management agreements) at least 100 MWs of solar generation assets or 250 MWs of power generation assets (of which at least 50 MWs were solar generation assets).

	The LLC Agreement will provide that any transfer of an interest in the Class A Member shall not require any consent of the Class B Member so long as Investor maintains an ownership interest in the Class A Member.
Voting	
Fundamental Decisions	Certain fundamental decisions to be agreed in connection with the execution of definitive documentation shall require the consent of at least 66% of Class A Shares and 66% of Class B Shares both prior to the later of the Flip Date and the expiration of the Class A Member's Deficit Restoration Obligation (the "Sunset Date") and after the Sunset Date.
Major Decisions:	Certain major decisions to be agreed in connection with the execution of definitive documentation shall require the consent of at least 51% of Class A Shares and 51% of Class B Shares prior to the Sunset Date. There will be no Major Decisions applicable following the Sunset Date.
Miscellaneous	
Guaranty: Scope	The obligations of the Sponsor affiliates under the ECCA, the LLC Agreement, MSA and O&M Agreement (if not direct obligations of Sponsor) shall be guaranteed by the Guarantor.
Indemnity: Cash Sweep	The Class A Member shall have the right to sweep 100% of the cash payable to the Class B Member for claims relating to breaches of the ECCA or the LLC Agreement if damages are not paid within 30 days of a claim being finally determined or not disputed until such claim is paid in full; provided that, in the event that (the Class B Member is disputing its liability for claim, 100% of the cash payable to the Class B Member shall be swept into an escrow account up to the amount of such claims until the Class B Member's liability for such claim is finally determined.
Removal of Managing Member	The Managing Member shall not be removed without "Cause," which shall be defined as (i) fraud, willful misconduct or gross negligence of the Managing Member, (ii) a material breach of the Managing Member's obligations under the LLC Agreement, and (iii) a "Bankruptcy" of the Managing Member; *provided, however*, that in the case of clause (ii), if such breach is curable, the Managing Member shall have the opportunity to cure such breach within 30 days of receiving written notice from the Class A Member of such breach; *provided, further*, that if such breach cannot be cured within such period, and the Managing Member is proceeding with diligence to cure such breach, the 30-day cure period shall be extended by an additional 60 days, for a total cure period of 90 days.

Continued

Project Credit Support	The Class B Member will provide the credit support required to be posted at the Final Funding Date under the Company's project documents. Thereafter, the Class B Member shall be required to maintain and replenish such credit support in exchange for an annual fee to be paid by the Company. Any draw of any credit supported provided by the Class B Member shall be treated as a member loan to the Company.
Purchase Option – Timing and Purchase Price	For a purchase price equal to the greater of (i) fair market value as then determined by a third-party appraiser if not mutually agreed to by the parties, (ii) Investor's projected book balance on Target Flip Date to be determined on the Final Funding Date and (iii) the amount which preserves achievement of the All-In Return, the Class B Member shall have the right to elect to buy the Class A Member's interest within 180 days after each of the following dates (x) the Flip Date and (y) the fifth anniversary of the Flip Date.

[a]NB: percentage allocations and dates in this section subject to adjustment to conform to the final model.
[b]Subject to due diligence and discussion.

3. Back leverage term sheet for a portfolio of residential solar projects

Private & Strictly Confidential **Execution Version**

[Sponsor]
Indicative Summary of Terms & Conditions
Up to $_____ MM of Senior Secured Term Loan and
Up to $___ million of Senior Secured Letter of Credit Facility
[Lender]
[Date]

*These Indicative Terms and Conditions (this "**Term Sheet**") are not a commitment or offer to lend money, extend, arrange or underwrite credit or borrow money. Any such commitment is expressly subject to receipt of final credit committee approvals with respect to the applicable Senior Secured Credit Facilities described below, satisfactory due diligence, finalization of satisfactory definitive documentation, no material adverse change, and satisfaction of relevant conditions precedent. This Term Sheet is intended as an outline only and does not purport to summarize or contain all the conditions, covenants, representations, warranties and other provisions which would be contained in definitive legal documentation. Indicative pricing presented in this Term Sheet is for information purposes only and represents the pricing Lender believes could be achieved, based upon current market conditions, with final pricing to be determined at time of launch based on then-existing market conditions. Lender retains the right in its sole discretion to determine, for any reason, not to participate in the proposed Senior Secured Credit Facilities. This Term Sheet is delivered to you with the understanding that you shall not disclose this term sheet or any of the terms and conditions set forth herein to any person, except (i) to your directors, officers, employees, accountants, advisors and legal counsel who require such information in order to evaluate the transactions contemplated hereby, in each case on a confidential need-to-know basis and (ii) where disclosure is required by law (in which case you agree to inform us promptly thereof, prior to such disclosure if possible). Federal law requires that all financial institutions obtain, verify and record information that identifies each entity establishing a credit relationship with the financial institution. As a result, Lender hereby notifies you that we will ask for information that identifies the Borrower and Guarantors listed below, including tax identification number, address, and documents evidencing legal incorporation, formation or existence. Lender may also request information about directors and executive officers of such parties.*

Borrower:	A special purpose limited liability company wholly-owned by the Sponsor (defined below) owning equity interests in the Portfolio (defined below).
Sponsor:	[___]

Continued

Residential Solar Portfolio:	The Sponsor, via the Borrower, is creating a vehicle to monetize long-term contracted revenues that are generated from its deployment of photovoltaic systems (the "Systems") installed across the United States (the "Portfolio"). Prior to completion of the Systems, ownership of the Systems comprising the Portfolio will be sold to tax equity funds (each a "Tax Equity Fund") in which third-party investors (the "Tax Equity Investors") receive an economic return from each System's federal tax benefits and operating cash flow. Long-term contracted revenues are primarily generated through leases (the "Leases") or power purchase agreements ("PPAs" and, together with the Leases, the "Host Customer Agreements") with homeowners (the "Homeowners") once such Systems have been "placed-in-service".
Subsidiary Guarantors:	The Borrower will own a collection of wholly-owned subsidiaries (each, a "Subsidiary Guarantor") that directly or indirectly own the equity interests in the Portfolio. Each Subsidiary Guarantor will cross-subsidize and cross-collateralize the obligations of the Borrower under the Senior Secured Credit Facilities.
Portfolio Installer:	[___]
O&M and Administrative Servicer:	[___]
System Equipment Providers:	System equipment providers, including in respect of solar panels and inverters (and related warranties and performance guarantees), to be reviewed by the Independent Engineer. An approved listing of providers to be agreed upon and included in the definitive documents.
Financial Closing Date:	The date when all conditions set forth in the Conditions Precedent to Closing and Initial Disbursement have been satisfied or waived. The Financial Closing Date is expected to occur on or prior to [___].
Sole Bookrunner:	[___]
Joint Lead Arranger(s):	[___], and other institutions to be determined.
Lender(s):	[___]
Administrative Agent:	[___]
Depositary Agent:	[___]
Collateral Agent:	[___]
Issuing Bank:	[___] or one of the other Lenders participating in the DSR LC Facility (defined below) on the Financial Closing Date and from time to time.

Guaranty:	The Sponsor shall have provided a guaranty (subject to due diligence) in respect of liabilities stemming from tax equity agreements and related payment/liquidity obligations guaranteed to the Tax Equity Fund operating companies by the Sponsor.
Senior Secured Credit Facilities:	Senior secured term loan (the "Term Loan"): An up to $[___] million senior secured credit facility funded as a term loan, which monetizes the Sponsor's long-term contracted revenues from the Portfolio. Debt Service Letter of Credit ("DSR LC Facility"): $[ ] million senior secured letter of credit facility. The DSR LC is being provided to meet the Borrower's obligations under the Debt Service Reserve (defined below). Amounts drawn under a letter of credit issued under the DSR LC Facility shall convert to a loan and be repaid in accordance with the Collections Account Waterfall.
Maturity Date:	The [___] (__th) anniversary of the Financial Closing Date.
Debt Sizing Parameters:	The Term Loan will be sized at each draw through the end of the Availability Period in a manner consistent with the following: Loan to Portfolio Value Test ("LPVT"): The ratio of loan amount divided by PV (__%) (defined below). PV (%): The present value of the Borrower's cash available for debt service under the Base Case Financial Model (to be defined) for the Portfolio discounted at a rate of [___] percent. The Term Loan will be sized to achieve a maximum level of debt equal to the lesser of (a) an up to approximately [___] LPVT or (b) a minimum and average debt service coverage ratio ("DSCR") for the Borrower equal to at least [___]x, each over the remainder of the Host Customer Agreements in the Portfolio assuming no contract renewals and a [___] year initial term.
Eligible Systems:	Includes all Systems and related cash flows based on current Host Customer Agreement that meet the following criteria, determined at the related draw: • [FICO requirements for Portfolio] • [Location density requirements for Portfolio] • The System shall be eligible for the investment tax credit under Section 48 of the Internal Revenue Code, precautionary fixture filings have been made and the System is installed in an approved jurisdiction. • Otherwise in accordance with the terms of this Term Sheet, including approval of the applicable System

Continued

	Equipment Providers and that the System has been "placed in service". The terms and conditions in the tax equity documentation relating to the Tax Equity Funds are subject to diligence and approval prior to signing of the loan documentation, including regarding subordination to, and cash diversion in favor of, the Tax Equity Investors.
Use of Proceeds:	Proceeds of the Term Loan will be used (a) to reimburse the Sponsor for capital costs associated with the deployment of a series of Systems comprising the Portfolio (each, a "Pool"), (b) to pay transaction costs related to the closing of the Senior Secured Credit Facilities, and (c) at the election of Borrower, to the extent not funded by an acceptable guarantee or letter of credit, to fund the Debt Service Reserve.
Availability:	The Term Loan will be available on a multi-draw basis in respect of each Pool no more than once quarterly in minimum draw amounts of $[___] million (or such lesser amount as may be available in connection with the final draw) for up to 18 months from the Financial Closing Date (the "Availability Period") provided that all conditions precedent to the Term Loan disbursement have been achieved. The DSR LC Facility shall be available from and after the Financial Closing Date to fund the Debt Service Reserve.
Interest Rate:	Term Loan: Three-month or six-month LIBOR plus an applicable margin of [___] bps per annum increasing by [___] bps on the [___] anniversary of the Financial Closing Date. Interest only payments shall be made quarterly. DSR LC Facility: Same as the Term Loan, to the extent drawn.
Letter of Credit Fee:	A letter of credit fee equal to the applicable margin under the DSR LC Facility multiplied by the average daily maximum aggregate amount available to be drawn under all issued letters of credit shall be payable quarterly in arrears.
Undrawn Commitment Fee:	[___] bps per annum payable quarterly in arrears during the Availability Period.
Scheduled Amortization:	Commencing prior to the expiry of the Availability Period, the Term Loan will amortize via quarterly payments made on each Scheduled Payment Date in accordance with the deemed amortization schedule described in Debt Sizing Parameters (as determined as at the end of the Availability Period) provided that amount amortized per the amortization schedule set at each funding date shall not be less than the amount established on each prior funding date. Any voluntary or mandatory prepayments made

	during the Availability Period shall reduce the Total Commitment by such amount.
Scheduled Payment Dates:	The last business day of the first month following each calendar quarter after the Financial Closing Date.
Distribution Trap:	Subject to cash flow waterfall guidelines (to be established), the Borrower shall be permitted to make Distributions quarterly (on Scheduled Payment Dates). If the Borrower's actual rolling DSCR is below [___]x on any Scheduled Payment Date, then the Borrower will be required to suspend distributions to the Sponsor until the Borrower is able to demonstrate for [___] successive quarters a DSCR equal to or greater than [___]x. If the Borrower's actual [ ] quarter rolling DSCR is below [___]x for [ ] successive Scheduled Payment Dates, amounts in the distribution trap account shall be applied to prepay the Term Loan in inverse order of maturity.
Depositary, Accounts, Collateral Agent Fees:	To be determined pursuant to a competitive bid process.
Interest Rate Hedging Requirements:	By no later than the end of the Availability Period, the Borrower will enter into and maintain a minimum of [___]% (up to a maximum of [___]%) of the floating rate exposure under the Term Loan through the date the Term Loan is shown to be fully repaid under the Base Case Financial Model. Immediately following the completion of the interest rate hedging transactions, the Base Case Financial Model and amortization schedule shall be updated to reflect the fixed rate obtained in connection with such hedging. In connection with such update, a debt amount shall be determined for each Scheduled Payment Date that is in compliance with the Debt Sizing Parameters (the amount for each Scheduled Payment Date being the "Target Debt Amount") and if, on the next Scheduled Payment Date, the Target Debt Amount is less than the principal outstanding under the Term Loan, then excess cash flow (as defined in the definitive documentation) shall be swept to repay the Term Loan on each Scheduled Payment Date until the principal outstanding under the Term Loan is less than or equal to the Target Debt Amount. Any mandatory prepayments made during such period shall also reduce the Target Debt Amount. Each hedge bank, provided that such hedge bank is a Lender under the Term Loan at the time the hedging transaction is entered into, shall share in the Collateral on a pari passu basis.

Continued

Mandatory Prepayments:	Subject to the due diligence and reasonable acceptance of tax equity arrangements, mandatory prepayments shall be subject to such tax equity arrangements and otherwise usual and customary for similar transactions, including, but not limited to the following: **1.** [___]% debt issuance; **2.** [___]% casualty proceeds; **3.** [___]% pro rata asset sale proceeds; Subject to permitted equity cures (discussed further below) and required Lender consent, the Sponsor shall have the right to make capital contributions into any Subsidiary Guarantor to enable such Subsidiary Guarantor to pay amounts due in connection with any call, put or purchase option in respect of the interest of the applicable Tax Equity Investor in a Pool without triggering any prepayment obligation in respect of the Senior Secured Credit Facilities.
Optional Prepayments:	The Senior Secured Credit Facilities may be prepaid, in whole or in part, at the Borrower's election at par without premium or penalty.
Applications of Prepayments:	Optional Prepayments shall be applied to the Term Loan and the DSR LC Facility, as directed by the Sponsor.
Debt Service Reserve:	A debt service reserve account shall be funded on the Financial Closing Date in an amount at least equal to the next [ ] months of scheduled interest, principal and periodic fees associated with the Term Loan and DSR L/C Facility plus, during an initial ramp-up period, a reserve for commitment fees and certain agency fees (the "DSRA Required Level"). On each Scheduled Payment Date and each disbursement date the Debt Service Reserve shall be funded to the DSRA Required Level. An acceptable guarantee or letter of credit (including a letter of credit issued under the DSR LC Facility) may be used to replace the funding obligation of the Borrower.
Inverter Reserve:	In the event that the Independent Engineer or the Lenders (acting in consultation with, and with the recommendation of, the Independent Engineer) become aware of a Serial Defect (to be defined) of inverters included in the Portfolio that is not covered by an acceptable manufacturer's warranty (as determined in consultation with the Independent Engineer), the Borrower shall establish an account, to be funded by the account waterfall, to reserve for the anticipated cost of replacement inverters as confirmed by the Independent Engineer.

Call/Put Option Reserve:	The Borrower shall fund a call/put option reserve with available cash on a quarterly basis (after payment of debt service but prior to any Distribution) up to an amount to be agreed in the definitive documentation.
Flip Reserve:	Shall be mutually agreed (if required) in respect of any Tax Equity Fund that has a floating flip date (i.e. timing for the "flip" in cash distribution percentages determined based on whether the applicable Tax Equity Investor has achieved its target yield).
Servicer Termination Event:	Subject to the rights of the Tax Equity Investors and pursuant to back-up servicing arrangements **(i)** the Back-Up Servicer (as defined below) or a transition manager may be appointed by the Lenders to act in respect of those Tax Equity Funds that are wholly-owned by the Sponsor or that do not have existing back-up servicing or transition management arrangements in place, **(ii)** the Borrower shall be required to maintain back-up servicing or transition management arrangements in respect of all Tax Equity Funds at all times during the term of the Term Loan and **(iii)** should the Borrower's [___] quarter rolling DSCR be less than or equal to [___]x, there are material uncured breaches under the O&M and Administrative servicing agreements or other trigger events to be agreed (consistent with the precedent documentation) occur, then, at the election of the Lenders, the O&M and Administrative Servicer shall be terminated and all authority, power, obligations and responsibilities of the O&M and Administrative Servicer automatically pass to the Back-Up Servicer or its designee upon appointment. Cure rights shall be agreed to provide O&M and Administrative Servicer with the opportunity to cure defaults triggering a Servicer Termination Event. The tax equity rights referenced above and the back-up servicing arrangements remain subject to the due diligence and reasonable acceptance of such arrangements by the Lenders.
Collateral:	Term Loan, DSR LC Facility and interest rate hedges: • A first priority security interest in the Borrower's assets, contracts and accounts; • A pledge of 100% of the Sponsor's membership interests in the Borrower and the Subsidiary Guarantors (the pledge of the membership interests in the Borrower to be provided by a passive holding company); and

Continued

	• To the extent permitted by the underlying tax equity fund partnership agreements, a first priority security interest in the Subsidiary Guarantors' assets, contracts and accounts, including a pledge of 100% of each Subsidiary Guarantor's direct equity interests in the Portfolio but excluding Excluded Property (defined below) and proceeds therefrom. The tax equity fund partnership agreements and restrictions therein remain subject to the due diligence and reasonable acceptance of such arrangements by the Lenders.
Collections Account Waterfall:	All Collections (defined below) of the Borrower shall be placed in a designated account of the Borrower that is pledged to the Lenders and applied in the following order of priority on each Scheduled Payment Date: **1.** Administrative Agency Fees and Depositary, Accounts and Collateral Agency Fees owing in respect of the Term Loan; **2.** For non-financed structures, contractual amounts due to the O&M and Administrative Servicer and Back-Up Servicer or transition manager (such amounts subject to the due diligence of the Lenders) and approved operating expenses subject to an approved yearly budget (with an agreed % variance contingency) and as otherwise approved by the Lenders); **3.** Term Loan and DSR LC Facility Undrawn Commitment Fees; **4.** Quarterly interest of the Term Loan and the DSR LC Facility and periodic interest rate hedge payments; **5.** Commencing on a Scheduled Payment Date occurring during the Availability Period, quarterly scheduled amortization payments and hedge termination payments; **6.** Repayment of all draws under the DSR LC Facility; **7.** Funding of the Debt Service Reserve up to the DSRA Required Level; **8.** Funding of the Inverter Reserve up to its then required level (if any); **9.** Funding of the Flip Reserve up to its then required level (if any); **10.** Funding of the Call/Put Option Reserve up to its then required level (if any); **11.** Application towards any applicable Mandatory Prepayments (other than as provided in priority 5 above); **12.** Application towards any Optional Prepayments; and **13.** Distributions to Sponsor (subject to agreed distribution conditions, including any Distribution Trap).

Collections:	Shall mean, without duplication (i) to the extent that the equity interests in any Portfolio entity are wholly-owned by the Borrower, the related (A) Rents (defined below), including all scheduled payments and prepayments under any Host Customer Agreement, (B) payments in respect of certain governmental incentives such as PBIs that are not Excluded Property, (C) payments relating to the Systems by lenders with respect to, or subsequent owners of, property where such Systems are installed pending assumption of a Host Customer Agreement relating to such System, (D) proceeds of the sale, assignment or other disposition of any Collateral, (E) insurance proceeds and proceeds of any warranty claims arising from installer or vendor warranties, in each case, with respect to any System, (F) reimbursements to any Person in respect of promotion appeasements or indemnified appeasements, (G) all recoveries including all amounts received in respect of litigation settlements and work-outs, (H) all purchase payments received from a Customer with respect to any System and (I) all other revenues, receipts and other payments to such Portfolio entity of every kind arising from their ownership, operation or management of the Systems, but excluding Excluded Property and proceeds thereof, and (ii) to the extent the equity interests in any Portfolio include tax equity financing, (A) all amounts received by the Borrower or the Subsidiary Guarantors as distributions with respect to its membership interests in a Portfolio entity, excluding Excluded Property and proceeds thereof and (B) amounts paid by the Sponsor to the Borrower, including under the Guaranty. "Rents" shall mean the monies owed to the Portfolio entity by the Homeowners pursuant to the Host Customer Agreements, including any lease payments under any Lease and power purchase payments under any PPA. "Excluded Property" shall mean (i) all cash proceeds from any upfront solar energy incentive programs, including proceeds pursuant to the California Solar Initiative (which are not subject to state income tax), or any other state or local solar power incentive program which provides incentives that are substantially similar to those provided under the California Solar Initiative (and which are similarly not subject to state income tax), (ii) all cash proceeds from any state income tax credit, including proceeds pursuant to the refundable Hawaii Energy Tax Credits and (iii) Marketable RECs and all revenues and proceeds of Marketable RECs. In no event shall any such amounts be considered part of Collections pursuant to the Base Case Financial Model.

Continued

	"Marketable RECs" shall mean a renewable energy certificate representing any and all environmental credits, benefits, emissions reductions, offsets and allowances, howsoever entitled, that are created or otherwise arise from a System's generation of electricity, including, but not limited to, a solar renewable energy certificate issued to comply with a state's renewable portfolio standard and in each case resulting from the avoidance of the emission of any gas, chemical, or other substance attributable to the generation of solar energy by a System (including renewable energy credits sold under a forward sale agreement).
Conditions Precedent to Closing and Initial Disbursement:	The Senior Secured Credit Facilities will contain usual and customary conditions precedent to the Financial Closing Date, including but not limited to: **1.** All Financing Documents shall have been executed and delivered, shall be in full force and effect, and shall be in form and substance reasonably satisfactory to the Lenders. **2.** All System servicing documentation (including back-up servicing arrangements) shall have been executed and delivered, shall be in full force and effect, and shall be in form and substance reasonably satisfactory to the Lenders and the Lenders shall have received any consents to collateral assignment required in respect thereof. **3.** Any tax equity investment documentation entered into in respect of the Pool is in a form and substance reasonably satisfactory to the Lenders. **4.** Customary resolutions, incumbency, corporate documents, and officers' certificates from the Borrower and the Sponsor shall have been delivered. **5.** Legal opinions in respect of enforceability, corporate and collateral security matters. **6.** The Borrower's accounts shall be established. **7.** The Lenders shall have received a satisfactory report from their insurance consultant, addressing the adequacy of the insurance coverage to be maintained and all required policies shall have been issued and shall be in full force and effect. **8.** The Lenders shall have received a satisfactory report on the Portfolio from the Independent Engineer. **9.** The Lenders shall have received a satisfactory report on the Portfolio from the Model Auditor. **10.** The Lenders shall have received customary reliance letters from agreed consultants allowing the

	Administrative Agent and the Lenders to rely on the underlying reports prepared by the consultants (which shall include a report from the Business Operations Consultant). **11.** The Administrative Agent shall have received (a) the final Portfolio budget and construction schedule, (b) the initial O&M budget through the end of [20__] (as reviewed by the Independent Engineer) and (c) the initial Base Case Financial Model, acceptable to the Lenders **12.** The Lenders shall have received such financial and other information concerning the Borrower, the Sponsor and the transaction as it shall have requested in order to complete each Lender's respective "Know Your Customer" processes. **13.** All indebtedness, if any, of the Borrower and the Subsidiary Guarantors shall have been discharged and terminated, and the Lenders shall have received reasonably satisfactory evidence thereof (including customary payoff letters and lien releases). Other conditions as necessary resulting from due diligence or specific to the Senior Secured Credit Facilities to be agreed, including in respect of the tax equity documentation.
Conditions Precedent To Each Distribution Subsequent to Closing and Funding:	Conditions to all draws (including the initial draw on the Financial Closing Date) shall be made in connection with the funding of a Pool and will be usual and customary for transactions of similar type, including but not limited to: **1.** All Host Customer Agreements for all Systems in the Pool are with qualifying Homeowners, shall have been executed, shall be in form and substance reasonably satisfactory to the Lenders and shall be in full force and effect, except as to materiality thresholds to be agreed. **2.** The Lenders shall have received and approved an updated Base Case Financial Model and amortization schedule in connection with the Pool which shows compliance with the Debt Sizing Parameters. **3.** The Lenders shall have received a certification from the Borrower in a form to be agreed which shall, among other things, contain certifications that the Systems included in the Pool have been "placed in service" (as such term or an equivalent term is defined in the tax equity documentation), including receipt of permission to operate from the local utility in respect of Systems included in the Pool.

Continued

	4. The warranties for all equipment used in the installation of the Systems in the Pool shall be in full force and effect, except as to materiality thresholds to be agreed, and shall be in form and substance satisfactory to the Lenders. **5.** All required permits and governmental approvals for the Systems in the Pool shall have been issued and shall be in full force and effect. **6.** Proper title; perfection of the security interests granted by the security documents, including filings and any consents to collateral assignment in respect thereof. **7.** Collateral (including applicable Systems) free and clear of liens (other than permitted liens); satisfactory lien releases of any lenders which have previously financed assets included in such Pool. **8.** Certification by the Borrower of compliance with the qualifications for purchase and/or contribution and/or lease of Systems under the applicable tax equity documentation of such Pool being financed, subject to the due diligence and reasonable acceptance of such arrangements by the lenders. **9.** The applicable reserves are fully funded to their required level or shall be fully funded from the proceeds of the disbursement. **10.** Representations and warranties of the Borrower and the Subsidiary Guarantors are true and correct in all respects (in respect of the Financial Closing Date) or all material respects (in respect of any subsequent disbursement). **11.** No default, Event of Default or Distribution Trap shall exist or be continuing. **12.** To the knowledge of the Borrower, no condemnation is pending or threatened, and no unrepaired casualty exists, with respect to any of the Systems included in the Pool, except as would not reasonably be expected to have a material adverse effect on the Borrower or the Portfolio. **13.** There is no breach of any provision of, or default under the terms of, the System servicing documentation (including back up servicing arrangements) or the tax equity documentation, in each case, except as materiality thresholds to be agreed.

	Other conditions as necessary resulting from due diligence or specific to the Senior Secured Credit Facilities to be agreed, including in respect of the tax equity documentation and any bring-downs in respect of the Pool and Pool documentation.
Representations & Warranties:	Usual and customary for financings of this kind (including as to exceptions, materiality thresholds, knowledge and other qualifications and "baskets"), including, but not limited to, the following: organization, requisite power and authority, qualification; equity interests and ownership; due authorization; no conflict with organizational documents, laws or agreements; binding obligation of credit documents; reasonableness of projections; the absence of any material adverse effect (definition to be agreed); environmental matters; disclosure of broker's fees; payment of taxes; title (including in respect of Systems); no default or Event of Default; governmental regulation (including with respect to energy regulatory matters, margin stock, etc.); employee matters; employee benefit plans; solvency; status as senior indebtedness; Patriot Act; intellectual property; absence of litigation; insurance; compliance with laws, including consumer protection, ERISA matters; consents and approvals (including governmental consents); no defaults under Host Customer Agreements; Host Customer Agreements and System servicing documentation (including back up servicing arrangements) in full force and effect; System warranties; Systems installed by qualified installers/dealers; access rights; tax and tax equity; customary disclosure representations; and priority and perfection of liens in the Collateral. The representations and warranties will be required to be reaffirmed in connection with each disbursement (including the disbursement on the Financial Closing Date).
Affirmative Covenants:	Usual and customary for financings of this kind (including as to exceptions, materiality thresholds, qualifications and "baskets"), including, but not limited to, the following: maintenance of corporate existence and rights; separateness and single-purpose covenants; PATRIOT Act reporting and other similar lender KYC requirements; servicing in accordance with prudent industry practice; maintenance of insurance; warranties; Systems installed by qualified installers/dealers; access rights; applicable law; Host Customer Agreements; Systems servicing documentation (including back up servicing arrangements), tax equity documentation; delivery of

Continued

	financial statements and related certificates (including regarding applicable DSCR tests); delivery and approval of annual budgets by the Independent Engineer; delivery of periodic Portfolio operating reports (including tracking of expected flip dates and reporting on inverter failure rates); delivery of notices of defaults, litigation and other matters; books and records and visitation rights; payment of taxes; continued perfection of security interests and filings; further assurances and additional collateral; environmental; payment and/or performance of contractual obligations; maintenance of properties; compliance with law, governmental regulation (including with respect to energy regulatory matters) and maintenance of permits; use of proceeds; and interest rate hedging.
Negative Covenants:	Usual and customary for financings of this kind (including as to exceptions, materiality thresholds, qualifications and "baskets"), including, but not limited to: incurrence of negative pledges; indebtedness (other than permitted indebtedness which shall not include any debt for borrowed money); liens (other than permitted liens); investments; guarantees; fundamental changes, acquisitions and dispositions (including downstream) unless permitted; creation of subsidiary interests or joint ventures (including tax equity investments); capital expenditures; amendments or waivers of organizational documents; amendment or waivers with respect to Host Customer Agreements, including warranties and performance guarantees (provided that the Subsidiary Guarantors may enter into payment facilitation agreements to amend or modify the electricity or lease rate, annual escalator or term of any Host Customer Agreement in respect of a defaulted solar asset subject to terms and conditions to be agreed in the definitive documentation, including all necessary approvals from any applicable Tax Equity Investors), System servicing documentation (including back up servicing arrangements) and changes to tax equity documentation which could reasonably be expected to be materially adverse to the Lenders; entering into material documents; sale and lease-back transactions; transactions with affiliates; speculative transactions; termination; dividends and other restricted payments; permitted activities; conduct of business and changes to the fiscal year; tax; accounts
Change of Control:	A "Change of Control" shall be deemed to have occurred if: **1.** The Sponsor ceases to (i) own at least [___]% of the equity interests in Borrower, and retain control of the management of the Borrower; or

	2. The Borrower ceases to directly or indirectly, own [___]% of the equity interests in the Subsidiary Guarantors and any of their subsidiaries holding the Portfolio assets (other than pursuant to a permitted tax equity transaction, if any).
Events of Default:	Usual and customary for financings of this kind (including as to exceptions, materiality thresholds, qualifications, notice rights, cure periods and "baskets"), including, but not limited to, the following: failure to make payments under the credit documents when due; breach of applicable Sponsor entity (including as servicer), the Borrower or the Subsidiary Guarantors under the definitive loan documents, tax equity documents or servicing documentation (including a failure by the Sponsor to pay under the Tax Equity Indemnity); representations or warranties incorrect when given; voluntary or involuntary bankruptcy of the Borrower, the Sponsor or any Guarantor; judgments and attachments relating to the Borrower or any Subsidiary Guarantor; ERISA; abandonment; impairment of title or security interests in Collateral; actual or asserted invalidity of subsidiary guarantees and other credit documentation; and Change of Control.
Equity Cure Rights:	Sponsor may equity cure the Borrower up to [___] in any period of [___] consecutive fiscal quarters.
Assignment and Participation:	Usual and customary for transactions of this type; provided that (i) so long as no default or Event of Default has occurred and is continuing, assignments and participations during the Availability Period may only be made to lenders that meet tangible net worth requirements to be agreed in definitive documentation and (ii) no assignment shall be made to a competitor (to be defined) of the Sponsor.
Governing Law:	[___]
Borrower's Counsel	[___]
Lenders' Counsel:	[___]
Independent Engineer:	[___]
Insurance Consultant:	[___]
Business Operations Consultant:	[___]
Back-Up Servicer:	[___]

Continued

Model Auditor:	[___]
Costs & Expenses:	Borrower shall reimburse the Lenders for all reasonable costs and expenses, including the legal fees, incurred by such parties in connection with the negotiation, execution and syndication of the Term Loan and the DSR LC, regardless of whether the Borrower satisfies the conditions precedent for the closing date and/or the funding date.

4. Sample confirm for a commodities hedge for a utility-scale wind project

Monthly Realized Revenue $\sum_{i=1}^{intervals}\left(P_i^N \times V_i^A\right)$

where,

Intervals shall reflect, for each Month, the total number of Settlement Intervals for that Month.

P_i^N shall be the published real-time price at the ERCOT node assigned to the Project for each Settlement Interval.
V_i^A shall be the actual volume generated by the Project for each Settlement Interval.
If Seller has failed to schedule or make deliveries of the WPP Hourly Quantities or NPP Hourly Quantities to the WPP Delivery Point or NPP Delivery Point (or, if the Alternative TA Hedge has been executed, the quantities sold by Seller under the transaction contemplated by the definitions thereof), during any Settlement Interval as required hereunder, all Settlement Intervals for such hour shall be excluded from the summation in determining the Monthly Realized Revenue.

Monthly Floating Obligation $\sum_{j=1}^{hours}\left(P_j^{EWI} \times V_j^W + P_j^{ENI} \times V_j^N + P_j^{ETA}\right)$

where,

hours shall reflect, for each Month, the total number of hours for that Month.

P_j^{EWI} shall be the arithmetic average of the published real-time prices for WPP Delivery Point for each hour.

V_j^W shall be the applicable WPP Hourly Quantity for each hour (expressed in MWh).

P_j^{ENI} shall be the arithmetic average of the published real-time prices for the NPP Delivery Point for each hour.

V_j^N shall be the applicable NPP Hourly Quantity for each hour (expressed in MWh).

P_j^{ETA} shall be $[____]/MWh; provided that if the Alternative TA Hedge has been executed, P_j^{ETA} shall equal the arithmetic average of the published real-time prices for each hour for the delivery point for the purchases made under the transaction contemplated by the definition thereof.

If Seller has failed to schedule or make deliveries of the WPP Hourly Quantities, NPP Hourly Quantities, or, if the Alternative TA Hedge has been executed, the quantities sold by Seller under the transaction contemplated by the definitions thereof, during any Settlement Interval as required hereunder, such hour shall be excluded from the applicable summation in determining the Monthly Floating Obligation.

Mismatch:	**Monthly Realized Revenue less the Monthly Floating Obligation.**
Monthly Payment of Tracking Account:	A payment and true-up mechanism will be performed at the end of each Month as follows: If the Mismatch for any Month is greater than 0, then Seller shall make a payment to Buyer in an amount equal to the Adjustment Amount. To the extent Seller makes a payment to Buyer, the Tracking Account Balance shall simultaneously be adjusted by adding the amount of the payment to the Tracking Account Balance. If the Mismatch for any Month is less than 0, then Buyer shall make a payment to Seller in an amount equal to the Adjustment Amount. To the extent Buyer makes a payment to Seller, the Tracking Account Balance shall simultaneously be adjusted by subtracting the amount of the payment from the Tracking Account Balance. Payments required under the preceding two paragraphs with respect to the Mismatch for a particular Month shall be due (i) if the Adjustment Amount is payable by Seller, on or before the later of the fifteenth (15th) Local Business Day of the Month following such Month, or the fifteenth (15th) day after Seller's receipt of an invoice from Buyer, and (ii) if the Adjustment Amount is payable by Buyer, on or before the later of the fifteenth (15th) day of the Month following such Month and the third Local Business Day after Buyer's receipt of the data necessary to calculate such amount; provided that, in either case, if such day is not a Local Business Day, then such payment shall be made on the next Local Business Day. Buyer shall prepare an invoice for the Adjustment Amount and all other amounts due with respect to each Month promptly after receipt of the report containing the settlement data necessary to calculate the Mismatch for such Month. Buyer acknowledges that such reports shall contain data, all or a portion of which, may not be final ERCOT settlement quality data and agrees that Buyer shall calculate the Mismatch and corresponding Adjustment Amount for each Month based on such data (regardless of whether it is final) when such data necessary for such

Mismatch:	**Monthly Realized Revenue less the Monthly Floating Obligation.**
Monthly Payment of	calculations for such Month has first been provided by (or on behalf of) Seller. Buyer shall reconcile and true-up such amounts with the final settlement amounts on a monthly basis, and any reconciliation shall be payable on the timeline set forth above with respect to monthly settlement of Adjustment Amounts (or with respect to the Month ending on the TA Termination Date, to be included in the applicable payments required to settle the Tracking Account Balance, as applicable). For the purposes of the foregoing, the parties agree that they shall include in such monthly reconciliation and true-up, any restatement or resettlement of data made by ERCOT (including due to correction of a Relevant Price by ERCOT) in the three (3) Months immediately following the Month to which such restatement or resettlement data relates, but not in any subsequent Months. Each invoice provided by Buyer for each Month shall contain such level of detail and supporting documentation as reasonably requested by Seller to support the amounts invoiced thereunder, and shall contain a summary of all amounts payable by the parties with respect to such Month, including with respect to the WPP Daily Payment and NPP Daily Payment obligations.

The Tracking Account Balance shall accrue interest each day at a per annum rate equal to the Three-Month LIBOR in effect on such day, plus [___] basis points for negative balances in the tracking account through and including [____________], and [___] basis points thereafter and such interest shall be added to the applicable Tracking Account Balance at the last day of each Month. Interest on negative Tracking Account Balances shall be calculated using negative numbers, and such negative numbers shall be added to the Tracking Account Balance.

Beginning on [____________], the Undrawn Tracking Account Balance shall accrue interest each day at a per annum rate equal to [___] basis points, and such interest shall be added to the applicable Tracking Account Balance at the last day of each Month. Interest on Undrawn Tracking Account Balances shall be calculated using negative numbers, and such negative numbers shall be added to the Tracking Account Balance. Upon the consummation of the Alternative TA Hedge, the Undrawn Tracking Account Balance shall cease to accrue interest on the effective date of the exercise and shall not thereafter accrue additional interest prior to the TA Termination Date (but in each case without limiting any default interest payable with respect thereto payable in accordance with the Agreement).

If the Tracking Account Balance is less than the Tracking Account Limit at the end of any Month ending prior to the TA Termination Date, then Seller shall pay to Buyer an amount equal to the absolute value of the difference between the Tracking Account Limit and the Tracking Account Balance on or before the later of the fifteenth (15th) Local Business Day of the Month following such Month, or the fifteenth (15th) day after Seller's receipt of an invoice from Buyer; provided that if such day is not a Local Business Day, then such payment shall be made on the next Local Business Day. Upon such payment by Seller, the Tracking Account Balance shall be increased by the amount of the payment.

If the Tracking Account Balance is less than zero as of the TA Termination Date, then Seller shall repay to Buyer an amount equal to the absolute value of such Tracking Account Balance (the "Seller Tracking Account Obligation") by one or any combination of the following (such election of methods being at Seller's discretion):

(a) making a single payment to Buyer on or before five (5) Business Days after the TA Termination Date; or

(b) making twelve (12) equal payments (unless if the TA Termination Date is [____________], in which case, eight (8) equal payments), with such payments due on each June 30, September 30, December 31, and March 31 thereafter until such amounts have been paid in full. If any such day is not a Local Business Day, then such payment shall be made on the next Local Business Day. The unpaid amount of the Seller Tracking Account Obligation shall accrue interest each day at a per annum rate equal to the Three-Month LIBOR in effect on such date plus [___] basis points through and including [____________], and [___] basis points thereafter and such interest shall be due and payable with each quarterly payment. On the date that is thirty-six (36) months (unless if the TA Termination Date is [____________], in which case, twenty-four (24) months) from the first quarterly payment date, all accrued but unpaid interest and the remaining balance of the Seller Tracking Account Obligation shall be due and payable by Seller to Buyer. The Seller Tracking Account Obligation may be prepaid in whole or in part without premium or penalty. Prepayments shall be applied first toward accrued but unpaid interest and then toward remaining payments on a pro rata basis. If Seller fails to make any payment under this paragraph when due and payable and such failure is not remedied on or before the second Local Business Day after notice of such failure is given by Buyer to Seller, the unpaid balance of the Seller Tracking Account Obligation and all accrued but unpaid interest thereon shall become immediately due and payable upon Buyer's demand.

(c) Notwithstanding anything in the foregoing clauses (a) or (b) to the contrary, if the scheduled date of payment for any amount due under the foregoing clauses (a) or (b) is not a Local Business Day, then such payment shall be made on the next Local Business Day.

On or before the fifth (5th) Business Day after the TA Termination Date, as applicable, Seller shall provide written notice to Buyer of which of the above repayment option(s) that it will utilize in connection with the repayment of the Seller Tracking Account Balance. If Seller fails to so provide such notice on or before such fifth (5th) Business Day after the TA Termination Date, as applicable), clause (a) above shall apply.

As used herein, the following terms shall have the following meanings:

"Adjustment Amount" means, with respect to any Month, (a) if the Mismatch for such Month is greater than 0, an amount equal to the lesser of (1) the Mismatch for such Month and (2) the difference of (x) Tracking Account Funding Cap − (y) Prior Tracking Account Balance, and (b) if the Mismatch for such Month is less than 0, an amount equal to the absolute value of the greater of (1) the Mismatch for such Month and (2) the difference of (x) Tracking Account Limit − (y) Prior Tracking Account Balance.

"Month" means a calendar month.

"Prior Tracking Account Balance" means the Tracking Account Balance at the end of the immediately preceding Month, provided that for the purposes of this definition, the Prior Tracking Account Balance shall not be a number less than the Tracking Account Limit and the initial Prior Tracking Account Balance shall be zero. For the avoidance of doubt, the Prior Tracking Account Balance at the end of the immediately preceding Period reflects the application of the Adjustment Amount for the immediately preceding Period.

"Settlement Interval" means the time period for which the ERCOT real time Energy market is settled, which as of the Trade Date is fifteen (15) minutes.

"Tracking Account Balance" means the Tracking Account Balance determined in accordance with these Tracking Account provisions; provided that the initial Tracking Account Balance shall be zero.

"Tracking Account Funding Cap" means $0.

	"Tracking Account Limit" means negative $[____] million; provided that if the Alternative TA Hedge is consummated, upon the effective date of such purchase transaction, the Tracking Account Limit shall be negative $[___] million.
Prepayments of Tracking	"Undrawn Tracking Account Balance" means (i) the Tracking Account Limit minus (ii) the lesser of (a) the Tracking Account Balance and (b) negative $[___] million, provided that (y) if the Alternative TA Hedge is consummated, the Undrawn Tracking Account Balance shall be zero and (z) in no event shall the Undrawn Tracking Account Balance be greater than zero (0).
Account Balance: Payments on non-	Seller may prepay a negative Tracking Account Balance or the Seller Tracking Account Obligations, in either case, in whole or in part, at any time and from time to time without premium or penalty.
Business Days	Notwithstanding any provision hereof to the contrary, any payment hereunder that otherwise would be due on a day that is not a Local Business Day instead shall be due on the first Local Business Day to occur thereafter.
Other Energy Terms:	Notwithstanding anything to the contrary set forth in the Agreement or in this Confirmation, at any time prior to the Project Completion Date, the foregoing "Monthly Realized Revenue," "Monthly Floating Obligation,"

Continued

	"Mismatch," and "Monthly Payment of Tracking Account" sections of this Confirmation shall not apply, and all payments in respect of this Confirmation shall be made in accordance with the provisions of the Power Annex. Notwithstanding anything to the contrary set forth in the Agreement: (i) any amounts with respect to a particular Month that Seller is obligated to pay Buyer pursuant to Section (c) (i) of the Power Annex, up to the lesser of (A) the sum of the absolute value of the Tracking Account Limit plus the Prior Tracking Account Balance, and (B) $[___] million, shall be payable on the applicable date described in the "Monthly Payment of Tracking Account" section of this Confirmation, and (ii) any amounts that Buyer is obligated to pay Seller pursuant to Section (c) (ii) of the Power Annex shall be payable on the applicable date described in the applicable "WPP Daily Payment" and "NPP Daily Payment" sections of this Confirmation. The Parties acknowledge and agree that Seller shall not be considered to have incorrectly reported any data for the purpose of Part 4(s) of the Schedule as a result of delivering data that is not final settlement quality data and that may be subject to adjustment after receipt of settlement quality data from ERCOT. With respect to this Transaction, the sole Market Disruption Events shall be "Price Source Disruption" and "Disappearance of Commodity Reference Price".
Changes in Delivery Point:	If, after the Trade Date, the Delivery Point of Energy is redefined, replaced or eliminated by ERCOT, including through the addition or subtraction of a material portion of the electrical buses from the Hub comprising the Delivery Point, and such change has a material adverse effect on one of the parties to this Transaction, then at the affected party's reasonable request, the parties shall promptly enter into negotiations to specify a new Delivery Point(s) that allocates the economic benefits and burdens of this Transaction between the parties in the same proportion as they were allocated on the Trade Date (in respect of Energy). For avoidance of doubt, a redefinition of the Delivery Point shall not include a circumstance whereby ERCOT replaces certain hub buses pursuant to its Process for Defining Hubs. A redefinition would include the replacement of all or substantially all of the hub buses that currently define the Delivery Point.

Income Tax Designation:	Party B intends the Transaction to be a hedging transaction for U.S. federal income tax purposes pursuant to Section 1221(a) (7) of the U.S. Internal Revenue Code of 1986, as amended, and Treasury Regulation Sections 1.1221-2 and 1.446-4; provided, however, that such intent shall have no bearing on Party A's tax treatment of the Transaction.

4. Account Details:
Payments to [Buyer]:

Account for payments:	**Bank: [________]** ABA #: [________] Account No.: [________]
Payments to [Seller]:	Reference: [________].
Account for payments:	The Revenue Account.

(signatures follow)

This Power Confirmation shall become effective and be binding on the parties hereto when executed and delivered by each party hereto.

Accepted and confirmed as of the date first above written:

[____________________]	[____________________]
("Buyer")	**("Seller")**
By: ____________________	By: ____________________
Name:	Name:
Title:	Title:

Signature page to the power confirmation

Annex A

WPP Hourly Quantity (MW per hour)

During the WPP Term, the WPP Hourly Quantity shall be set forth in the table below opposite the applicable calendar month; otherwise the WPP Hourly Quantity shall be zero.

MW/h	7 × 16	7 × 8		MW/h	7 × 16	7 × 8
18-Apr				23-Oct		
18-May				23-Nov		
18-Jun				23-Dec		
10-Jul				24-Jan		
18-Aug				24-Feb		
18-Sep				24-Mar		
18-Oct				24-Apr		
18-Nov				24-May		
18-Dec				24-Jun		
19-Jan				24-Jul		
19-Feb				24-Aug		
19-Mar				24-Sep		
19-Apr				24-Oct		
19-May				24-Nov		
19-Jun				24-Dec		
19-Jul				25-Jan		
19-Aug				25-Feb		
19-Sep				25-Mar		
19-Oct				25-Apr		
19-Nov				25-May		
19-Dec				25-Jun		
20-Jan				25-Jul		
20-Feb				25-Aug		
20-Mar				25-Sep		
20-Apr				25-Oct		
20-May				25-Nov		
20-Jun				25-Dec		
20-Jul				26-Jan		
20-Aug				26-Feb		
20-Sep				26-Mar		
20-Oct				26-Apr		
20-Nov				26-May		

20-Dec				26-Jun		
21-Jan				26-Jul		
21-Feb				26-Aug		
21-Mar				26-Sep		
21-Apr				26-Oct		
21-May				26-Nov		
21-Jun				26-Dec		
21-Jul				27-Jan		
21-Aug				27-Feb		
21-Sep				27-Mar		
21-Oct				27-Apr		
21-Nov				27-May		
21-Dec				27-Jun		
22-Jan				27-Jul		
22-Feb				27-Aug		
22-Mar				27-Sep		
22-Apr				27-Oct		
22-May				27-Nov		
22-Jun				27-Dec		
22-Jul				28-Jan		
22-Aug				28-Feb		
22-Sep				28-Mar		
22-Oct				28-Apr		
22-Nov				28-May		
22-Dec				28-Jun		
23-Jan				28-Jul		
23-Feb				28-Aug		
23-Mar				28-Sep		
23-Apr				28-Oct		
23-May				28-Nov		
23-Jun				28-Dec		
23-Jul				29-Jan		
23-Aug				29-Feb		
23-Sep				29-Mar		

For the purposes of the foregoing:

7 × 16 shall be HE 0700 through HE 2200 CPT, Monday through Sunday, including NERC Holidays, or any successor period in accordance with ERCOT protocols.

7 × 8 shall be HE 0100 through HE 0600 and HE 2300 through HE 2400 CPT, Monday through Sunday, including NERC Holidays, or any successor period in accordance with ERCOT protocols.

Annex B

NPP Hourly Quantity (MW per hour)

Prior to the month of [____________], the NPP Hourly Quantity shall be zero. Commencing the month of [____________] through and including the month of [____________], the NPP Hourly Quantity shall be set forth in the table below opposite the applicable calendar month. Commencing the month of [____________] and thereafter, the NPP Hourly Quantity shall be zero.

Month	7 × 16	7 × 8
Jan		
Feb		
Mar		
Apr		
May		
Jun		
Jul		
Aug		
Sep		
Oct		
Nov		
Dec		

For the purposes of the foregoing:

7 × 16 shall be HE 0700 through HE 2200 CPT, Monday through Sunday, including NERC Holidays, or any successor period in accordance with ERCOT protocols.

7 × 8 shall be HE 0100 through HE 0600 and HE 2300 through HE 2400 CPT, Monday through Sunday, including NERC Holidays, or any successor period in accordance with ERCOT protocols.

Annex C

TA Hourly Quantity (MW per hour)

Prior to the month of [____________], the TA Hourly Quantity shall be zero. Commencing the month of [____________] through and including the month of [____________], the TA Hourly Quantity shall be set forth in the table below opposite the applicable calendar month:

Month	7 × 16	7 × 8
Jan		
Feb		
Mar		
Apr		
May		
Jun		
Jul		
Aug		
Sep		
Oct		
Nov		
Dec		

If the TA Termination Date is [____________], for [____________] through [____________]:

Month	7 × 16	7 × 8

For the purposes of the foregoing:

7 × 16 shall be HE 0700 through HE 2200 CPT, Monday through Sunday, including NERC Holidays, or any successor period in accordance with ERCOT protocols.

7 × 8 shall be HE 0100 through HE 0600 and HE 2300 through HE 2400 CPT, Monday through Sunday, including NERC Holidays, or any successor period in accordance with ERCOT protocols.

Index

Note: 'Page numbers followed by "f" indicate figures, "t" indicates tables'.

L

M

N

O

Printed in the United States
By Bookmasters